Lecture Notes in Energy

Volume 38

Lecture Notes in Energy (LNE) is a series that reports on new developments in the study of energy: from science and engineering to the analysis of energy policy. The series' scope includes but is not limited to, renewable and green energy, nuclear, fossil fuels and carbon capture, energy systems, energy storage and harvesting, batteries and fuel cells, power systems, energy efficiency, energy in buildings, energy policy, as well as energy-related topics in economics, management and transportation. Books published in LNE are original and timely and bridge between advanced textbooks and the forefront of research. Readers of LNE include postgraduate students and non-specialist researchers wishing to gain an accessible introduction to a field of research as well as professionals and researchers with a need for an up-to-date reference book on a well-defined topic. The series publishes single and multi-authored volumes as well as advanced textbooks.

More information about this series at http://www.springer.com/series/8874

Ricardo Guerrero-Lemus
Les E. Shephard

Low-Carbon Energy in Africa and Latin America

Renewable Technologies, Natural Gas and Nuclear Energy

Springer

Ricardo Guerrero-Lemus
Departmento de Física
Universidad de La Laguna
La Laguna
Spain

Les E. Shephard
Department of Civil and Environmental Engineering
University of Texas at San Antonio
San Antonio, TX
USA

ISSN 2195-1284 ISSN 2195-1292 (electronic)
Lecture Notes in Energy
ISBN 978-3-319-52309-5 ISBN 978-3-319-52311-8 (eBook)
DOI 10.1007/978-3-319-52311-8

Library of Congress Control Number: 2017930945

Printed on acid-free paper

This Springer imprint is published by Springer Nature
The registered company is Springer International Publishing AG
The registered company address is: Gewerbestrasse 11, 6330 Cham, Switzerland

To Inés, Claudia (pichi-pichi) and mami
To Darlene—831!

Preface

Africa and Latin America are comprised of some of the world's most prosperous nations and some of the world's poorest. With more than 20% of the global population, these nations all strive to enhance their economic prosperity and to build a social fabric and a business community that allows their citizens' opportunities for success in the future. For many nations, in these regions, any goals beyond basic sustenance represent a marked improvement in the standard of living and basic services, but all nations recognize the inextricable link between economic prosperity and energy consumption and the challenges associated with building a secure energy future that fuels their long-term economic growth.

This book is intended to serve as an introduction and initial source of information for students, researchers, and other professionals interested in the energy sectors for nations that comprise both Africa and Latin America (Fig. 1) with a specific focus on low-carbon energy systems. This book coalesces information that is often difficult to find in the published literature to provide the most current material on how the energy sector is evolving in these countries and the challenges they face in moving from a disaggregated, nonstandard energy sector framework to a fully integrated, yet distributed sector. The most important up-to-date numerical data related to energy production, capacity, efficiencies, production costs, etc., are exposed in 14 chapters, 208 figures, and 52 tables, integrated in terms of units and methodology. We have attempted to rely on the recent (2014–2016) technical peer-reviewed literature in our assessments of each technology and the role they play in these nations, but for many countries, this information is often limited and for some nearly nonexistent. As such, we have also relied on government, non-government, and trade organization publications where necessary to supplement insights gained from the refereed literature.

This book begins with an assessment of the current energy situation and trends in Africa and Latin America and the significant constraints on meeting their future energy needs with current practices. These constraints include social, political, regulatory, financial, technical, economic, and policy considerations and challenges. We begin by examining the current energy trends in Africa and Latin America and the constraints that current practices place on meeting future energy

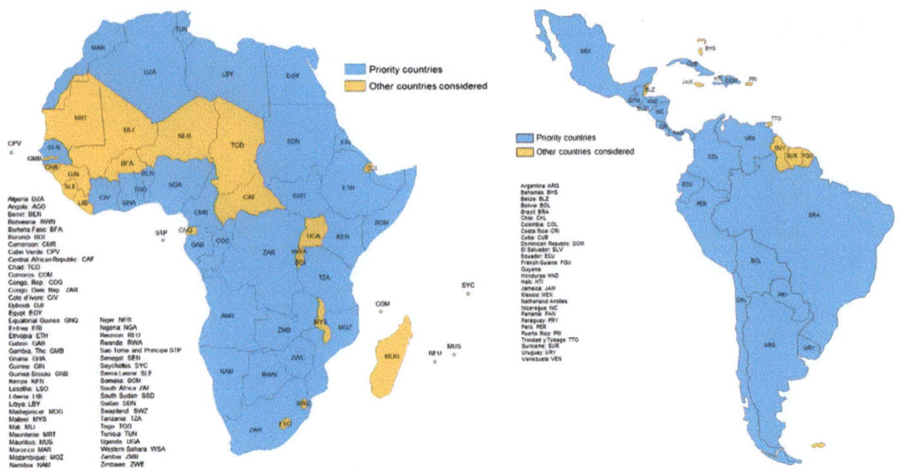

Fig. 1 African and Latin American priority countries and other countries considered in this book

needs. Later chapters present a more detailed description and analyses of each low-carbon energy technology and the role they play in countries that comprise these two regions. These chapters are supported by a large number of illustrations and data summary tables to offer valuable insights into the topics and technologies discussed. We have integrated 94 "Case examples" from the refereed literature in each of the chapters that identify specific examples of technology developments and deployments or a synthesis of the challenges, successes, and deliberations related to specific technologies and/or the complementary capability that has arisen as a result of access to low-carbon energy resources (e.g., ethanol gel stoves). Our case examples incorporate experiences from nearly every nation in these two regions and are intended in part to serve as "models for success" that may be emulated elsewhere within African and Latin American countries.

This book is intended to provide a basis for understanding the energy context for both Africa and Latin America by serving as a resource to help define strategies that accelerate the deployment of indigenous low-carbon energy technologies in a manner that enhances long-term economic prosperity. The authors enjoy "real-world experience" in teaching energy concepts and principles in "emerging" countries, and this book summarizes much of the information we use in the classroom interactions with our students. Both of our universities draw significantly upon students from African and Latin American countries, and our cities serve as gateways to these regions for trade, commerce, and education. Also, we plan to use this book as our resource for teaching classroom and online courses in the coming years in our respective universities. The authors will be available for readers to discuss any data or analysis published in the book (rglemus@ull.edu.es), and the readers will be encouraged to propose any additional and recognized content that they consider can enrich future editions. The readers who collaborate in the enrichment of future edition content will be mentioned in the acknowledgements of the edition where this content is added.

Priority countries for this book were identified based on the available reliable data on the energy sector.

African Countries

Algeria, Angola, Benin, Botswana, Cameroon, Congo, Democratic Republic of Congo, Cote d'Ivore, Egypt, Eritrea, Ethiopia, Gabon, Ghana, Kenya, Libyan Arab Jamahiriya, Morocco, Mozambique, Namibia, Nigeria, Senegal, South Africa, Sudan (covering South Sudan), United Republic of Tanzania, Togo, Tunisia, Zambia, Zimbabwe, and other African countries briefly considered (Burkina Faso; Burundi; Cape Verde; Central African Republic; Chad; Comoros; Djibouti; Equatorial Guinea; Gambia; Guinea; Guinea-Bissau; Lesotho; Liberia; Madagascar; Malawi; Mali; Mauritania; Mauritius; Niger; Reunion; Rwanda; Sao Tome and Principe; Seychelles; Sierra Leone; Somalia; Swaziland; Uganda; and Western Sahara).

Latin American Countries

Argentina, Bolivia, Brazil, Chile, Colombia, Costa Rica, Cuba, Dominican Republic, El Salvador, Ecuador, Guatemala, Haiti, Honduras, Mexico, Nicaragua, Panama, Paraguay, Peru, Uruguay, Venezuela, and other Latin American countries briefly considered (Antigua and Barbuda; Aruba; Bahamas; Barbados; Belize; Bermuda; British Virgin Islands; Cayman Islands; Dominica; Falkland Islands; French Guyana; Grenada; Guadeloupe; Guyana; Jamaica; Martinique; Montserrat; Netherlands Antilles; Puerto Rico; St. Kitts and Nevis; Saint Lucia; Saint Pierre et Miquelon; St. Vincent and the Grenadines; Suriname; Trinidad and Tobago; and Turks and Caicos Islands).

To discuss regional energy figures (mainly supply, capacities, and production), we use the IEA and US EIA Statistics Databases. We consider these sources very rigorous, but the methodology employed produces 2-year delayed data with respect to present. To compensate this drawback, in many chapters, more updated estimations, provided by global and prestigious associations related to the specific technology, are referred.

This book would not have been possible without the selfless support of many that believe as we do that we must improve the economic prosperity of global citizens everywhere and that energy is key to a prosperous future. Brooke L.E.S. Fontenot-Amedee has been gracious with her time and insight on information technology, and The Good Shephard Foundation has provided financial and moral support from the onset. Also Prof. José Manuel Martínez-Duart and Prof. Antonio Lecuona have provided significant content to this book.

The University of La Laguna, the University of Texas System, and the University of Texas at San Antonio have continuously encouraged collaborative research opportunities on renewable energy between our universities. Dr. Alfonso "Chico" Chiscano, MD has dedicated his life to the spirit of collaboration between San Antonio and the Canary Islands and has continuously nourished this relationship over decades.

We also want to make special mention to our image designer, Aneliya Stoyanova, and to Oyinkansola Adeoye, who has contributed along with many others technical support.

La Laguna, Spain
San Antonio, TX, USA
December 2016

Ricardo Guerrero-Lemus
Les E. Shephard

Contents

Acronyms

AC	Alternating current
ACSR	Aluminum conductor steel reinforced
AD	Anaerobic digestion
AHWR	Advanced heavy water reactor
BMP	Biochemical methane potential
BWR	Boiling water reactor
CANDU	CANada Deuterium Uranium
CAPP	Central African Power Pool
CC	Combined cycle
CCGT	Combined cycle gas turbine
CCS	Carbon capture and storage
CFL	Compact fluorescent light
CHP	Combined heat and power
COD	Chemical oxygen demand
DM	Dry matter
DSO	Distribution system operator
DU	Depleted uranium
EAPP	East African Power Pool
EIA	US Energy Information Administration
EPA	US Environmental Protection Agency
ETC	Evacuated tube collector
FBR	Fast breeder reactor
FPC	Flat plate collector
GCR	Gas-cooled reactor
GCV	Gross calorific value
GDP	Gross domestic product
GHG	Greenhouse gases
HLW	High-level waste
HTC	Hydrothermal carbonization
HTGR	High-temperature gas-cooled reactor

HV	High voltage
HVAC	High-voltage alternating current
HVDC	High-voltage direct current
IAEA	International Atomic Energy Agency
ICE	Internal combustion engine
IEA	International Energy Agency
IGBT	Insulated-gate bipolar transistor
ILW	Intermediate level waste
IMF	International Monetary Fund
IPP	Independent power producer
IR	Inferred resources
IRENA	International Renewable Energy Agency
ISO	Independent system operator
LCA	Life cycle analysis
LCOE	Levelized cost of electricity
LED	Light-emitting diode
LHV	Lower heating value
LLW	Low-level waste
LNG	Liquid natural gas
LPG	Liquified propane gas
LRMC	Long-run marginal cost
LV	Low voltage
lwg	Live weight gain
LWGR (RBMK)	Light water gas reactor
LWR	Light water reactor
MOX	Mixed oxide
MSR	Molten salt reactor
MSW	Municipal solid waste
MV	Medium voltage
NEP	National energy plan
NG	Natural gas
NGO	Nongovernmental agency
NGV	Natural gas-fueled vehicles
NORM	Naturally occurring radioactive material
OCGT	Open cycle gas turbine
OTEC	Ocean thermal energy conversion
PHWR	Pressurized heavy water reactor
PPA	Power purchase agreement
PWR	Pressurized water reactor
RAR	Reasonably assured resources
RDF	Refused derived fuel
RepU	Reprocessed uranium
RES	Renewable energy sources
RPS	Renewable purchase standards
RTO	Regional transmission organization

SAPP	South African Power Pool
SHC	Solar heating and cooling
SMR	Small modular reactor
SRF	Solid refuse fuel
SRMC	Short-run marginal cost
STP	Standard temperature and pressure
SWH	Solar water heating
SWT	Small wind turbine
SWU	Separative work unit
TPEC	Total primary energy consumption
TPES	Total primary energy supply
TRE	Tradable renewable energy
TSO	Transmission system operator
UMA	Arab Maghreb Union
USEC	United States Enrichment Corporation
VAT	Value-added tax
VLLW	Very low-level waste
WAPP	West African Power Pool
WtE	Waste to energy

Countries

AGO	Angola
ARG	Argentina
ATG	Antigua and Barbuda
BDI	Burundi
BEN	Benin
BFA	Burkina Faso
BHS	Bahamas, The
BLZ	Belize
BMU	Bermuda
BOL	Bolivia
BRA	Brazil
BRB	Barbados
BWA	Botswana
CAF	Central African Republic
CHL	Chile
CIV	Cote d'Ivoire
CMR	Cameroon
COD	Congo, Dem. Rep
COG	Congo, Rep
COL	Colombia
COM	Comoros
CPV	Cabo Verde
CRI	Costa Rica

CUB	Cuba
CUW	Curacao
CYM	Cayman Islands
DJI	Djibouti
DMA	Dominica
DOM	Dominican Republic
DZA	Algeria
ECU	Ecuador
EGY	Egypt, Arab Rep
ERI	Eritrea
ETH	Ethiopia
EU	European Union
GAB	Gabon
GHA	Ghana
GIN	Guinea
GMB	Gambia, The
GNB	Guinea-Bissau
GNQ	Equatorial Guinea
GTM	Guatemala
GUY	Guyana
HND	Honduras
HTI	Haiti
JAM	Jamaica
KEN	Kenya
LBR	Liberia
LBY	Libya
LSO	Lesotho
MAR	Morocco
MDG	Madagascar
MEX	Mexico
MLI	Mali
MOZ	Mozambique
MRT	Mauritania
MUS	Mauritius
MWI	Malawi
NAM	Namibia
NER	Niger
NGA	Nigeria
NIC	Nicaragua
PAN	Panama
PER	Peru
PRY	Paraguay
RWA	Rwanda
SDN	Sudan
SEN	Senegal

SLE	Sierra Leone
SLV	El Salvador
SOM	Somalia
SSD	South Sudan
STP	São Tomé and Principe
SUR	Suriname
SWZ	Swaziland
SYC	Seychelles
TCD	Chad
TGO	Togo
TTO	Trinidad and Tobago
TUN	Tunisia
TZA	Tanzania
UGA	Uganda
URY	Uruguay
USA	United States
VCT	St. Vincent and the Grenadines
VEN	Venezuela, RB
ZAF	South Africa
ZMB	Zambia
ZWE	Zimbabwe

List of Case Examples

Chapter 1
Executive Summary

Ricardo Guerrero-Lemus and Les E. Shephard

Abstract Energy consumption in Africa and Latin America has grown at a rate greater than the total energy consumption worldwide since 1980, consuming nearly 10% of the total global energy and ~20% of global renewable energy, largely biomass. Other significant sources of renewables include hydropower and geothermal. Both regions have diverse energy resources distributed non-uniformly between nations. Coal, oil and natural gas production is restricted to a few nations but is often used for electricity production across several countries. Natural gas reserves are prolific in parts of both regions and are likely to contribute to expanded electricity production in future decades particularly if investments in energy infrastructure occur as suggested.

Keywords Africa · Latin America · Energy · Electricity · Renewable technologies · Natural gas · Nuclear energy · Carbon emissions

Energy consumption in Africa and Latin America has grown at a rate greater than the total energy consumption worldwide since 1980, consuming nearly 10% of the total global energy and ~20% of global renewable energy, largely biomass. Other significant sources of renewables include hydropower and geothermal. Both regions have diverse energy resources distributed non-uniformly between nations. Coal, oil and natural gas production is restricted to a few nations but is often used for electricity production across several countries. Natural gas reserves are prolific in parts of both regions and are likely to contribute to expanded electricity production in future decades particularly if investments in energy infrastructure occur as suggested.

In terms of production, Africa and Latin America represent more than 23% of the global renewable energy produced, with most (~21%) of this energy provided by traditional biofuels and waste (36% of total global production) to support daily sustenance needs for heating and cooking rather than electricity production characteristic of nations with mature economies. These regions possess tremendous potential for significant growth of renewable energy resources derived from access to large land areas (more than 37% of the combined global surface area), the availability of abundant renewable resources (i.e., wind, water and solar), regional

R. Guerrero-Lemus and L.E. Shephard, *Low-Carbon Energy in Africa and Latin America*, Lecture Notes in Energy 38,
DOI 10.1007/978-3-319-52311-8_1

commitments to cut global emissions by 2030 and a growing recognition by some nations that low carbon development can enhance economic prosperity while reducing potential impacts on climate change. Considerations that influence future energy decisions for African and Latin American nations include life cycle CO_2, land requirements, water consumption, surface area, population and population density, cost and the ease of business transactions including financing. As the ease of transactions within the business environment improves there is likely to be greater diversification of energy sources and expansion of entrepreneurial activities.

Moreover, Africa and Latin America have diverse energy reserves that vary between nations but seldom achieve capacity levels necessary for the long-term sustainment, or more importantly growth, of an individual national or regional economy. Current demand for energy in many nations is low (i.e., energy consumption per capita) so existing fossil fuel reserves are often adequate to meet existing needs and projections for growth in the future. An energy paradox exists for Africa and Latin America driven largely by socioeconomics that suggests current energy resources are adequate for sustainment, but substantial energy resource development is necessary to improve regional GDP. Comparisons of GDP and CO_2 emissions per capita for the different countries in Africa and Latin America show that only a few nations have been capable of increasing their GDP per capita while at the same time reducing their CO_2 emissions (Figs. 1.1 and 1.2). However, both regions have substantial opportunities to increase their GDP based on the expanded penetration of low carbon indigenous energy technologies into their infrastructure (Fig. 1.3).

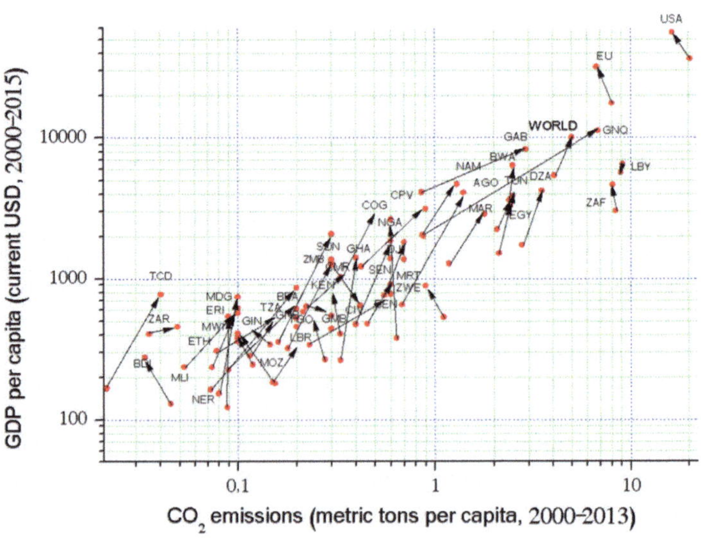

Fig. 1.1 Changes in CO_2 emissions (2000–2013) and GDP (2000–2015) for African countries

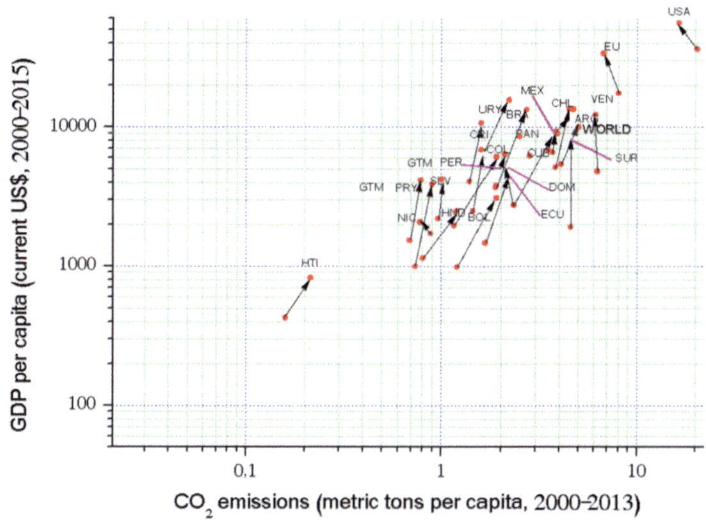

Fig. 1.2 Changes in CO_2 emissions (2000–2013) and GDP (2000–2015) for Latin American countries

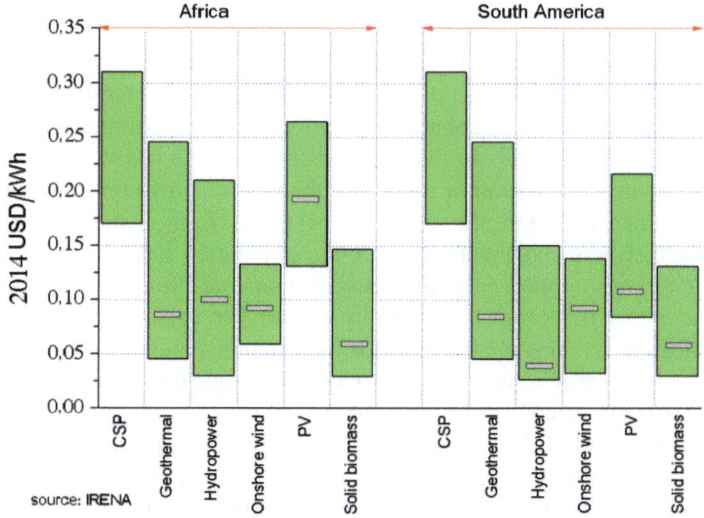

Fig. 1.3 Variations in cost for traditional renewable energy resources located in Africa and Latin America

- **Energy infrastructure** has a significant impact on energy consumption in both Latin America and Africa. Electric grid integration in both Africa and Latin America is limited by variations in frequency, sub-regional power pools that lack interconnections, geographic barriers and political differences (Fig. 1.4).

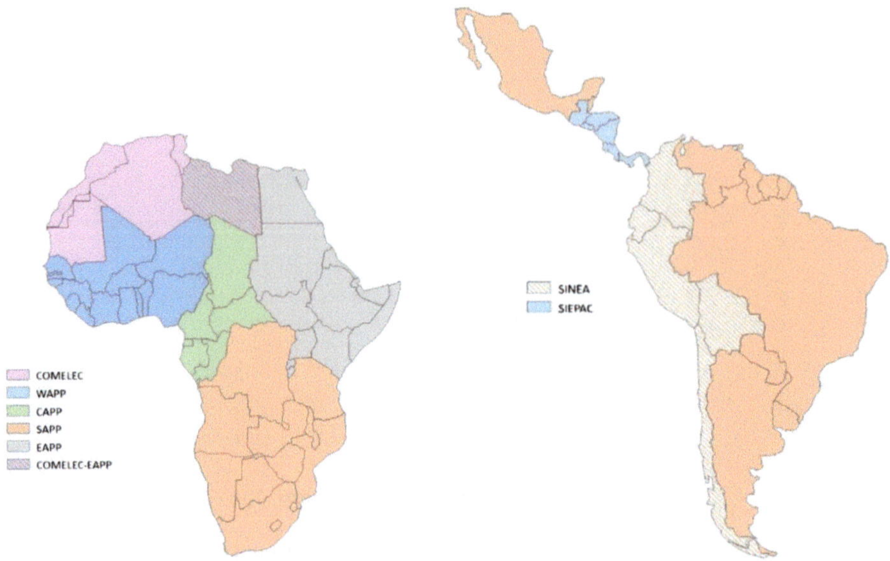

Fig. 1.4 Interconnections of the electric grid in Africa and Latin America limits the sharing of electricity between nations

Significant efforts in both regions are directed at improving the connectivity between nations but still less than 8% of power crosses international borders in any African region. Sub-Saharan Africa contains one of the least electrified regions in the world with an overall electrification rate below 50% and with 17 nations having electrification rates less than 20%. This contrasts significantly with North Africa and the majority of urban Latin America (Haiti is an exception), where more than 99% of the population has access to electricity. Electricity consumption in both regions remain well under the global average (21 and 60% of the world electricity output per capita, respectively). Rural electrification remains challenging for both regions, and those nations that rely heavily on biomass typically lack a modern energy infrastructure of any type, hence access to electricity is very limited.

- **Electric grid reliability** remains problematic in many nations where proper maintenance is limited, power theft is prevalent and power outages frequent, limiting the availability of power in many countries with the concomitant impact on local economies. In many countries the traditional grid extension model requires a multidimensional approach covering regulation, finance, economic development and social dimensions to analyze the business case of each electrification project.
- **De-centralized and off-grid electricity supply** is related to scale and to variations in daily power demand (e.g., lighting load 3–4 h after sunset). The base load for many nations with decentralized generation approaches zero, further reducing the financial viability of the system until the local economy develops to

some minimal level that can sustain a base load. The increase and improvement of grid interconnections can be the determinant factor for reaching the optimal low carbon energy scenario in both regions, but political trust and fair rules between countries are needed. Thus, the uneven geographic distribution of high-quality resources demonstrate that regional collaboration and grid interconnection will be necessary to promote low-cost clean wind, solar, hydropower and geothermal energy to all countries.

- **The electricity mix** in Africa and Latin America is highly varied but is typically tied directly to indigenous resources controlled by individual nations. While much of the electricity is generated using antiquated fossil fuel technology, it should be recognized that Latin America is the most decarbonized region in the world because of the abundance of hydropower.
- **Biomass** persists today as the predominant fuel for many of the less developed African and Latin American countries (Figs. 1.5a and 1.5b). Nearly 38% of the world population relies on biomass for cooking, largely in rural areas, but in Africa this value rises to a 67% with much of this contribution from sub-Saharan Africa. In Latin America, 15% of the population relies on biomass as it represents an affordable energy source for many households. Biomass utilization in Africa varies significantly between Maghreb countries with almost no reliance on traditional biomass for energy consumption, and Sub-Saharan countries, where most countries have a strong dependence on traditional biomass. These variations are attributed to lower GDP per capita, the dominance of rural areas, ease of access to forest and agricultural waste and the limited availability of fossil fuels in sub-Saharan countries. In Latin America, where GDP per capita is higher, the reliance on traditional biomass is much lower than in Africa. Very limited amounts of biomass are used for electricity production in either region although some electricity output from biomass occurs in most Latin American countries and can be significant in Brazil and Chile. The use of biomass is estimated to contribute to more than 577,000 premature deaths in Africa and 74,000 in Latin America's low- and middle-income countries. Exposure is

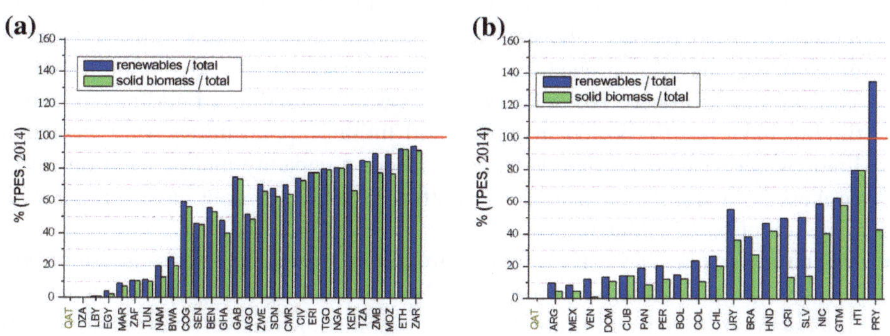

Fig. 1.5 Variations in the total primary energy supply (TPES) for both African and Latin American countries highlighting the significant contributions of biomass and traditional renewables to total energy supply

particularly high among women and young children, who spend the most time near the domestic hearth, with many premature deaths attributed to ischemic heart diseases, strokes and chronic obstructive pulmonary diseases in adults, and acute lower respiratory infections in children under 5 years.

- **Solar photovoltaic** (PV) production in Africa and Latin America combined comprise less than 3% of the total global photovoltaic energy production in spite of the uniform distribution of solar resources across both regions. While the PV capacity is far below many other global regions, there are some countries experiencing rapid growth. South Africa has invested significantly in solar PV because of recent blackouts and lack of supply and has more than 920 MW of ground mounted solar installed. Other nations are investing in solar supply chain capabilities (i.e., Mozambique has developed a solar panel manufacturing capability near Maputo and Rwanda has recently commissioned the largest solar PV facility in the sub-Sahara region with an 8.5 MW facility being constructed). In Latin America new solar PV capacity is being driven largely by utility-scale solar deployments. Chile added 396 MW of solar PV in 2014 alone because of high electricity prices and high solar irradiation in some areas, and is now considered a leader among all Latin American nations relative to solar PV.
  Reported prices of PV systems vary widely and depend on many factors including system size, location, customer type, connection to an electricity grid and technical specifications. Economies of scale, lack of technical expertise, trading barriers and other considerations increase the final cost of electricity in these regions. The levelized cost of electricity (LCOE) ranges from USD 0.131–0.264/kWh in Africa with an average value of USD 0.190/kWh, and USD 0.084–0.216/kWh in South America with an average value of USD 0.110/kWh (Fig. 1.3). Increased solar PV modularity, reduced costs of both large utility-scale and smaller residential-scale systems, limited operation and maintenance costs, and conversion efficiencies independent of power capacities will likely contribute to future broader scale PV adoption in both Africa and Latin America.
- **Solar thermal energy** is also underdeveloped in most areas. Thus, Africa and Latin America have very limited Concentrating Solar Power (CSP) with 2 and 8.6 GWh, respectively. Solar thermal energy for heating water is utilized locally in both regions with greater concentrations in Brazil, Mexico, South Africa, Tunisia and Morocco. No solar cooling systems for direct building air conditioning have been reported in Africa and limited opportunities are reported in Latin America with no detailed information on capacity and/or location. Solar cookers hold significant promise in both regions but to date have limited wide scale adoption in either region.
- **Hydropower** plays an extraordinarily important role in the energy future of both Africa and Latin America. More than 20% of total global hydropower generation occurs in Africa and Latin America combined. Moreover, more than 90% of existing renewable electricity in both regions is produced with hydropower and the capacity is expected to increase over time as several nations are expected to

pursue new hydropower development or expansion of existing capacity. Hydropower production in some countries exceeds 85% of demand, with more than 100% of electricity demand in Paraguay produced with hydropower. Two dams, Itaipu in Brazil and Guri in Venezuela, currently have capacities that exceed 10 GW. The production of electricity from hydropower in African countries depends on the availability of the resource and the existing power demand. Several African countries obtain almost all electricity from hydropower whereas others have almost no capacity because of limited or no hydro-resource or a GDP that limits electricity demand. It is estimated that 92% of the existing total hydropower capacity in Africa is untapped. Small hydropower capabilities are likely to become more pervasive over time as environmental concerns, drought and climate change limit efficiency and public acceptance of larger facilities.

- **Marine energy** plays a very marginal role in renewable electricity production as it has virtually no role in Africa or Latin America. As technology evolves and costs decrease, marine energy may play a role in select future locations. However, it is largely believed that more traditional renewable resources will be adopted much earlier in the energy development cycle.
- **Geothermal power** provides a small amount of electricity worldwide (nominally 0.31%) and similarly in Africa (0.27%) and Latin America (0.64%). New geothermal capacity is being developed in several nations located proximal to tectonic plate boundaries or "hot spots" across both regions in part because of capacity factors that can exceed 90% and a competitive cost structure in both Africa and Latin America. On the other hand, heat energy obtained from underground areas at nearly constant temperature of about 60–70 °C, and injected to heat pumps is not used in Africa and Latin America. However, there are options to use this geothermal heat to locate specific industries and spas in hotels. A significant barrier to continued growth of geothermal energy in both regions is the need to connect the resource with demand centers through the national grid.
- **Wind energy** is an established technology for producing low cost electricity in many regions around the world with abundant wind resources. Africa and Latin America contribute less than 5% of the total global wind production and no offshore capacity has been located in these regions. In Latin America wind resources are being developed in 12 nations and have become a significant contributor to electricity production in Nicaragua and Panama ($\sim$8 to 9%). The largest wind farm ($\sim$270 MW and expanding) is located in Panama. As importantly, the manufacturing of wind turbine blades, towers, and balance of plant components has become an important contributor to local economies in Brazil and Mexico and is contributing to a nascent industry throughout the region. Future expansion of wind resources in both regions will likely require significant investments in transmission as the best wind resources are located distant from existing urban demand centers. In Africa, most wind production occurs in South Africa, Morocco and Egypt with limited development elsewhere, mostly along coastal regions. There are no offshore wind deployments in

either Latin America or Africa. While wind generally has strong public support in both regions, there is a growing resistance in some communities opposing large-scale wind production for several reasons: (i) the perception of adverse impacts to natural and cultural heritage; (ii) the perception of negative strong effects that the wind farm development would have on local productive activities; (iii) local community distrust toward the wind developing company, local authorities and environmental regulators; and (iv) the threat to life projects of the local people. Sharing benefits from wind energy with the local communities and avoiding environmental impact and damage are the best recipes to increase wind energy in Africa and Latin America.

- **Liquid biofuels** (mainly bioethanol and biodiesel) are produced from biomass through chemical and biological processes, and are primarily used in the transportation sector and as an alternative cooking fuel option in Africa. In the transportation sector, biofuel production is limited largely to Brazil and to a lesser extent Colombia and Argentina, with almost no production ongoing in Africa. Bioethanol can be directly used in some specific internal combustion engines (i.e., flex vehicles), with E85 serving as a standard for Brazilian vehicles. In both Africa and Latin America, diesel engines are preferred over gasoline, although Brazil is the second largest producer of bioethanol in the world. Efforts to produce ethanol in Africa (largely Ethiopia) have been impacted severely by delays, water limitations and falling gasoline prices globally.
  Ethanol and ethanol gels (manufactured by mixing ethanol with cellulose) are emerging as preferred cooking fuel options in Africa. Ethanol gel may be preferred because it is clean-burning fuel that does not spill and is denatured to prevent accidental ingestion. Several ethanol gel stoves are now available, incorporating single or double burners, which accommodate custom pots and have the ability of utilizing ethanol with water contents as high as 50%. Mozambique has developed the first supply chain to distribute ethanol and efficient ethanol stoves. Stable policies, a balance between subsidies to fossil fuels and biofuels, and sharing profits from biofuels cultivation and transformation with local communities are key elements to increase the share of biofuels in the transportation sector.

- **Waste-to-energy** provides very limited amounts of electricity or heat on a global basis and has virtually no contribution in either Africa or Latin America except in Mexico, which has begun producing electricity in small amounts (i.e., $\sim$140 GWh) annually. There is a wide variety of municipal solid waste (MSW) production rates between countries in Africa related to GDP per capita, and variations in national policies (some nations have very few policies on MSW and very limited data on MSW collection rates) to minimize the environmental impact of human activity, mainly carbon and methane emissions, or to utilize MSW as a potential energy resource in the future. In Latin America the average values of MSW production rates are significantly higher but are less related to GDP per capita than other socio-economic factors that are difficult to identify. Total collection rates for both urban and rural areas are available for most of Latin America indicating collection rates exceed 70%. Landfill gas and biogas projects are increasing in frequency within Africa (e.g., South Africa and Kenya) and Latin America (e.g., Colombia, Brazil and Bolivia). However, instead of carbon capture and storage in landfills, which generally occurs at the beginning of the resource flows, the principle of carbon capture, transformation and reuse adding waste-to-energy paths in the cycle could be further developed and applied throughout the whole socio-industrial metabolism.

- **Natural gas** is playing an increasingly prominent role in the development of electricity generation systems in Africa and Latin America, as it is versatile, supports both baseline and peaking power generation demands, provides proven reserve generation capacity that supplements non-dispatchable power (i.e., wind and solar), has lower GHG emissions and uses less water than coal-based generation and has lower capital costs and shorter new plant construction times than nuclear. However, Africa consumes much less gas than it produces on a country-by-country basis in part because of limited infrastructure investment. In Latin America the gas pipeline infrastructure is more highly developed with gas being transported between countries and with all countries that produce natural gas using it to generate electricity. Unconventional production of natural gas is associated with a number of socio-political, environmental and economic issues that have limited drilling in some areas and created local concerns about impacts on water supply, excessive road traffic, accelerated deterioration of local and regional roads, impacts on air quality and enhanced methane emissions. Natural gas can play a key role in the development of the future low-carbon energy future in Africa and Latin America if social and environmental impacts can be mitigated and the extension of the power grid are achieved.
- **Nuclear energy** plays a role in both of these regions. Today, there are nine operational nuclear power plants in Africa and Latin America that provide limited power in absolute terms and relative to their national power mix. South Africa has two operational nuclear power plants with a total of 1830 MW capacity. South Africa began working on nuclear power in the early 1960s as part of the Atoms for Peace project and on nuclear weapons in the early 1970s. In 1989 South Africa became the only nation in the world to voluntarily dismantle their nuclear weapons. Recently, South Africa has discontinued research on a new power reactor design (i.e., Pebble Bed Reactor) and continues to defer decisions on additional expansion of nuclear power. In Latin America, Argentina has two operating reactors and one under construction and is contemplating further expansion. Brazil has two operating reactors and one under construction. Mexico has two operating reactors in Laguna Verde. Brazil is the only nation currently producing uranium for nuclear fuel manufacturing. The capacity and connectivity of the electric grid in Africa (and potentially some parts of Latin America) limits the further development of nuclear power using traditional nuclear power plants. Several nations in both regions are considering small modular reactors (SMRs) as an option (modular capacity is variable—

nominally 25–250 MW). Argentina is finishing the construction and testing of the world's first civilian modular reactor that is expected to go critical in 2017. However, there are many reasons to expect a very limited nuclear future in most African and Latin American countries.

In conclusion, the challenges associated with the development of a low carbon energy future for both Africa and Latin America are substantial, but not insurmountable. Nations in both regions are utilizing indigenous resources to meet their existing needs but limitations in the availability of power markedly impacts many nations to expand their economies and compete in a regional or global market. The challenge is even more pervasive for those nations whose populations are largely rural. The lack of an interconnected grid where electric power is shared across national borders precludes economic expansion. New technology may preclude the need to invest in traditional infrastructure for many nations. Several nations have been successful in developing a robust energy system and can serve as models for those that aspire to extend their existing capacity. As the ease of transactions within the business environment improves there is likely to be greater diversification of energy sources and expansion of entrepreneurial activities. Investments in these nations is essential, our future world depends on it!

Chapter 2
General Description

Ricardo Guerrero-Lemus and Les E. Shephard

Abstract Africa and Latin America possess tremendous potential for significant growth of renewable energy. Much of this potential is derived from access to large land areas with more than 37% of the combined global surface area, the availability of abundant renewable resources (i.e., wind, water and solar), commitments to cut global emissions by 2030 and a growing recognition by some nations that low carbon energy development enhances economic prosperity. Moreover, Africa and Latin America produce more than 23% of the global renewable energy, but much of this energy is provided by traditional biofuels (36% of total global production) to support daily sustenance needs (i.e., heating and cooking) rather than electricity production. An energy paradox exists for Africa and Latin America as they have diverse and often abundent energy reserves that vary between nations but they seldom achieve generation levels necessary for growth of national or regional economies. In this chapter a general description of both regions is provided. This description incorporates updated key parameters that should influence present and future energy decisions by governments, business sector, researchers and professionals including primary energy supply, reserves, carbon emissions versus GDP, water consumption, land consumption, easiness of doing business, etc. As the ease of transactions within the business environment improves there is likely to be greater diversification of energy sources and an overall expansion of entrepreneurial activities.

Keywords Africa · Latin america · Energy · Electricity · Renewable technology · Power plants · CO_2 · Carbon emissions · Carbon capture · CCS · Coal · Oil · Natural gas · Combustion · Energy storage · Water · Fossil fuels

2.1 Introduction

The global potential for the production of renewable energy resources is huge and exceeds the world total primary energy supply (TPES) based on conventional energy resources for each primary renewable energy resource (Fig. 2.1). The renewable energy resource with the largest potential (considering only the surface

R. Guerrero-Lemus and L.E. Shephard, *Low-Carbon Energy in Africa and Latin America*, Lecture Notes in Energy 38,
DOI 10.1007/978-3-319-52311-8_2

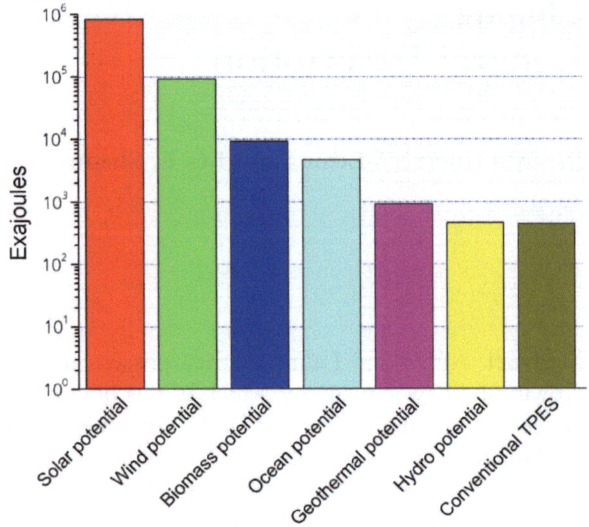

Fig. 2.1 Global technical potential for the different renewable energy resources, compared to the world conventional annual TPES in 2014 [4]

land above the sea level) is solar energy, followed by wind energy. Much of the technical potential for renewables is located in Africa and Latin America, as these regions represent 22 and 15%, respectively, of the world surface area (i.e., 134,324,741 km^2) [1], and comprise 15.2 and 8.5% of the world's population (i.e., 7.35 billion inhabitants) (2015) [2]. Africa represents 5.64% of the 572 Ej (2014) world total primary energy supply (TPES: indigenous production + imports – exports − international marine bunkers − international aviation bunkers ± stock exchanges) and Latin America 6.31% [3].

In Africa, the actual percentage of renewable energy used is huge compared to the rest of the world (Table 2.1) and in Latin America it is substantial. Renewable energy in both of these region is primarily associated with biofuels and the basic sustenance needs of heating and cooking rather than the growth of modern renewable technologies for producing electricity and reflects the limitations on the deployment of energy infrastructure throughout these regions. Fossil fuels play a significant role in each region, nuclear energy is present in both regions and hydropower is substantial in parts of Latin America (Table 2.1).

In relation to the proven reserves and average annual consumption of the different fossil fuel resources [5] and uranium [6], the most updated statistics indicate that proven reserves of fossil fuels are adequate in both regions for meeting the current demand and average consumption growth (Table 2.2).

Energy consumption per capita (Table 2.3) for both regions is well below global averages, reflecting the significant gap between these regions and the most advanced regions in the world relative to energy consumption per capita. This situation can be considered as an opportunity to move Africa and Latin America to a more sustainable energy supply system based on low carbon technologies.

Table 2.1 Main TPES (EJ) indicators for Africa, Latin America and the World in 2012 (last data available) [3]

TPES 2014 (Mtoe)	World	Africa	Latin America
Coal	164	5	2
Crude oil	182	5	15
Natural gas	121	5	9
Nuclear	27	0.2	0.3
Hydro	14	0.4	3
Geothermal, solar, etc.	8	0.2	0.4
Biofuels and waste	59	15	6
Total	575	32	36
% world renewables/total	14	49	26

Table 2.2 Proven reserves of oil, coal and natural gas, known recoverable reserves of uranium, consumption, increase in consumption and proven reserves in year terms [5, 6]

Proven reserves	World	Africa	Latin America
Proven reserves oil (billion barrels, 2015)	1663	127	340
Proven reserves coal (million short tonnes, 2014)	1,071,560	14,640	16,785
Proven reserves natural gas (trillion cubic feet, 2015)	6950	604	291
Known recoverable uranium (thousand tonnes, 2013)	5903	1253	276
Oil consumption (thousand barrels per day, 2013)	91,195	3601	6628
Coal consumption (million short tonnes, 2011)	8186	221	50
Natural gas consumption (billion cubic feet, 2013)	121,357	4573	4698
Uranium production (tonnes, 2013)	59,673	10,505	198
% consumption oil (2004–2013)	1.31	2.79	3.49
% consumption coal (2003–2012)	4.20	1.76	4.19
% consumption natural gas (2003–2013)	2.61	5.91	3.93
% production uranium (2003–2013)	5.30	5.87	-3.13
Reserves oil (years)	39	48	51
Reserves coal (years)	44	77	65
Reserves natural gas (years)	36	26	31
Reserves uranium (years)	36	37	–

Table 2.3 Fossil fuel consumption per capita (2014)

	World	Africa	Latin America
Consumption per capita oil (barrel/year)	4,62	1,20	3,93
Consumption per capita coal (short tonnes/year)	1,14	0,20	0,08
Consumption per capita natural gas (thousand cubic feet/year)	16,84	4,17	7,63

2.2 Carbon Emissions and Climate Change

If we consider the thermodynamic balance of a planet at a constant temperature, the amount of absorbed energy as solar radiation must equal the amount of energy emitted back to space at longer wavelengths (infrared). On Earth, re-emitted radiation reaches 239 W/m^2. According to thermodynamics, a body emitting energy with this power density would have a mean temperature of −18 °C. However, the average temperature on Earth is larger due to the presence of greenhouse gases in the atmosphere, which absorb and re-emit infrared radiation while keeping the lower atmosphere and the Earth's surface warm (Fig. 2.2) [7].

The increase in global energy consumption associated with increased economic development in recent decades is also related to the increase in annual CO_2 emission rates (Fig. 2.3) [5]. Global economic recessions related to the economic crises in 1974, 1980–82, 1990 and 2008–09 are readily apparent as small reductions in annual CO_2 emission rates (Fig. 2.3).

Carbon is emitted but also absorbed on a global scale. A global carbon budget published in the literature (Table 2.4) [8] suggests that fossil fuels and cement are increasing their shares in global CO_2 emissions while established forests are decreasing their role as CO_2 sinks (although the overall effect of deforestation is to warm the planet, replacing the trees with crops or grassland makes the ground paler and more reflective, and particles created from sulphur oxide reflects light into space [9]). Consequently, the atmosphere is increasing in its share of the carbon budget resulting in an increase in atmospheric CO_2 content.

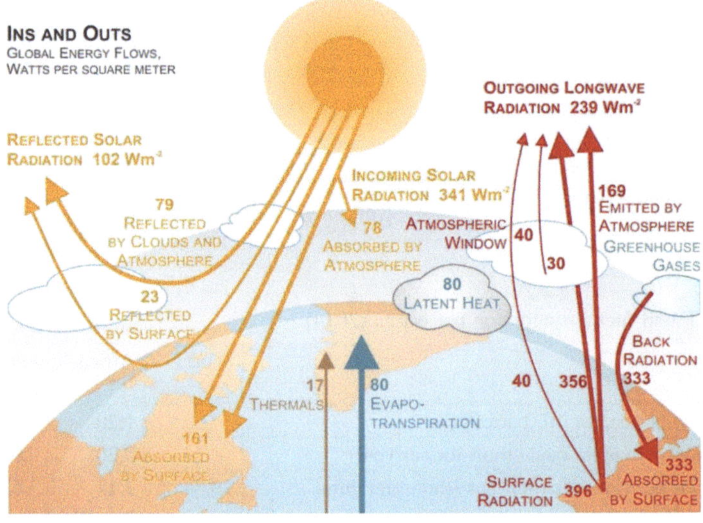

Fig. 2.2 Description of the thermodynamic balance on Earth [5]

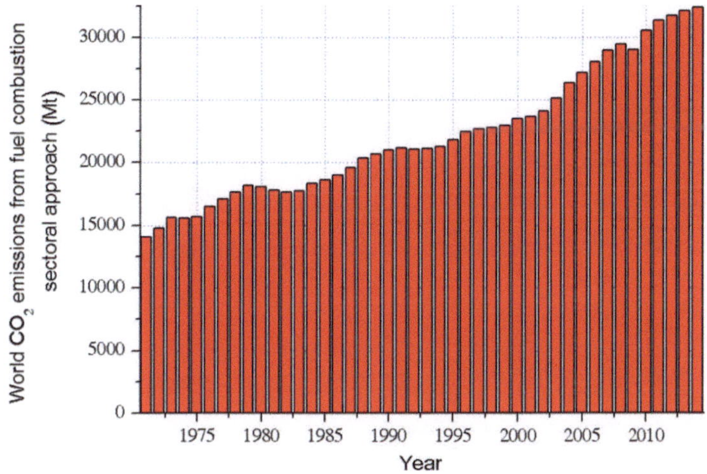

Fig. 2.3 Evolution of the world's annual CO_2 emission rates in the period 1978–2014 [3]

Table 2.4 Global carbon budget decomposed in terms of sources and sinks, and calculated for the periods 1990–1999 and 2000–2007 [8]

Pg CO_2/year	1990–1999	2000–2007
Sources (CO_2 emissions)		
Fossil fuel and cement	6.5 ± 0.4	7.6 ± 0.4
Land-use change	1.5 ± 0.7	1.1 ± 0.7
Total sources	8.0 ± 0.8	8.7 ± 0.8
Sinks (CO_2 absorption)		
Atmosphere	3.2 ± 0.1	4.1 ± 0.1
Ocean	2.2 ± 0.4	2.3 ± 0.4
Terrestrial (established forest)	2.5 ± 0.4	2.3 ± 0.5
Total sinks	7.9 ± 0.6	8.7 ± 0.7
Global residuals	0.1 ± 1.0	0.1 ± 1.0

Increasing CO_2 emissions to the atmosphere is causing the average CO_2 levels in the atmosphere to rise very significantly, from the 280 ppm in the pre-industrial era to above 400 ppm currently measured (Fig. 2.4) [10].

The increasing CO_2 levels in the atmosphere is reducing the Earth's radiation of heat into space and, consequently, producing an increase in the average global temperature [11, 12]. Under these conditions, temperature increases of around 0.15 °C per decade are estimated [12]. Global warming is not only associated with direct changes in weather conditions and subsequent food availability [7, 13], but also with an increase in extreme weather events [14, 15] and even civil conflicts [16].

Experience over the last several decades shows that the implications of projected global mean temperature changes tends to underestimate regional (and national level) impacts because the global changes are much smaller than the expected changes on average and extreme regional temperatures occur over most land areas.

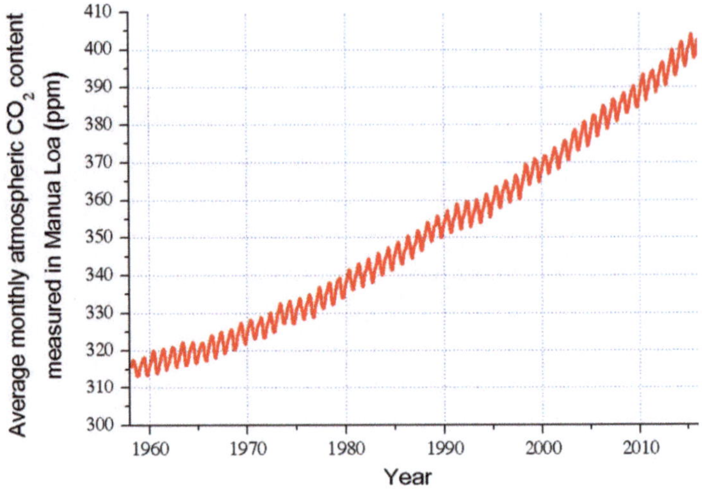

Fig. 2.4 Average monthly atmospheric CO_2 content measured at Mauna Loa Laboratory (Hawaii) [10]

Thus, models show these enhanced changes in many areas of Africa and Latin America [17].

Technological advances in the energy sector will help reduce atmospheric CO_2 emissions. For example, the replacement of coal fired power plants by natural gas combined cycle power plants (Fig. 2.5), or nuclear power plants, and carbon capture and sequestration techniques can reduce emissions. However, it is obvious that

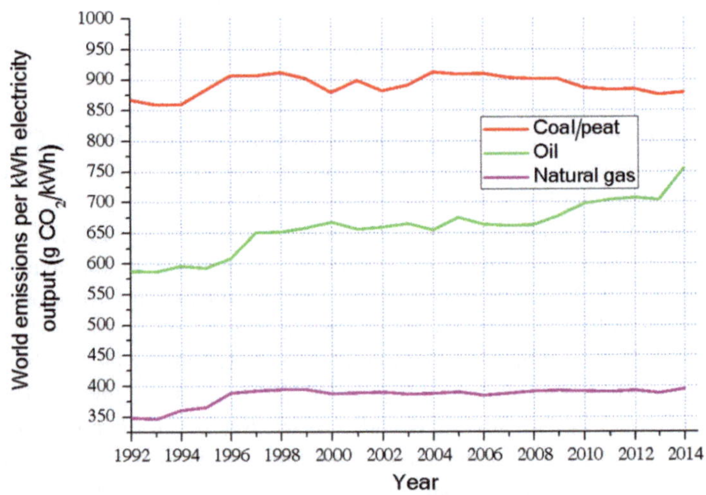

Fig. 2.5 Evolution of CO_2 emissions per kWh of electricity produced by conventional plants fired with coal, oil or natural gas [3]

significant reductions in CO_2 emission rates from the energy sector is a complex challenge and will require a systems solution that includes a combination of improving energy efficiency, replacing conventional technologies with renewables, and significant investments in infrastructure, education and outreach, among others.

The Kyoto Protocol has identified several measures to encourage large industries to reduce their CO_2 emission rates. One of the most popular measures has been the creation of a market for carbon emission permits to incentivize more efficient energy consumption and to improve the technologies used in industrial processes. However, the price evolution of the emission permits in the most important market, the EU Emission Trading Scheme, has been below initial expectations (Fig. 2.6) [18], generating very little incentive to reduce CO_2 emissions or to introduce new technologies to capture and store CO_2. However, some estimates indicate that this market has facilitated the reduction in CO_2 emissions by a few percentage points per year compared to a scenario without any permit market [19].

A new policy recently approved to reduce global carbon emissions is derived from the Paris Climate Change Conference on November 2015. After this summit, most African and Latin American countries have submitted their Intended Nationally Determined Contribution (INDC) to cut global GHG emissions by 2030 to the United Nations (Table 2.5) [20]. However, most targets are directly or indirectly tied to foreign support. For less developed countries, limited confidence exists in the data available to support key decisions on the policies to be implemented.

South Africa is currently deliberating the introduction of a carbon tax which has become quite controversial because of the large financial implications for the national utility (Eskom) and the mining sector [21]. Conversely, Zambia has imposed a carbon tax on all motor vehicles since 2006 [21].

Fig. 2.6 Evolution of the daily CO_2 emission permits price in the EU Emission Trading Scheme [18]

Table 2.5 Intended national determined contributions in Africa and Latin America to cut global GHG emissions by 2030 [20]

Country	%	Reference
Angola	35	BAU_{2005}
Benin	3.5	2016
Botswana	15	2010
Burkina Faso	6	BAU_{2030}
Burundi	3	BAU_{2030}
Cameroon	35 (C)	2010
Cabo Verde	20 (C)	BAU_{2030}
CAR	5 (C)	BAU_{2030}
Chad	18.2	BAU_{2030}
Congo	48 $(C)_{2025}$	BAU_{2025}
Côte d'Ivoire	28	2012
DR Congo	17 (C)	2000
Djibouti	40	BAU_{2030}
Eq. Guinea	20 (C)	2010
Eritrea	39.2	BAU_{2030}
Ethiopia	64	BAU_{2030}
Gabon	50	BAU_{2025}
Gambia	45.4	BAU_{2030}
Ghana	15	BAU_{2030}
Guinea	13 (C)	1994
Kenya	30 (C)	BAU_{2030}
Lesotho	10	BAU_{2030}
Liberia	15	BAU_{2030}
Madagascar	14 (C)	BAU_{2030}
Malawi	30 (C)	2015
Mali	31.6 (C)	BAU_{2030}
Mauritania	2.7	BAU_{2030}
Morocco	32 (C)	BAU_{2030}
Namibia	89 (C)	BAU_{2030}
Niger	3.5	BAU_{2030}
Nigeria	20	BAU_{2030}
Tanzania	10–20 (C)	BAU_{2030}
Togo	11.14	2010
Tunisia	41 (C)	2010
Uganda	22	BAU_{2030}
Zambia	47 (C)	BAU_{2030}
Zimbabwe	33 (C)	BAU_{2030}
Argentina	15	BAU_{2030}
Brazil	43	2005
Chile	30	2007
Colombia	20	BAU_{2030}

(continued)

Table 2.5 (continued)

Country	%	Reference
Costa Rica	25	2012
Dominican R	25 (C)	BAU_{2030}
Ecuador	20.4 (C)	BAU_{2030}
Guatemala	11.2	2005
Haiti	5	BAU_{2030}
Honduras	15 (C)	BAU_{2030}
Mexico	25	BAU_{2030}
Paraguay	10	BAU_{2030}
Peru	20	BAU_{2030}
Venezuela	20 (C)	BAU_{2030}

BAU Business as usual; *C* Conditional to international support

2.3 Low Carbon Development Concept

The *low carbon development* concept is an explicit strategy to increase GDP per capita while simultaneously decreasing the carbon emissions related to the consumption of energy. A comparison of GDP per capita and CO_2 emissions per capita [1] for the different countries in Africa and Latin America (Figs. 2.7 and 2.8) shows a logarithmic relationship between GDP and CO_2 emissions. For most countries since 2000, as GDP increases, CO_2 emissions increase. However, some countries have been capable of increasing their GDP per capita and reducing their CO_2

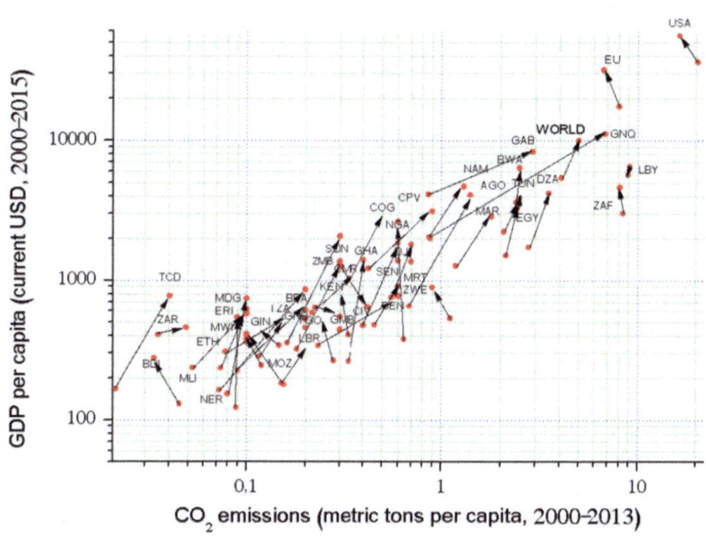

Fig. 2.7 GDP per capita (2000–2015) versus CO_2 emissions per capita (2000–2013) in the African countries from 2000 to 2013 [1]

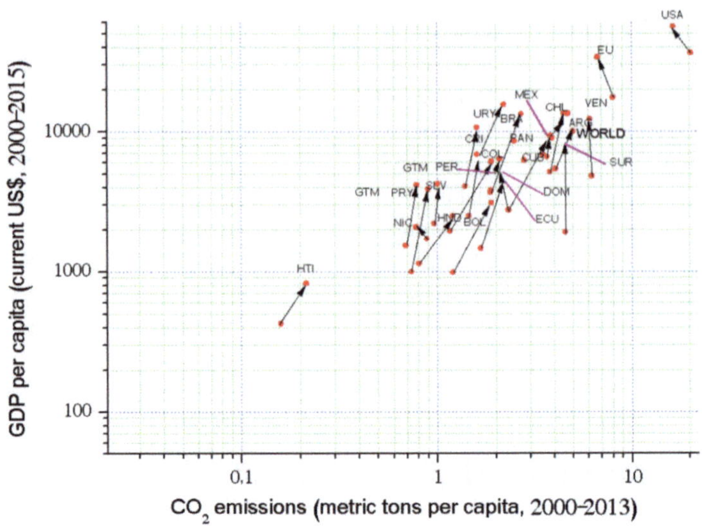

Fig. 2.8 GDP per capita (2000–2015) versus CO_2 emissions per capita (2000–2013) for Latin American countries from 2000 to 2013 [1]

emissions (e.g., Burundi, Eritrea and Zimbabwe in Africa and the Dominican Republic, Suriname and Venezuela in Latin America), in part because their economies are very limited.

In contrast to the most common evolution of GDP per capita and CO_2 emissions per capita in Africa and Latin America, the trends in the European Union (EU) and USA show low carbon development dynamics. Thus, it is possible to grow in terms of GDP per capita in parallel to a lower consumption of fossil fuels for countries showing high GDP values. This behaviour is encouraging not only because it means that increases in living standards can be reached without increasing global CO_2 levels, but also because this growth is mostly based on indigenous energy resources. Africa and Latin America, both regions with under average energy consumption per capita (Table 2.3), have a huge opportunity to increase their GDP mainly based on the penetration of low carbon indigenous energy technologies.

The principal environmental impacts associated with the development of the energy sector in emerging nations include increasing CO_2 emissions, water consumption and land use. Since the impacts from each technology are affected differently by external constraints (e.g., location, energy mix, variations in the prices of raw materials, etc.) there are not well established methodologies for calculating specific impacts. However, general guidance is available for most renewable technologies that can be used to estimate relative environmental impacts.

The production of electricity with any renewable technology produces CO_2 emissions when the entire technology life cycle is considered (Fig. 2.9) [22]. The manufacture of renewable energy technologies requires energy, and today this energy is usually supplied by a power grid where fossil fuels play a major role.

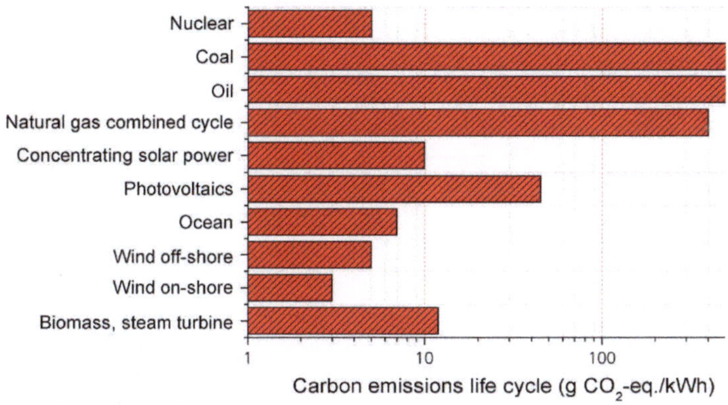

Fig. 2.9 CO_2 emissions per kWh of electricity produced from different renewable energy technologies considering the full lifecycle of these systems [22]

Many of the materials that integrate the renewable energy technologies into power plants (cement, steel, aluminum, etc.) are also produced with energy supplied from power grids or heating systems where fossil fuels play an important role. In the long term, life cycle CO_2 emissions will decrease as the contribution of renewable energy to total energy consumption increases in those regions where the manufacturing, construction, operations, maintenance and disposal of renewable power plants become increasingly free of carbon emissions.

The largest CO_2 emitters per kWh produced are coal-fired power plants, well above the emissions from renewable technologies. Recent studies also suggest that crude-oil extracted from oil sands by new procedures increases air pollution with respect to established estimates as observed in the crude extracted from the Canadian oil sands [23].

Another important environmental impact is water consumption per kWh produced by different energy technologies. Water production can require large amounts of energy in many locations for water transport, distribution and treatment. As observed in Fig. 2.10 [24], water consumption for wind energy and photovoltaics is almost negligible but is substantial for other renewable technologies, mainly hydropower and geothermal. Conventional power plants are also important water consumers, with coal fired power plants the largest water consumer. This result, added to the fact that coal fired power plants are the larger carbon emitters of all power plant technologies (Fig. 2.10), makes coal the leading technology in terms of environmental impact. Recent studies conclude that the energy return from water invested for the most water-efficient fossil fuel technology is one or two orders of magnitude greater than the most water-efficient biomass technologies [25]. Hence, the development of biomass energy technologies could produce or exacerbate water shortages around the globe, and this effect should be further analysed.

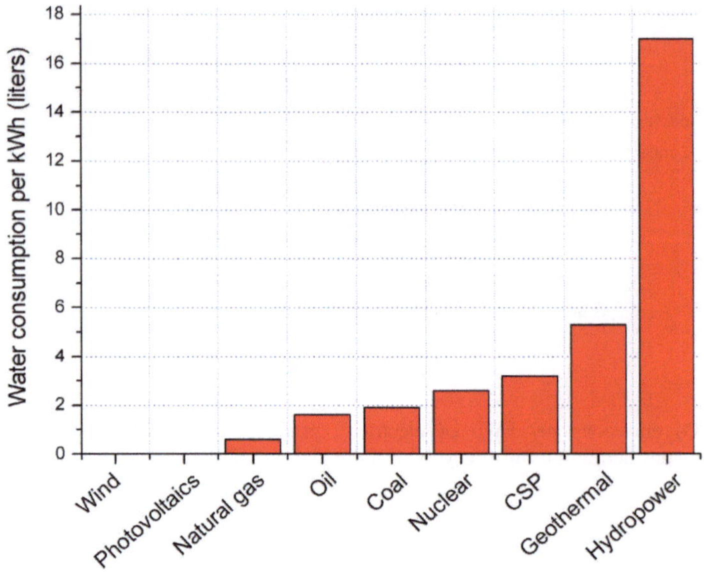

Fig. 2.10 Water consumption per kWh for different conventional and renewable energy technologies [26]

Finally, land utilization is also an important environmental parameter to be considered. It is important to recognize that there exists a significant gap in terms of power density per unit area between fossil fuels and renewable technology (Fig. 2.11) [27]. A coal mine or oil field, for instance, yields five to 50 times more

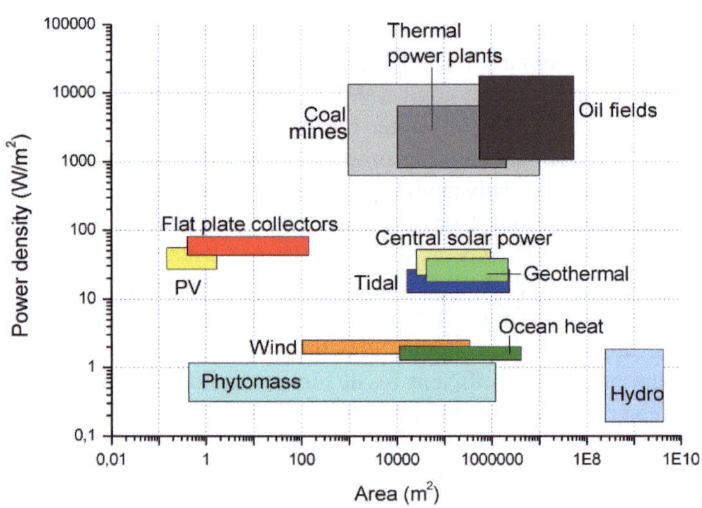

Fig. 2.11 Power density per unit area for different conventional and renewable energy resources and technologies [27]

power per square meter than a solar facility, 10–100 times more than a wind farm, and 100–1000 times more than a biomass plant. Even if the energy needed to extract, transport, and process coal is considered, it still yields 50 times more energy per unit of land than ethanol from corn and 10 times more than ethanol from sugarcane.

To achieve a low carbon energy development process it is necessary to move from coal and oil based energy utilities to lower carbon utilities that rely on natural gas, nuclear energy and renewables. Other indirect impacts and cost, mainly related to the use of water and land occupancy, should be also considered.

2.4 Main Country Indicators

The disparities between African and Latin American countries is not only evident in GDP and CO_2 emissions per capita (Figs. 2.6 and 2.7) but also in terms of population (2015), surface area and density of inhabitants per km^2 (Table 2.6) [1].

Another important issue for consideration is the ease of doing business in Africa and Latin America, particularly when foreign investment and technology is required for low carbon development. The World Bank every year publishes a report called Doing Business that can be considered the most important and influential tool for first approximation in investment decisions (Table 2.7) [28]. In 2015 the World Bank analysed 189 countries for ease of doing business and found that many of the most difficult countries to do business with are located in Africa. Some Latin American countries are also also included as very difficult countries in terms of doing business on the World Bank list.

In addition to the relationships highlighted above, there is also an interesting logarithmic correlation between the ease of doing business and electricity consumption per capita (Fig. 2.12) for the countries considered by the IEA statistics [3] (does not include countries involved in internal conflicts). We can conclude from these relationships that entrepreneurial activities increase as the electrification of a country increases. Chapter 3 provides more detail on this relationship.

2.5 Carbon Capture and Storage Systems

2.5.1 Overview

Carbon capture and storage (CCS) is being proposed as a key technology to avoid CO_2 emissions to the atmosphere from fossil fuel combustion related to electricity power plants. According to the latest available statistics (for the year 2014 [29]), 23,816 TWh of electricity production worldwide, 40.8% is obtained from coal,

Table 2.6 Ranking of African and Latin American countries in terms of population. Surface (S) area and population density (D) also included [3]

Country Name	Population (×1000)	S: km^2 (×1000)	D: h/km^{-2}
Africa	1,183,919	29,823	37
Nigeria	182,201	924	193
Ethiopia	99,391	1104	87
Egypt, Arab Rep.	91,508	1001	83
Congo, Dem. Rep.	77,266	2345	30
South Africa	54,957	1219	44
Tanzania	53,470	947	54
Kenya	46,050	580	78
Sudan	40,234	1879	21
Algeria	39,667	2382	17
Uganda	39,032	242	161
Morocco	34,378	447	75
Mozambique	27,978	799	33
Ghana	27,409	239	111
Angola	25,022	1247	18
Madagascar	24,235	587	40
Cameroon	23,344	475	48
Cote d'Ivoire	22,702	322	65
Niger	19,899	1267	15
Burkina Faso	18,106	274	64
Mali	17,600	1240	13
Malawi	17,205	118	142
Zambia	16,212	753	20
Zimbabwe	15,603	391	37
Senegal	15,129	197	74
Chad	14,037	1284	10
Guinea	12,609	246	49
South Sudan	12,340	644	18
Rwanda	11,610	26	459
Tunisia	11,108	164	67
Burundi	11,179	28	377
Benin	10,880	115	92
Somalia	10,787	638	17
Togo	7305	57	123
Eritrea	6536	118	56
Libya	6278	1760	4
Central African Republic	4900	623	8
Congo, Rep.	4.559	342	13
Liberia	4.397	111	39

(continued)

Table 2.6 (continued)

Country Name	Population (×1000)	S: km^2 (×1000)	D: h/km^{-2}
Mauritania	3.984	1031	4
Namibia	2.348	824	3
Lesotho	2.098	30	69
Botswana	2.039	582	4
Gambia, The	1.909	11	169
Guinea-Bissau	1.746	36	48
Gabon	1.711	268	6
Swaziland	1.268	7	73
Mauritius	1.261	2	618
Djibouti	886	3	38
Equatorial Guinea	778	8	28
Cabo Verde	504	4	125
Sao Tome and Principe	198	1	206
Latin America	591,564	20,203	30
Brazil	207,847	8516	24
Mexico	127,017	1964	63
Colombia	48,229	1141	43
Argentina	43,417	2780	15
Peru	31,377	1.285	24
Venezuela, RB	31,108	912	34
Chile	17,948	756	24
Guatemala	16,343	108	146
Ecuador	16,144	256	62
Cuba	11,390	109	102
Bolivia	10,725	1.098	10
Haiti	10,711	27	377
Dominican Republic	10,528	48	216
Honduras	8075	112	73
Paraguay	6639	407	17
El Salvador	6127	21	303
Nicaragua	6082	130	47
Costa Rica	4807	51	97
Panama	3929	75	52
Uruguay	3432	176	19
Guyana	767	214	4

21.6% from natural gas and 4.3% from oil, while the remaining electricity is derived from from nuclear, hydro and renewables. CO_2 can be captured and stored from other centralized emission sources, mainly in the fuel transformation sector and from plants located in the industrial sector.

Table 2.7 Easiness of doing business ranking in African and Latin American countries in 2015 [29]

Africa	No.	Latin America	No.
Mauritius	32	Mexico	38
Rwanda	62	Chile	48
Morocco	68	Peru	50
Botswana	72	Colombia	54
South Africa	73	Costa Rica	58
Tunisia	74	Panama	69
Zambia	97	Guatemala	81
Namibia	101	El Salvador	86
Swaziland	105	Uruguay	92
Kenya	108	Dominican Republic	93
Ghana	114	Paraguay	100
Lesotho	114	Honduras	110
Uganda	122	Brazil	116
Cabo Verde	126	Ecuador	117
Egypt, Arab Rep.	131	Argentina	121
Mozambique	133	Nicaragua	125
Tanzania	139	Suriname	156
Malawi	141	Bolivia	157
Cote d'Ivoire	142	Haiti	182
Mali	143	Venezuela, RB	186
Burkina Faso	143		
Ethiopia	146		
Togo	150		
Gambia, The	151		
Burundi	152		
Senegal	153		
Zimbabwe	155		
Benin	158		
Sudan	159		
Niger	160		
Gabon	162		
Algeria	163		
Madagascar	164		
Guinea	165		
Sao Tome and Principe	166		
Mauritania	168		
Nigeria	169		
Djibouti	171		
Cameroon	172		
Congo, Rep.	176		

(continued)

Table 2.7 (continued)

Africa	No.	Latin America	No.
Guinea-Bissau	178		
Liberia	179		
Equatorial Guinea	180		
Angola	181		
Chad	183		
Congo, Dem. Rep.	184		
Central African Republic	185		
South Sudan	187		
Libya	187		
Eritrea	188		

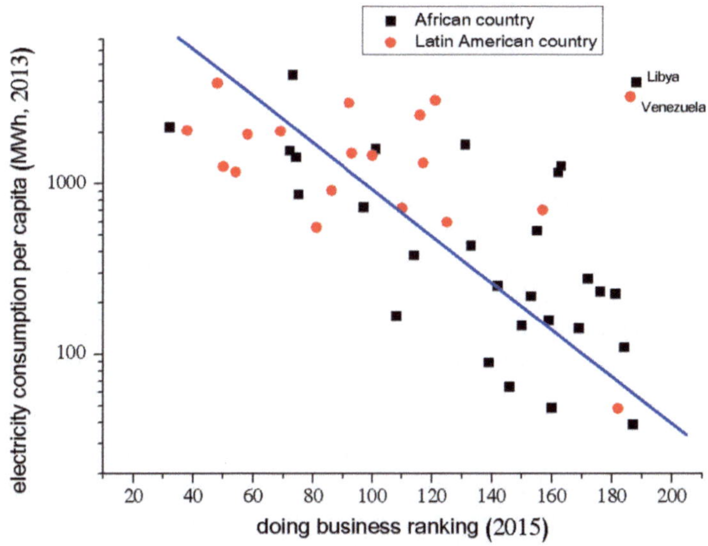

Fig. 2.12 Electricity consumption versus easiness of doing business in African and Latin American countries [3, 28]

The capture of carbon dioxide from centralized sources involves several steps:

(a) Capture of the carbon dioxide emitted from power plants, industrial plants, etc., including the transformation and compression of the emitted gases.
(b) Transportation of the CO_2 gas to a sink by means of pipes, vessels, appropriate vehicles, etc.
(c) Injection of the CO_2 in underground geological formations, such as oil or gas deposits that are already empty, stored in the oceans, etc.

In 1972, the first large-scale integrated project (LSIP) in Val Verde (Texas, USA) started operation. 15 large installations are currently operating worldwide,

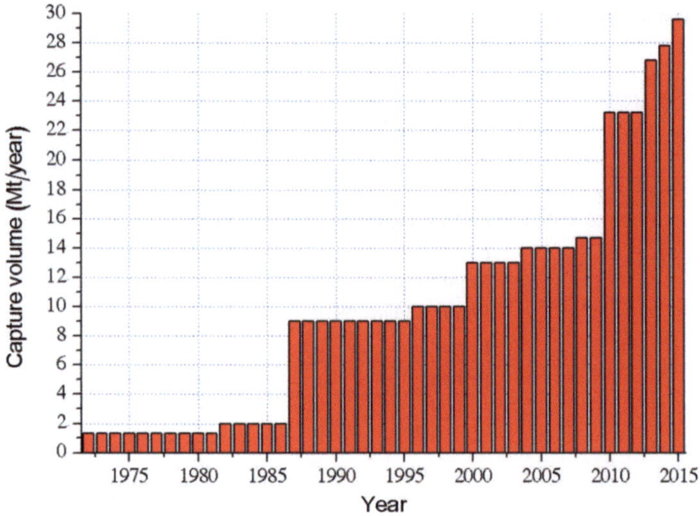

Fig. 2.13 Evolution of the world capture capacity for large CCS plants in operation (1972–2015) [30]

with most of them dedicated to enhanced oil recovery but also injecting into geological formations at different depths [30]. Ten of these plants operate with pre-combustion carbon capture technology, one with post-combustion carbon capture technology (explained below) and four with industrial separation processes, capturing about 30 Mt of carbon dioxide per year (Fig. 2.13).

The total CO_2 capture volume of all 15 projects in operation is about 0.1% of the annual CO_2 emissions from fuel combustion worldwide (last data available) [3]. However, between 80 and 120 Mt of CO_2 is sold commercially each year for a wide variety of applications (mostly for enhanced oil recovery, but also for chemical solvents, for decaffeination of coffee, carbonation of soft drinks, manufacture of fertilisers, refrigerants, plastics, etc.) and other applications can be valuable in future (enhanced algae cultivation, CO_2 concrete curing and bauxite residue carbonization) [26].

LSIPs are defined as those projects which involve the capture, transport and storage of CO_2 at a scale of: (i) not less than 800,000 tonnes of CO_2 annually for a coal-based power plant; and (ii) not less than 400,000 tonnes of CO_2 annually for other emission-intensive industrial facilities (including natural gas-based power generation).

2.5.2 *State of the Art*

There are four main technologies for CO_2 capture: (i) post-combustion capture, (ii) pre-combustion capture, (iii) oxy-fuelling, and (iv) chemical looping. The specific application of these technologies will vary depending on local and regional climatic conditions, specific fuel types and local variations.

The post-combustion capture of CO_2 from the exhaust gases of the combustion process, which has a CO_2 content between 2% and 25% [31], is usually performed at low pressure. The three main post-combustion processes are: (i) absorption, by the uptake of CO_2 into the bulk phase of another material (for example, dissolving CO_2 molecules into a liquid solution), the solvent is conducted to another compartment and heated to release the CO_2; (ii) adsorption, involving the selective uptake of CO_2 onto a solid surface surface, which is subsequently regenerated by lowering the pressure and/or increasing temperature to liberate the absorbed CO_2 (a claimed advantage of adsorption is that the regeneration energy should be lower relative to that of the absorption solvents); and (iii) membranes, by selectively permeating the flue gas through the membrane material (claiming low energy capture processes).

In the pre-combustion process the fuel is previously gasified producing a synthesis gas (syngas: a mixture of H_2 and CO). CO_2 removal from coal gasification syngas is a mature commercial process widely practiced throughout the world, mainly for removing CO_2 from natural gas to meet the purity standards. Then, the gas is cleaned from impurities that could damage the turbines, and the CO is mixed with steam to react and produce carbon dioxide and hydrogen. In this mix, the CO_2 is separated by an absorption process and prepared for transport and storage. The remaining gas, rich in oxygen, is mixed with additional oxygen and used to power a hydrogen gas turbine, which only produces water vapour as the exhaust. This turbine cycle with hydrogen is thermally more efficient for electricity production than the current ones [8]. Current commercially available pre-combustion CO_2 capture processes are based on the use of physical or chemical solvents. Typically all the solvents can accomplish greater than 90% CO_2 removal, under lower pressure than current post-combustion technology. Currently, there are a number of processes under development to demonstrate the feasibility of using pre-combustion in IGCC technology ("integrated gasification combined cycle") from coal and natural gas.

In the oxy-fuelling combustion process, fuel is burned in a pure oxygen environment instead of air, thereby avoiding the need of CO_2 separation in the exhaust gases as the content is up to about 90% [3]. If permitted by regulations, the raw, dehydrated flue gas may be stored directly without the need for further purification. Otherwise, the flue gas impurities (predominantly oxygen, nitrogen and argon) may be removed. One advantage of this technique is to facilitate the retrofitting of existing facilities, adding the ability to capture CO_2. The first electricity production plant using this technology became operational at Schwarze Pumpe in 2008. Oxyfuel combustion may be employed with solid fuels such as coal, petroleum coke, and biomass, as well as liquid and gaseous fuels.

The chemical looping combustion is still being developed to be applied in CCS, but its future looks promising, since it would consume much less energy than post-combustion technology and consequently improve the efficiency of the power plant compared to oxy-fuelling technologies. The chemical looping technique uses tiny particles of metal oxides as oxygen carriers, which make a loop between the combustion reactor and an air reactor. Thus, by reacting with fluids in a fluidized bed reactor (combustion reactor), the metal oxide particles introduce oxygen and again become metal. The oxygen and the fuel produce a mixture of CO_2 and water vapour in the reactor, which, after water condensation, the gas released is composed of nearly pure CO_2 that can be subsequently sequestered. Thus, the combustion takes place without any contact of fuel with air and, consequently, the exhaust contains no nitrogen content, making it easier to capture CO_2. The metal particles produced are transported to the other fluidized bed reactor where they react with air (air reactor) and re-oxidize, also releasing heat that can be used to increase the overall efficiency.

Transportation of CO_2 occurs largely in pipelines. There exists nearly 6,000 km of pipelines for pumping 50 Mtpa CO_2 worldwide, mainly in U.S. For geological storage under the seabed, the technology is more complex, but a plant with these characteristics is already operating (Snøhvit, Norway). In this facility, there are multiple lines originating from different CO_2 emissions plants and converging into a single larger pipe on the coast, where it is pressurized and connected to the geological location in the ocean where the CO_2 is stored.

Finally, geological storage of CO_2 can be produced by pumping the gas underground using several procedures, as illustrated in Fig. 2.14. One approach (process 1) is to store CO_2 in previous deposits of oil or gas that are already empty. CO_2 can also be used in so-called enhanced oil recovery. For this procedure, the injected CO_2 allows the extraction of oil not directly extracted by impulsion (process 2). CO_2 can also be injected in saturated sedimentary rock deposits of salt water (process 3). These rocks consist of alternating layers of water, silt, clay, etc. The CO_2 is trapped in the pores of the sand layers, in the capillaries of the clay layers, adsorbed on carbonaceous rocks, etc. It is also planned to store CO_2 in non-extractable coal veins (process 4) or to extract natural gas from these coal veins (process 5). Finally, there is the option for storage in basalt rocks, tar sands or in underground cavities (process 6). The storage of CO_2 in the subsurface of the ocean, usually at depths of 700 m or more, is also being considered.

In general, the technologies for CO_2 storage in deep geological formations are already highly developed, since they have been transferred from the exploration and extraction technologies for oil and natural gas. On the other hand, a storage capacity of 1,000–10,000 Gt CO_2 in salt deposits has been estimated, 600–1,200 Gt CO_2 in oil and gas fields, and 3–200 Gt CO_2 in coal veins to extract methane [32]. These magnitudes can be compared with 32.4 Gt CO_2 (Fig. 2.3) emitted by combustion of coal, oil and gas in 2014 (latest figures available) [3].

There is a significant effort in monitoring stored CO_2, using various technologies: (i) atmospheric techniques that utilize an infrared laser that is strongly absorbed by CO_2, so that the signal loss determines the CO_2 concentration; (ii) various near subsurface techniques that include geochemical analysis at shallow subsurface depths, measurement of CO_2 concentration in situ where gas is extracted and introduced into a chamber and topographic measurements of the electrical conductivity of materials below the surface; and (iii) deep subsurface techniques (in particular, geochemical)

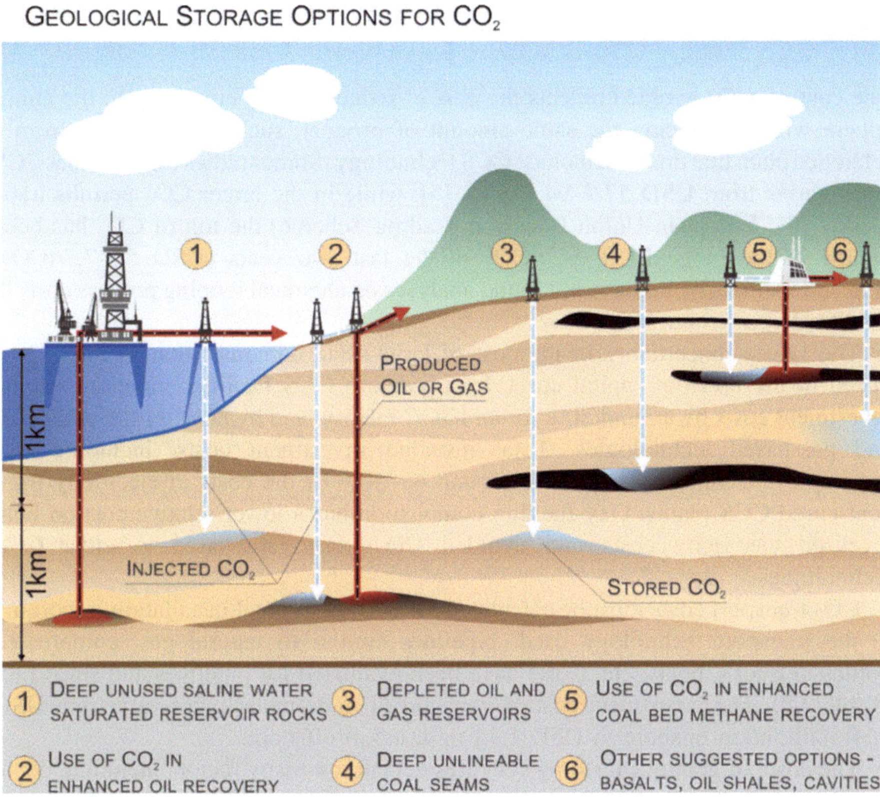

Fig. 2.14 Scheme of the different procedures for geological storage of CO_2 in the earth [10]

quite similar to those used in oil and natural gas fields, in which CO_2 migration can be studied using chemical tracers or measuring the variation of the signals from time-delay seismic analysis.

The IEA [33] published a study that shows the evolution of the thermal efficiency of various power plants based on coal and natural gas using different CCS technologies. These studies found efficiency losses ranging from 6.0 to 10.9%. In post-combustion technology, these efficiency losses are mainly due to the need to use steam to regenerate the solvent, and purification and compression processes of CO_2. In pre-combustion technology, the efficiency losses are attributed to the exothermic reactions losses that result from fuel conversion in the process prior to combustion. In oxy-fuelling combustion technology, the main efficiency losses are attributed to the need to feed the oxygen production unit. In chemical looping technology studies of these characteristics have not been currently detected.

2.5.3 Costs

The cost of CO_2 avoided reflects the cost of reducing CO_2 emissions to the atmosphere while producing the same amount of product, such as electricity, from a reference plant that does not include CCS technology. Some studies estimate that CCS costs varies from USD 57.9–94.0/tCO_2 [5], while in the larger CO_2 permits trade market (the European Union Emission Trading Scheme) the ton of CO_2 has been valued in the range EUR 5–7/tCO_2 in the last two years (USD 5.5-7.7/tCO_2) (Fig. 2.15). Finally, we have not found analyses of chemical looping process costs in the literature.

The largest uncertainty in the cost of large-scale demonstration plants is produced in the up-front capital costs. Incorporating CCS facilities increases capital investment costs by around 30% for an IGCC facility, and by 80% for the other coal and gas-based technologies. Total installed investment costs, including carbon capture technology, represent around 45–50% of the costs of electricity from coal-based CCS plants. Oxy-fuelling combustion has a lower relative cost on both levelised electricity costs and avoided CO_2 costs, compared to other CCS technologies.

CO_2 transport costs must be calculated for each individual installation, according to the transport technology used: pipelines similar to natural gas, compressed bottles in trucks, boats, etc. Costs have been evaluated for pipelines and ships [30] (Table 2.8). The investment cost of transport (1 m diameter pipe) vary from USD 0.80 million/km onshore to USD 1.34 million/km offshore.

The costs of geological storage cavities depend on many factors including depth below ground, the location (usually the location "offshore" is more expensive, etc.)

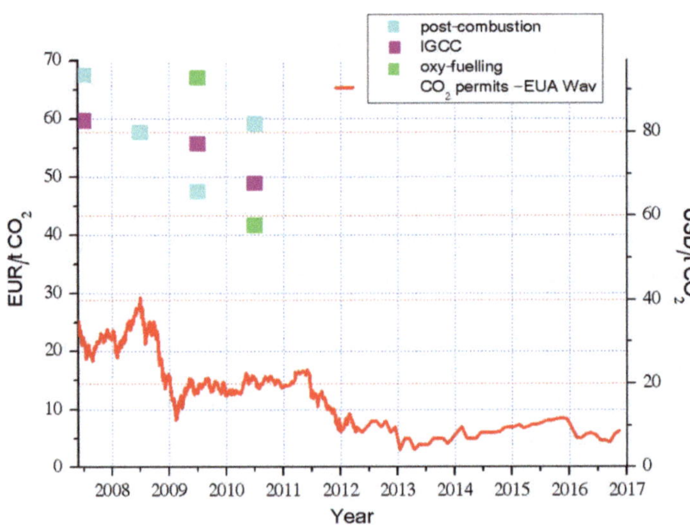

Fig. 2.15 CCS costs per ton of CO_2 for different technologies and evolution of the European Union Emission Trading Scheme for CO_2 permits

Table 2.8 Transport cost estimates for large-scale networks of 20 million tons per annum [30]

USD/t	180 km	500 km	750 km	1500 km
Onshore pipe	2.1	5.2	7.4	n/a
Offshore pipe	4.7	8.4	11.4	22.7
Ship (liq. incl.)	15.5	17.0	18.4	22.4

(Table 2.9). The CO_2 storage costs are highly variable, between USD 2.1–32.1/tCO_2 and can typically represent between 5 and 10% of the total cost of CCS [30]. Some estimations conclude that storage cost varies between USD 4.2–19.5/t CO_2 (Table 2.9).

The total CCS costs could be financially compensated in full and even produce an income if it is focused to value added applications. In this sense, CCS can be beneficial from an economic point of view for enhanced oil recovery (EOR). Thus, EOR activities have an estimated cost of 38–43 USD/tCO_2, and may recover a 5–23% more oil from oil fields [31], depending on reservoir conditions and oil—CO_2 miscibility. Also, other activities can add value to CO_2 capture, such as the manufacture of chemicals, especially synthetic gases.

The CCS costs can be translated to an increase of electricity costs. Thus, the additional costs due to CCS in electricity production range from USD 0.023–0.085/kWh, which would represent an increase of the electricity bill up to USD 8.5 cents per kWh with respect to current prices [3, 34].

Future CO_2 emission costs will depend on which technologies are more effective to capture, transport and storage of CO_2, the size of the CCS market, fuel prices, etc. The IEA estimates [35] that by 2030 a cost of USD 37.5/tCO_2 for CCS technologies, and the additional costs of electricity varying between USD 0.011–0.032/kWh. From these results, the International Energy Agency believes that it will be necessary to introduce an incentive of 50 USD/tCO_2 to facilitate the introduction of CCS in the energy sector [35]. However, as it has been exposed above, the evolution of prices for CO_2 emission permits in the European Climate Exchange market shows that the technology is still far from being attractive enough to be applied in power generation.

Thus, if we consider the data published by the IEA on CO_2 emissions per kWh in electricity production from natural gas, oil and coal power plants and they are adjusted to the daily price of the future of CO_2 emission rights [18] (Fig. 2.16), we find that, after the severe drop in the prices of those rights in the second half of 2008, prices do not cover the lowest costs estimates for CCS, including coal plants.

Table 2.9 Cost estimates for different storage cavities

USD/t CO_2	On-shore	Off-shore
Depleted oil and gas fields (legacy wells)	4.2	8.4
Depleted oil and gas fields (no legacy wells)	5.6	13.9
Deep saline formation	7.0	19.5

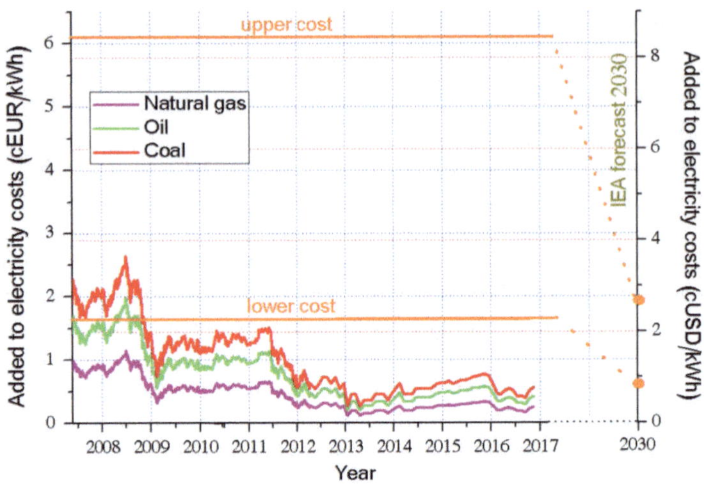

Fig. 2.16 Evolution of the additional costs to electricity production from coal, oil and natural gas, considering the variations in CO_2 emission permits in the European Union Emission Trading Scheme, and compared with current estimates for future costs of CCS exposed by the IEA

Also, if the current market price of future CO_2 permits remains relatively constant, it will be even more profitable in 2030 to purchase permits than to incorporate CCS technology into power plants.

Carbon taxes and emission trading schemes (ETS) have been implemented or scheduled for implementation in about 40 countries, mostly in the European Union, and 20 cities, mostly in North America and and East Asia, covering about a 12% of global emissions [26]. The combined value of carbon pricing instruments in 2015 was estimated at just under USD50 billion globally, of which almost 70% is attributed to ETSs and the reminder to carbon taxes. The existing carbon prices vary significatively, from less than USD1/tCO_2 in Mexico and Poland to USD130/tCO_2 in Sweden, but 85% of emissions covered are priced at <USD10/CO_2 [26]. In Africa, only South Africa is discussing a carbon tax. In Latin America, only the Chilean Parliament has approved (Sept 2014) the implementation of a national carbon tax that will put a price on emissions (equivalent to about USD 5/tCO_2eq) from 2017 onward, and applied to stationary sources with a thermal input capacity greater than 50 MW. Also the Mexican Ministry of Energy announced in February 2014 the possible development of an ETS in the energy sector, and Brazil is considering a ETS or carbon tax.

2.5.4 Carbon Emissions from CCS Based Power Plants

The IEA [8] published a study that shows the amount of CO_2 emitted, captured or avoided by power plants with different technologies for electricity production.

Therefore, adding CCS increases CO_2 emissions per kWh in power plants, since the thermal efficiency for these plants decreases (except in some cases where replaced technology is outdated and inefficient). Pre-combustion CCS technologies and post-combustion capture about 85–90% of emitted CO_2, while oxy-fuelling plants capture rate is about 90–97%. Also consider that each technology captures CO_2 with diverse purity, and variations in purity affect the value of the CO_2 produced.

Future technological trends are established according to the major challenges that CCS has for a principal role in mitigating climate change. In this respect, the biggest challenges lies in cost reductions and scaling by a factor of ten or more CCS facilities, as well as lower energy demand, especially in CO_2 capture processes.

2.6 Conclusions and Future Perspectives

Africa and Latin America possess tremendous potential for significant growth of renewable energy resources to support more than 23% of the 2015 global population. Much of this potential is derived from access to large land areas with more than 37% of the combined global surface area, the availability of abundant renewable resources (i.e., wind, water and solar), commitments to cut global emissions by 2030 and a growing recognition by some nations that low carbon development can enhance economic prosperity while reducing potential impacts on climate change. Also, Africa and Latin America produce more than 23% of the global renewable energy, with much of this energy provided by traditional biofuels and waste (36% of total global production) to support daily sustenance needs for heating and cooking rather than electricity production characteristic of nations with mature economies.

Africa and Latin America have diverse energy reserves that vary between nations but seldom achieve generation levels necessary for the long-term sustainment, or more importantly growth, of an individual national or regional economy. Current demand for energy in many nations is low (i.e., energy consumption per capita) so existing fossil fuel reserves are often adequate to meet existing needs and expected average growth in consumption. Coal, oil, natural gas and uranium are produced in both Africa and Latin America and nuclear power provides electricity to both. An energy paradox exists for Africa and Latin America driven largely by socioeconomics that suggests current energy resources are adequate to sustain the economy, but substantial energy resource development is necessary to improve regional GDP. Comparisons of GDP and CO_2 emissions per capita for the different countries in Africa and Latin America show only a few nations have been capable of increasing their GDP per capita while at the same time reducing their CO_2 emissions. Both regions have substantial opportunity to increase their GDP based on the expanded penetration of low carbon indigenous energy technologies into their infrastructure.

Three significant attributes of energy production that reflect environmental stewardship for Africa and Latin America include CO_2 emissions, water consumption and land utilization. As the relative contribution of renewable energy sources to total energy consumption increase, the life cycle CO_2 emissions produced from manufacturing, construction, O&M and disposal of renewable power plants will decrease. Water consumption per kWh for wind energy and photovoltaics is almost negligible, but can be substantial for other renewable technologies, mainly hydropower and geothermal. Land requirements may also vary depending on energy technology. Power density per surface occupied is a consideration that may reflect the impacts locally and regionally on extensive deployment of low CO_2. Other considerations that will influence future energy decisions for each African and Latin American nation include surface area, population and population density, and ease of business transactions. As the ease of transactions within the business environment improves there is likely to be greater diversification of energy sources and expansion of entrepreneurial activities.

Collectively, energy, environmental, economic and business parameters suggest that Africa and Latin America are regions with their own peculiarities, but each region is evolving to achieve higher living standards. An evolution to a low carbon energy future for these countries will require access to resources and technology and to finance and international trade. This future is possible and necessary for the global community.

References

1. World Bank statistics (2016). http://data.worldbank.org/indicator/AG.SRF.TOTL.K2. Accessed 6 Nov 2016
2. World Bank statistics (2016). http://data.worldbank.org/indicator/SP.POP.TOTL/countries?display=default. Accessed 6 Nov 2016
3. International Energy Agency database (2016) http://data.iea.org/. Accessed 6 Nov 2016
4. Nitsch F (2007) Technologische und energiewirtschaftliche Perspektiven erneuerbarer Energien. Deutsches Zentrum für Luft—und Raumfahrt
5. Energy Information Administration (2016) US Department of Energy. http://www.eia.gov/cfapps/ipdbproject/IEDIndex3.cfm?tid=5&pid=57&aid=6. Accessed 6 Nov 2016
6. World Nuclear Association (2016). http://www.world-nuclear.org/info/Nuclear-Fuel-Cycle/Uranium-Resources/Supply-of-Uranium/. Accessed 6 Nov 2016
7. Lobell DB et al (2011) Nonlinear heat effects on African maize as evidenced by historical yield trials. Nat Climate Change 1:42–45
8. Pan Y et al (2011) A large and persistent carbon sink in the world's forests. Science 333:988–993
9. Economist (2015) Supermodels. The science of climate change. Special report: climate change. Available 28 Nov 2015
10. Earth System Research Laboratory. Global Monitoring Division. National Oceanic and Atmospheric Administration. U.S. Department of Commerce. ftp://aftp.cmdl.noaa.gov/products/trends/co2/co2_mm_mlo.txt. Accessed 6 Nov 2016
11. IPCC (2008) Fourth Assessment report: climate change 2007 (AR4)
12. Met Office Hadley Centre. UK Government
13. Lobell DB, Schlenker W, Costa-Roberts J (2011) Climate trends and global crop production since 1980. Science 333:616-620

14. Barriopedro D et al (2011) The hot summer of 2010: redrawing the temperature record map of Europe. Science 332:220–224
15. Shiermeier Q et al (2011) Extreme measures. Nature 477:148–149
16. Hsiang SM, Meng KC, Cane MA (2011) Civil conflicts are associated with the global climate. Nature 476:438–441
17. Seneviratne SI, Donat MG, Pitman AJ, Knutti R, Wilby RL (2016) Allowable CO_2 emissions based on regional and impact-related climate targets. Nature 529:477–483
18. SENDECO_2. http://www.sendeco2.com/es/precios-co2. Accessed 30 Nov 2015
19. The Economist (2009) Good policy and bad. Especial report on climate change and the carbon economy. Available 3 Dec 2009
20. INDC United Nations. http://www4.unfccc.int/submissions/INDC/Submission%20Pages/submissions.aspx. Accessed 30 Nov 2015
21. Climatescope (2015) The clean energy country competitiveness index. The multilateral investment fund, UK Department of International Development, Power Africa and Bloomberg New Energy Finance
22. Krewitt W (2009) External costs of energy technologies. New energy externalities developments for sustainability. Brussels, 16–17 Feb 2009
23. McLinden C (2012) Oil-sand pollution quantified. Research highlights. Nature 483:126
24. Cho A (2010) Energy's tricky tradeoffs. Science 329:786–787
25. Mulder K, Hagens N, Fisher B (2010) Burning water: a comparative analysis of the energy return on water investment. Ambio 39:30–39
26. World Bank (2015) State and trends of carbon pricing
27. Kerr RA (2010) Do we have the energy for the next transition? Science 329:780–781
28. World Bank (2016) Doing business 2015 report. http://data.worldbank.org/indicator/IC.BUS.EASE.XQ. Accessed 30 June 2016
29. International Energy Agency (2016) Key world energy statistics
30. Global CCS Institute. http://www.globalccsinstitute.com/projects/large-scale-ccs-projects. Accessed 30 Nov 2015
31. The Economist (2010) The clouds of unknowing. Available 18 Mar 2010
32. IPCC (2005) Carbon dioxide capture and storage: Special report of the intergovernmental panel on climate change. Cambridge University Press, Cambridge, UK and New York
33. IEA Greenhouse Gas R&D Programme (2007).Capturing CO_2
34. FutureGen Alliance. www.futuregenalliance.org/. Accessed 30 Nov 2015
35. IEA (2008) Energy technology perspectives 2008, Chapter 7

Chapter 3
Current Energy Context in Africa and Latin America

Ricardo Guerrero-Lemus and Les E. Shephard

Abstract This chapter shows that energy consumption in Africa and Latin America has grown at a rate greater than the total energy consumption worldwide since 1980, consuming nearly 10% of the total global energy and ~20% of global renewable energy, largely biomass. Other significant sources of renewables include hydropower and geothermal. Both regions have diverse energy resources typically distributed irregularly within and between nations. Coal, oil and natural gas production is restricted to a few nations but is often used for electricity production across several countries. Natural gas reserves are prolific in parts of both regions which is likely to contribute to expanded electricity production in future decades particularly if investments in energy infrastructure occur as suggested. Moreover, energy infrastructure has a significant impact on energy consumption in both Latin America and Africa. However, those regions of Latin America and Africa that rely heavily on biomass typically lack a modern energy infrastructure. The presence of oil and gas pipelines connected to refineries, port facilities and production fields are indicators of trading partners internal to the regions and/or to other regions where demand may be higher (e.g., Maghreb to Europe). Electric grid integration in both Africa and Latin America is limited by variations in frequency, sub-regional power pools that lack interconnection, geographic barriers and political differences. Significant efforts in both regions are directed at improving the connectivity between nations but still less than 8% of power crosses international borders in any African region. Finally, energy efficiency and incentives to renewable energies are emerging as priority policies within several African and Latin American countries as they try to decouple economic growth and carbon emissions, and fundamentally alter the increasing trajectory of energy consumption. All these data and the analysis embedded in this chapter shows that the evolution to a low carbon energy future is nascent but on course in these regions.

Keywords Energy · Africa · Latin America · Electricity · Energy efficiency · Energy policy · Energy security · Carbon emissions · Renewables · Power consumption · Wind · Solar PV · Energy subsidies · Biomass · Hydropower · Biofuel · Interconnections · Energy demand · Geothermal · Water

R. Guerrero-Lemus and L.E. Shephard, *Low-Carbon Energy in Africa and Latin America*, Lecture Notes in Energy 38,
DOI 10.1007/978-3-319-52311-8_3

3.1 Introduction

Total primary energy consumption (TPEC) has evolved continuously from 1980 for Africa, Latin America and worldwide (Fig. 3.1) [1]. In 1980, Africa consumed 6.8 quadrillion Btu and Latin America 13.9 quadrillion Btu (2.4 and 4.9% of total global consumption, respectively). In 2014, (last data available) Africa consumed 18.2 quadrillion Btu and Latin America 35.6 quadrillion Btu (3.38 and 6.62% respectively). Hence, not only is energy consumption growing in Africa and Latin America, but it is growing at a faster rate than on the global scale.

If we consider the share of renewable electricity (Fig. 3.2) [1, 2] in 1980, Africa consumed 60.2 GWh (3.4% of the global consumption) and Latin America 223.1 GWh (12.7% of global consumption). In 2014 (last data available) Africa consumed 139.1 GWh (2.6% of global consumption) and Latin America 842.2 GWh, (15.7% of global consumption). Both regions have experienced increases in renewable electricity net consumption in absolute terms but Africa has experienced a decrease in its global share in part because of the significant growth in worldwide renewables since 2003 (Fig. 3.2).

If we consider the share of dry natural gas (Fig. 3.3) [1, 2], in 1980 Africa consumed 0.7 trillion Btu, a 1.4% of the global consumption and Latin America 2.0 trillion Btu, a 3.7%. In 2014 (last data available) Africa consumed 4.8 trillion Btu, (4.0% global consumption) and Latin America 7.0 trillion Btu (5.9%). That means not only a consolidated upward tendency that it is growing faster than in a global scale, but also a sharper increase in consumption compared to TPEC (Fig. 3.1).

Nuclear energy production in Africa (Fig. 3.4) [1, 2] was 3.9 GWh in 1984, 0.33% of the global consumption, and in Latin America nuclear power generated 2.2 GWh in 1980, 0.32% of the global consumption. In 2014 (last data available) Africa produced 13.8 GWh, and Latin America 30.8 GWh, 0.54% and 1.21% respectively of global consumption. Thus, the data shows a very small share of global nuclear electricity net generation in Africa and Latin America and this share is almost constant in Africa and slightly increasing in Latin America.

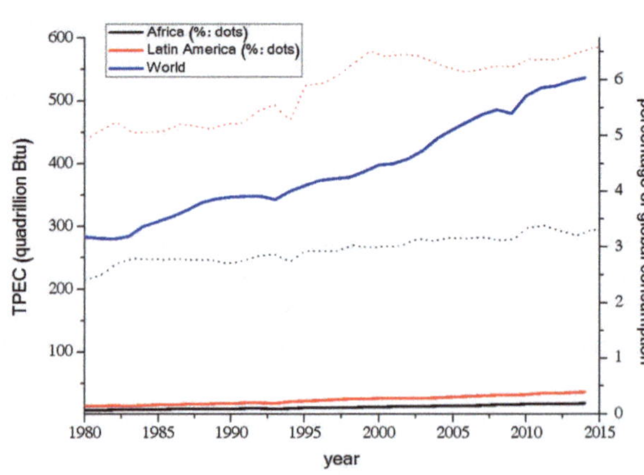

Fig. 3.1 Evolution of the total primary energy consumption (solid line) for Africa, Latin America and worldwide from 1980 to 2014 (last data available), and as a percentage of global consumption (dotted line) [1, 2]

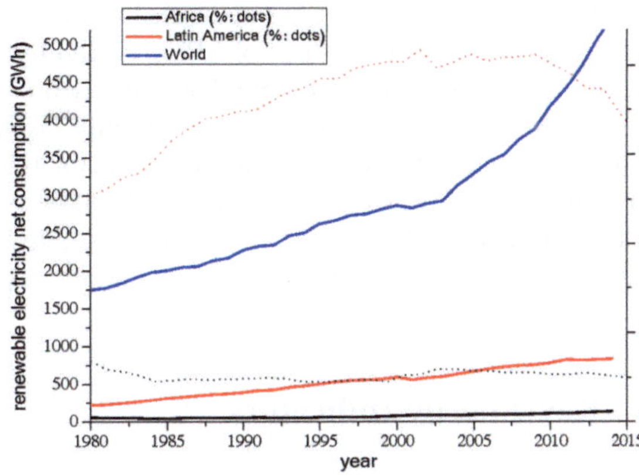

Fig. 3.2 Evolution of renewable electricity net consumption (solid line) for Africa, Latin America and worldwide from 1980 to 2014 (last data available), and as a percentage of global consumption (dotted line) [1, 2]

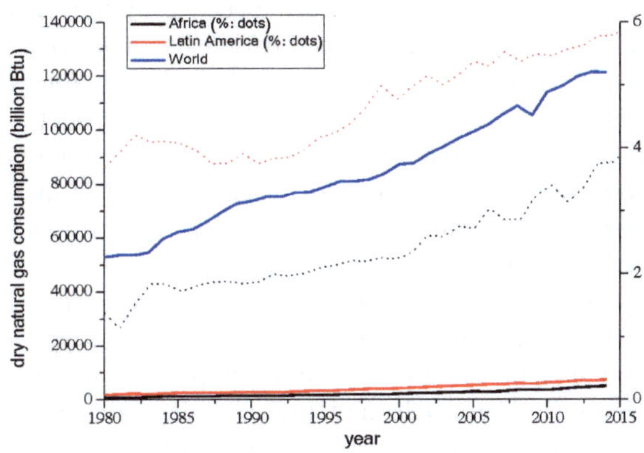

Fig. 3.3 Evolution of the dry natural gas consumption (solid line) for Africa, Latin America and worldwide from 1980 to 2014 (last data available), and as a percentage of global consumption (dotted line) [1, 2]

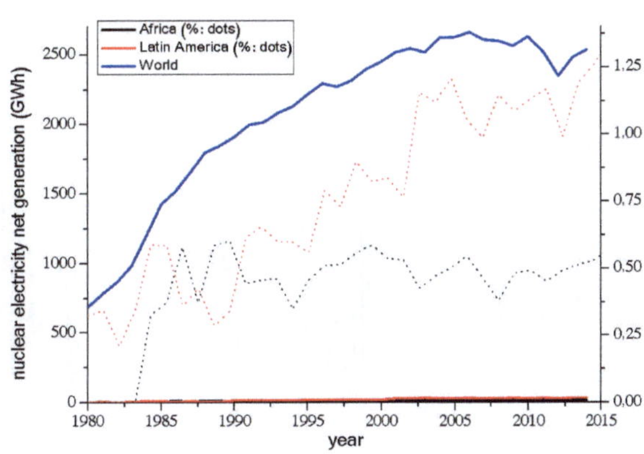

Fig. 3.4 Evolution of the nuclear electricity net generation (solid line) for Africa, Latin America and worldwide from 1980 to 2014 (last data available), and as a percentage of global consumption (dotted line) [1, 2]

Therefore, energy consumption is growing in Africa and Latin America faster than on a global scale but the growth in low carbon renewables lags behind global expansion. Natural gas consumption is growing in both regions. In Africa the global share of renewable electricity has decreased from 1980 to 2014, its share of nuclear energy is almost flat and its share of TPEC is slightly growing. The Latin American share of global renewable electricity has decreased significantly since 2010, and its share of nuclear energy and TPEC are slightly rising. These regional tendencies should be reversed if a low carbon energy sector is intended to be deployed. The following sections provide a more detailed analyses of the current energy sector and how to plan this future low carbon scenario, key energy indicators country by country, energy infrastructure and socioeconomic aspects related to energy.

3.2 Key Energy Indicators

In terms of energy resources, Africa and Latin America own massive fossil fuel reserves that are irregularly dispersed in different countries [1, 2]. Considering the countries with data available, Africa has low coal resources (Fig. 3.5) except in South Africa and a low dependence on coal for producing electricity except in South Africa and Botswana, where coal production is well matched with consumption. In contrast, Latin America has significant coal reserves that are homogenously distributed in South America, a lack of coal resources in Central America and limited coal dependence for producing electricity in most Latin American countries.

Considering those countries where data is available, Africa has important oil reserves (Fig. 3.6), mainly in Nigeria and Libya, and a low power dependence on oil except in small underdeveloped countries. Latin America has large oil reserves in Venezuela and high oil dependence for producing electricity in all Central American countries except Costa Rica (Costa Rica has an abundance of renewable resources).

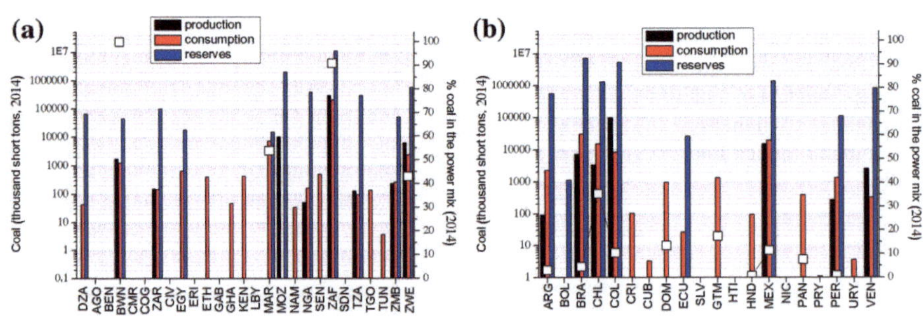

Fig. 3.5 Coal production, consumption, reserves [1] and percentage (*open square*) of coal-based electricity production [2] in 2012 in **a** Africa and **b** Latin America

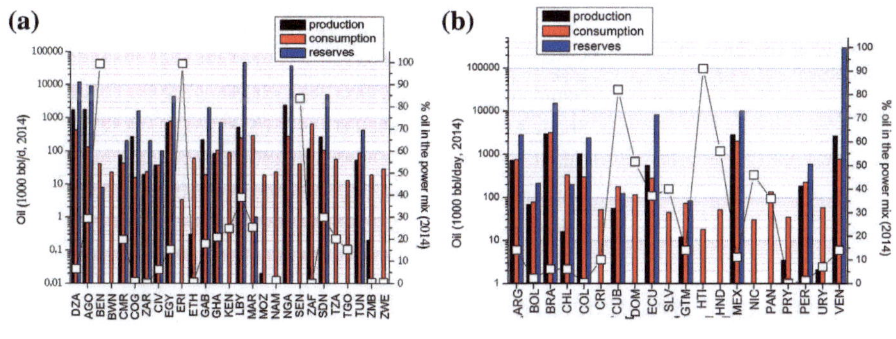

Fig. 3.6 Oil production, consumption, reserves [1] and percentage (*open square*) of coal-based electricity production [2] in 2012 in **a** Africa and **b** Latin America

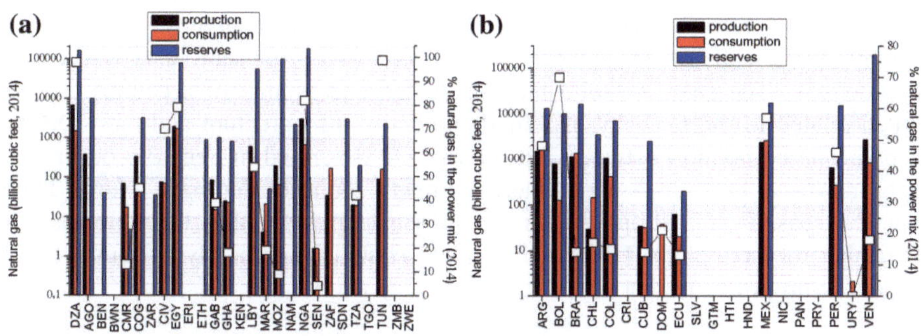

Fig. 3.7 Natural gas production, consumption, reserves [1] and percentage (*open square*) of coal-based electricity production [2] in 2012 in **a** Africa and **b** Latin America

Also considering those countries with data available, Africa has huge natural gas resources in many countries (Fig. 3.7) and a high dependence for producing electricity in most of them, as the production is quite well matched with consumption levels. In contrast, Central America shows very low and non-uniformly distributed natural gas reserves. In contrast, South America shows large natural gas reserves, mainly in Venezuela, and low national dependency for producing electricity except in Argentina, Bolivia, Mexico and Peru.

In Africa and Latin America, uranium reserves are located in a small number of countries with all except Brazil located in Africa (Table 3.1) [3]. However, as uranium must be enriched and appropriately encapsulated for fuelling a nuclear reactor, and nuclear technology is not easily available, there exists only a very soft interdependence between uranium reserves and opportunities to exploit this resource, compared to the technologies associated with fossil fuels.

Table 3.1 Known recoverable resources of uranium in 2013 [3]

Country	Tonnes U	Percentage of world (%)
Niger	404,900	6.9
Namibia	382,800	6.5
South Africa	338,100	5.7
Brazil	276,100	4.7
Botswana	68,800	1.2
Tanzania	58,500	1.0

3.3 Energy Infrastructure

3.3.1 Renewable Energy Supply

It is important to recognize that the countries worldwide with the most TPES based on renewable energy resources are located in Africa and Latin America (Fig. 3.8) [2]. This fact is mainly attributed to an underdeveloped economy where energy supply is basically solid biomass, and electricity, if present, is produced by hydropower. The share of solid biomass in the energy mix of a country is very much related to the maturity of a modern energy infrastructure and the GDP per capita. The correlation between TPES and GDP described in Sect. 3.6 occurs largely in Africa. In Latin America there are many countries that have a large TPES but it is based on hydropower and geothermal energy rather than biomass. In Paraguay for example, a TPES above 100% exists that is mainly attributed to the role of hydropower in that country.

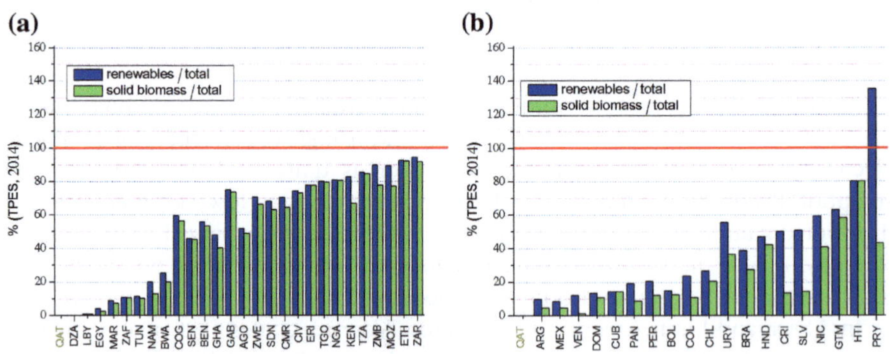

Fig. 3.8 Percentage of total primary energy supplied by renewable resources and only by solid biomass in 2014, **a** in Africa, and **b** Latin America [2]

3.3.2 Fossil Fuel Pipelines

The oil and gas infrastructure in Africa and Latin America was examined by analysing the distribution of oil and gas pipelines across the two regions as they are connected to refineries and oil and gas fields (Fig. 3.9) [4]. The concentrations of pipelines can be easily related to areas producing oil and natural gas (Figs. 3.5 and 3.6), but also to the supply of cities and transportation to ports. Then, the amount of oil and gas pipelines in an area where no oil and gas resources are available shows development of the area. Pipelines in operation crossing country borders can also show the level of commercial partnership between countries and regions. Thus, in Africa there is observed an increasing commercial activity based on the supply of natural gas from Maghreb to Europe, and also some agreements between inner and maritime countries for transporting the fuel to port areas (e.g. Sudan and South Sudan). In Latin America we can also observe countries supplying fuel to neighbours (e.g. natural gas from Bolivia to Brazil).

3.3.3 Power Grid

The other basic infrastructure for the energy sector is the power grid. A first approximation about power grid interconnections is summarized in this chapter.

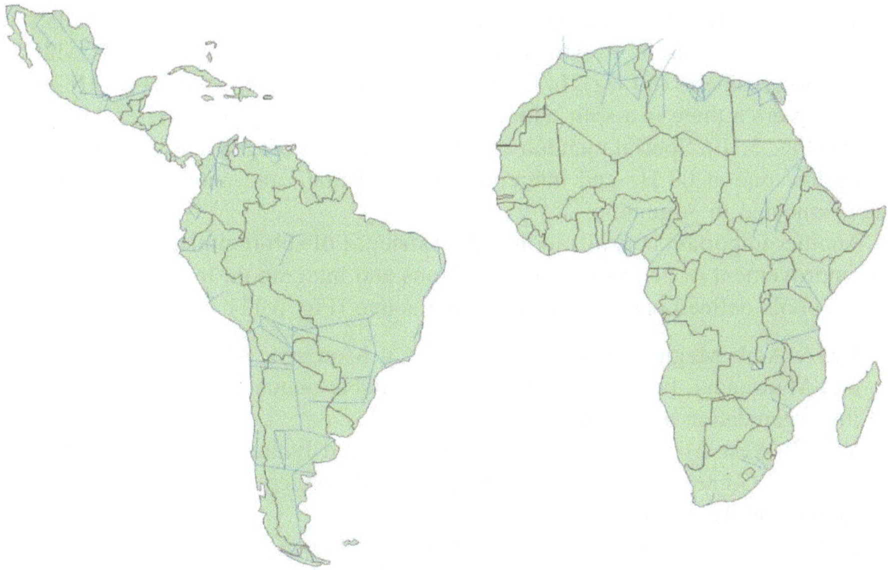

Fig. 3.9 Oil and gas pipelines in operation or under construction in Africa and America [4]

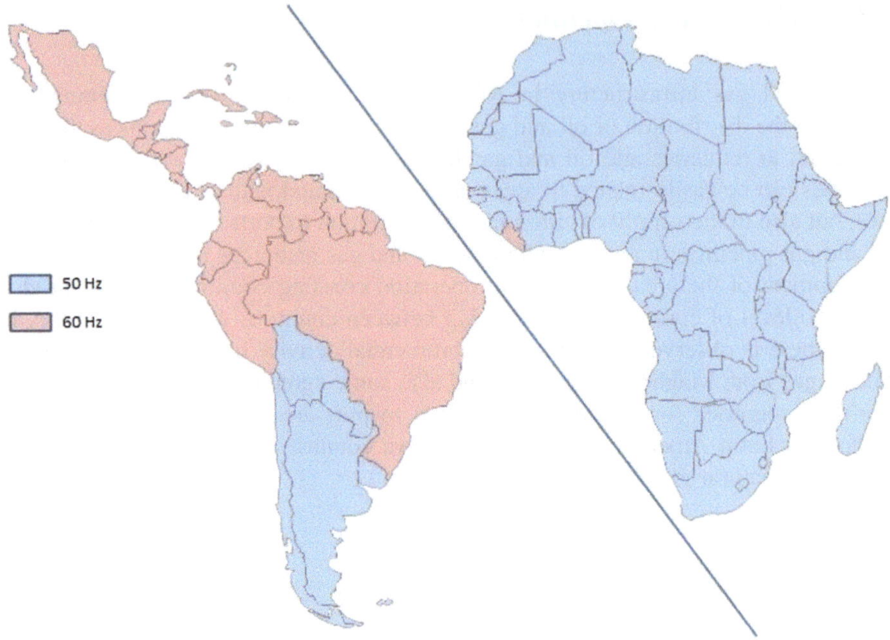

Fig. 3.10 Frequency of the national power grids

Also Chap. 4 is devoted to large, small and isolated power grids. A first consideration is that the grids can operate at 50 and 60 Hz, introducing difficulties to regional grid integration. In Africa the entire power grid operates at 50 Hz with the exception of Liberia (60 Hz). In Central America the grid operates at 60 Hz but in South America there is a split with some nations operating at 50 Hz and others at 60 Hz (Fig. 3.10). Special mention should be made of Paraguay, as this country has two grids, one at 60 Hz and interconnected to Brazil, and another at 50 Hz and interconnected to Argentina.

Another important consideration is the existence of sub-regional power pools for the improvement of national interconnections and integration. In Africa there are 5 power pools defined and most of them are active (Fig. 3.11):

- UMA Arab Maghreb Union (COMELEC): Morocco, Algeria, Libya, Tunisia and Mauritania. Founded in 1975, last available annual report of 2009 [5] but the organization is currently active.
- WAPP West African Power Pool: Benin, Côte d'Ivoire, Burkina Faso, Ghana, Gambia, Guinea, Guinea Bissau, Liberia, Mali, Niger, Nigeria, Senegal, Sierra Leone and Togo. Founded in 1999.
- EAPP Eastern African Power Pool: Egypt, Sudan, South Sudan, Djibouti, Ethiopia, Eritrea, Somalia, Uganda, Kenya, Rwanda and Burundi. Established in 2005 [6].

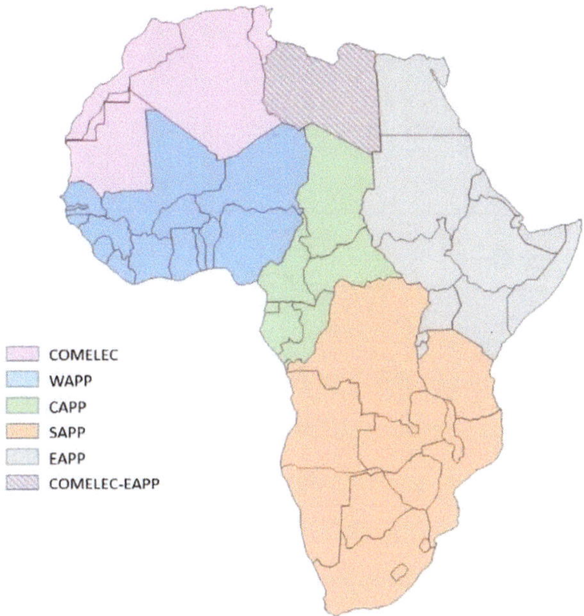

Fig. 3.11 Power pools in Africa

- CAPP Central African Power Pool: Chad, Central African Republic, Cameroon, Equatorial Guinea, Gabon and Congo. Under definition of a plan [7].
- SAPP Southern African Power Pool: DR Congo, Tanzania, Angola, Zambia, Malawi, Mozambique, Zimbabwe, Botswana, Namibia, Swaziland, Lesotho and South Africa. In operation.

However, less than 8% of power crosses international borders in any African region, despite the capacity shortages in many countries. The SAPP has the highest levels of trade, with 7.5% of power crossing international borders where the electricity exchange from Cahora Bassa in Mozambique to South Africa, and the return from South Africa to Mozal (aluminium smelter in Mozambique), represents a significant proportion of this amount. In CAPP and EAPP, less than 1% of power crosses international borders [8]. The lack of regional trade is not surprising, given Africa's chronic capacity shortage. Only countries with excess capacity could be expected to sell power across borders. For a time, South Africa had excess capacity, and the SAPP flourished, but since the energy crisis in 2008, exports of electricity from South Africa have slightly declined.

Currently these national grids are not all interconnected but most power pools have plans for future interconnections (Fig. 3.12). There are also some countries (e.g., DR Congo) considered in more than one interconnection plan. Libya should play an important role in the connection of NE and NW Africa, but at this time it is not defined because of internal problems. Some large grids are connected in DC (e.g., South Africa–Mozambique) or expected to be connected to reduce losses over

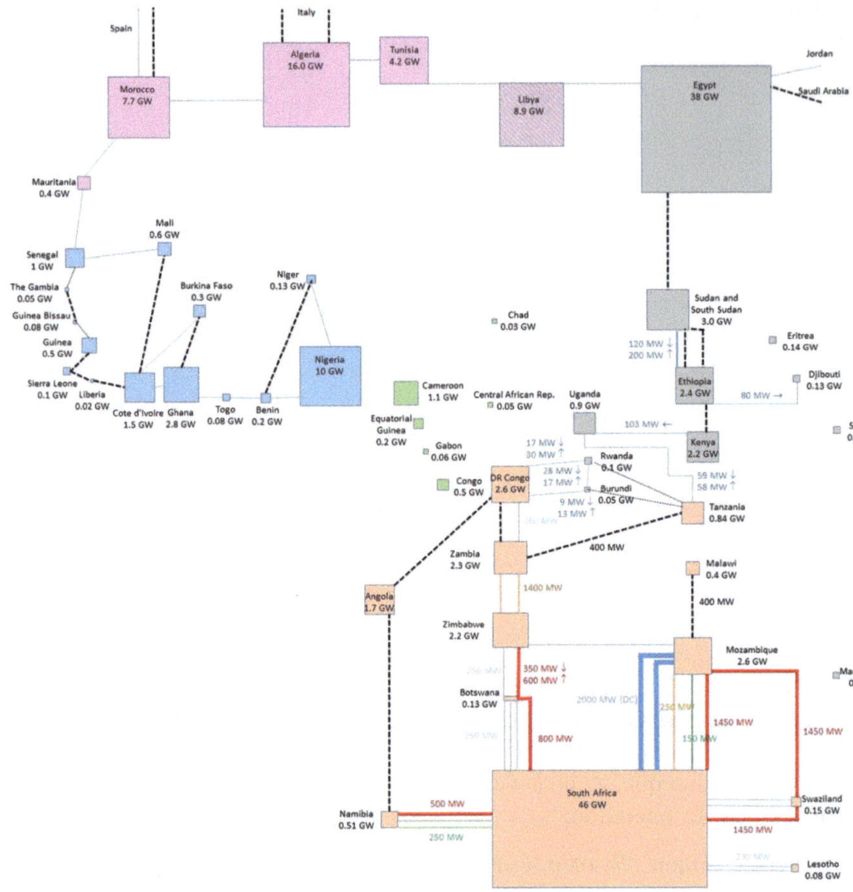

Fig. 3.12 Power grids in Africa (showing 2014 power capacities [1]). Dashed lines represent future interconnections

long distances. A transmission project to interconnect the West African Power Pool countries of Sierra Leone, Guinea, Liberia and Ivory Coast was approved in 2013 and is expected to be completed by 2017 [9]. An increase in interconnections between North Africa and Europe is expected in the future if instability in the Maghreb region diminishes.

In Latin America a regional operator only exists between Guatemala, El Salvador, Honduras, Nicaragua, Costa Rica and Panama and a regional grid called SIEPAC. This regional grid of 1800 km, 230 kV and 300 MW, was finished in 2014 and today plans exist for a second circuit (Fig. 3.13) [10]. Guatemala and Mexico are also connected by a new (2010) 400 kV transmission line of 225 MVA and 103 km, and a future connection between Panama and Colombia is scheduled. Also an agreement (SINEA) is expected to be successful between Colombia, Bolivia, Ecuador, Peru and Chile for starting regional interconnections to grow in the area [11].

Fig. 3.13 Power pools in Latin America

The interconnections between the Latin American countries (Fig. 3.14) show that Mexico is integrated into the North America power grid with USA and Canada and there is an isolated power grid in Baja California (Mexico) that could be connected to the main national grid in future. The SIEPAC is now starting into operation and the 300 MW capacity of the interconnecting grid (expected to increase to 600 MW in future) is significant for a regional 13.2 GW capacity. Also, it is important to mention the important effect that the Yacyretá (1.8 GW, 1998) and Itaipu (14 GW, 1984) hydropower plants between Paraguay-Argentina and Paraguay-Brazil, respectively, have related to interconnections between these countries, and also Salto Grande (1.8 GW, 1979) for increasing the interconnections between Argentina and Uruguay. There are also substantial advances in other Latin American countries, but geographical barriers and political concerns based on past conflicts can introduce difficulties for further interconnections. For example, because of a natural gas supply crisis between Argentina and Chile, a 700 MVA interconnection between Chile and Argentina established in 1999 is not working. There are also interconnection problems between Argentina and Brazil, between Colombia and Ecuador, between Colombia and Venezuela, and slow negotiations for a 400 MVA interconnection between Chile and Peru. Chile has 4 different,

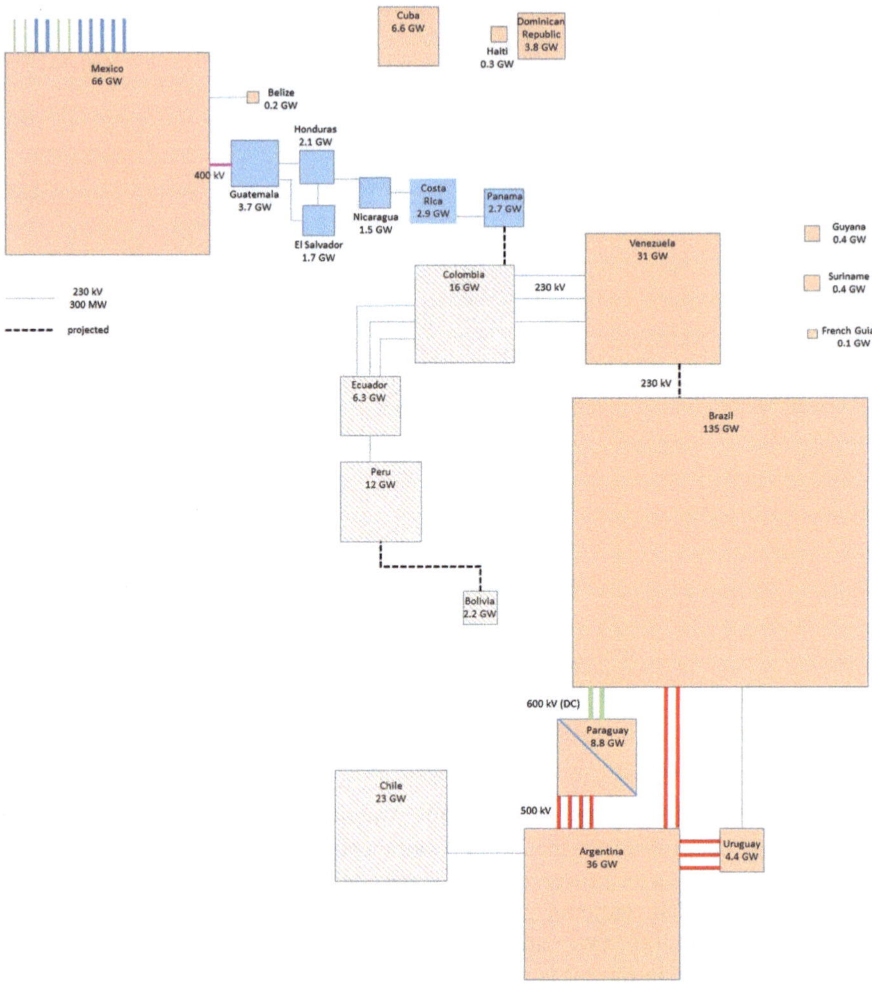

Fig. 3.14 Power grids in Latin America (showing 2014 power capacities [1]). Dashed lines represent future interconnections

independent grids placed in different regions of the country and that could be appropriately connected to other power grids: (i) Sistema Interconectado del Norte Grande (3698.7 MW); (ii) Sistema Interconectado Central (12,147.1 MW); (iii) Sistema Eléctrico de Aisén (40.2 MW); and (iv) Sistema Eléctrico de Magallanes (98.8 MW).

There are currently 12 interconnection projects in Latin America being planned that can substantially improve the regional integration: (i) Inambari hydropower complex (5 plants) between Perú and Brazil, adding a capacity of 20 GW, but the project depends on the agreement to export electricity to Brazil; (ii) Cachuela Esperanza hydropower plant between Bolivia and Brazil, adding 0.8 GW capacity, but the project depends on the agreement to export electricity to Brazil; (iii) 0.3 GW Colombia-Panama HVDC (614 km, 55 km on the seabed) interconnection; (iv) 2 GW Brazil-Argentina interconnection; (v) 0.2 GW swap Paraguay-Argentina-Chile from Yacyretá to Atacama; (vi) 230 kV SIEPAC II repowering 1800 km SIEPAC I; (vii) wheeling Chile by Argentina, transporting hydropower from Aysén to Santiago; (viii) 0.5 GW Brazil-Uruguay 500 kV interconnection (420 km); (ix) 2 GW Argentina-Paraguay-Brazil 500 kV interconnection (321 km) to connect Yacyretá and Itaipú; (x) 100 MW Peru-Ecuador interconnection that currently exists but not used; (xi) 125 MW Bolivia-Peru 230 kV interconnection (215 km); and (xii) 180 MW Bolivia-Chile 230 kV interconnection (150 km) [12].

3.4 Energy Regulations and Jobs

3.4.1 National Targets and Incentives

A number of countries are attempting to increase the amount of investment in the renewable energy sector by approving new regulations and objectives. The IEA/IRENA Joint Policies and Measures Database [13] has information about policies in renewable energies for 16 countries in Africa (Algeria, Botswana, Burkina Faso, DR Congo, Egypt, Ethiopia, Ghana, Kenya, Libya, Madagascar, Mali, Morocco, Nigeria, Senegal, South Africa, Tanzania, Tunisia, Uganda and Zimbabwe) and 9 countries in Latin America (Argentina, Bolivia, Brazil, Colombia, Chile, Dominican Republic, Ecuador, Mexico, Nicaragua, Peru and Uruguay). The primary targets for promoting change are summarized in Table 3.2. Many countries have defined goals for increasing the penetration share of RES, mostly in the electricity sector. Many of these countries also have national plans for defining the strategy to reach the targets and have approved incentives (mostly feed-in tariffs) to support targets. Also, there are sectoral programs, mostly for rural electrification and energy supply, and those related to promote the renewable resources of the country considered. There are many programs devoted to biofuels and blending in countries where important resources exist or the potential to build this sector exist. A number of programs devoted to net metering, CHP, lighting, and the role of women in promoting renewable energy are also being considered. Details on many of these programs are provided in later Chapters of this book.

3.4.2 Investment Climate and Jobs in Clean Energies

A database [14] that ranks Asian, African and Latin American countries in terms of investment climate and policies for clean energy investment shows Brazil, Chile,

Table 3.2 Main RES targets and sectorial policies defined in African and Latin American countries [13]

Country	Plan	Reach	Target	Sectoral plans
Algeria	2016	2030	27% of electricity generation from RES (22 GW)	PV feed-in tariff, efficiency, National Fund
Botswana	2009	2016		PV pilot, PV rural electrification, Energy Master Plan
Burkina Faso	2013		Sectoral Policy of Energy	Solar energy
DR Congo				Diversification of energy companies
Egypt	2008	2020	20% of total electricity generation from RES	Feed-in tariff for wind and solar, solar plan, RES Strategy
Ethiopia	2011	2015	Renewable capacity 8–10 GW	
Ghana	2006	2020	10% of RES in energy mix	Net metering, feed-in tariff for RES, Strategic Plan
Kenya	2011	2031	Geothermal 5.5 GW, wind 3 GW, PV 0.5 GW, small hydro 0.3 GW	Solar, biodiesel
Libya	2007	2020	10% of total energy generation coming from RES	
Madagascar				Rural electrification, strategic planning
Mali	1996	2025	15% of the energy generation from RES	Rural electrification by solar energy, new energy and women, jatropha, Action Plan
Morocco	2009	2020	42% of electricity generation capacity from RES	Self generation, National Agency
Nigeria	2011	2025	23% of electricity generation capacity from RES	Biofuels blending, RES tariffs
Senegal	2010		Renewable Energy Law	Biogas, biofuels, rural electrification and domestic fuels
South Africa	2009	2030	10% of electricity generation capacity from RES	Fuel blending, Electricity Plan, solar, households
Tanzania				Rural, Energy Program, electricity rules
Tunisia	2008	2010	10% of electricity generation capacity from RES	Connection to grid, solar plan
Uganda	2007	2017	61% of RES in gross final energy consumption	RES feed-in tariffs, solar, rural energy, hydro and CHP
Zimbabwe				Biofuels blending, rural electrification, Regulatory Act
Argentina	2007	2016	8% of RES in national electricity mix	Biofuels, wind energy

(continued)

Table 3.2 (continued)

Country	Plan	Reach	Target	Sectoral plans
Bolivia				Rural electrification
Brazil	2010	2019	Capacity targets: hydro 116.7 GW; small hydro 7 GW; biomass 8.5 GW; wind 6 GW	Ethanol export tax credit, wind energy, biomass, biodiesel and ethanol mandates, Light for All, rural electrification
Colombia	2010	2020	Grid connected regions: 6.5% non connected regions: 30% of RES in total power consumption	Flex fuel vehicles, biofuels, rural electrification fund
Chile	2013	2025	20% of total electricity generation from non-hydro RES	Access to energy, distributed PV, solar thermal, geothermal
Dominican Rep.	2007		Tax relief for RES	
Ecuador	2013	–	Feed-in tariff for RES (wind, solar, ocean, bioenergy, geothermal, hydropower)	Biodiesel blending
Mexico	2012	2026	35% of non-fossil fuel generation	National plan, energy sustainability fund, wind energy
Nicaragua	2013	2027	Indicative Electricity Generation Plan for multiple RES	Clean cookstoves, fuelwood and charcoal, rural electrification, geothermal
Peru	2008	2013	5% of total power demand from RES (excluding hydropower plants bigger than 20 MW)	Electric social compensation, biofuel production and commercialisation
Uruguay	2010	2015	15% of electricity demand met RES	Photovoltaics, wind energy, solar thermal, RES microgeneration, biomass, biodiesels and alcohols

South Africa, Kenya and Uganda with highly favorable investment climates. The ranking (Table 3.3) evaluates: (i) enabling framework (40%); (ii) financing and investment (30%); (iii) value chains (15%); and (iv) GHG management (15%). This ranking is quite similar to the one related to the ease of doing business, presented in Chap. 2, but there are some variations mostly related to countries with important renewable energy resources as these try to promote and attract foreign investment.

Currently, most stimulus to international investment on clean energy technologies in Africa and Latin America is produced by auctions promoted by national governments and utilities. Mexico's first-ever electricity auction garnered investment commitments estimated at $2.6 bn USD. The government was forced to re-run the auction after a flawed bid from Gestamp Wind was found to be too low,

Table 3.3 Ranking of African and Latin American countries in terms of investment climate and policies for clean energy investment (2015) [14]

Ranking	African country	Score	Ranking	LA country	Score
4	South Africa	1.91	2	Brazil	2.12
6	Kenya	1.74	3	Chile	1.97
9	Uganda	1.68	7	Mexico	1.72
12	Nigeria	1.58	8	Uruguay	1.69
17	Rwanda	1.41	14	Honduras	1.50
23	Tanzania	1.22	15	Costa Rica	1.49
26	Ethiopia	1.17	16	Peru	1.44
33	Malawi	1.01	18	Guatemala	1.40
34	Zambia	0.99	19	Colombia	1.39
35	Liberia	0.89	20	Argentina	1.39
36	Senegal	0.86	21	Panama	1.31
40	Sierra Leone	0.79	27	Nicaragua	1.14
41	Mozambique	0.77	29	Bolivia	1.04
42	Côte d'Ivore	0.71	30	El Salvador	1.03
43	Zimbabwe	0.70	31	Ecuador	1.03
47	Botswana	0.59	32	Dominican Rep.	1.02
49	Cameroon	0.56	45	Haiti	0.64
50	DR Congo	0.55	52	Paraguay	0.49
			54	Venezuela	0.40

and impacted results. The auction was the country's first private auction since policy makers decided in 2013 to allow private companies to compete for power contracts. Acciona, Enel Green Power, SunPower Systems and Recurrent Energy were among the 11 companies that won rights to generate and sell 2085 MW of clean power to the state-owned Comision Federal de Electricidad. As many as 18 projects were ultimately awarded, with solar making up 1691 of the megawatts contracted. The remaining 394 MW will come from wind resources, according to Energy Minister Pedro Joaquin Coldwell [15].

Separately, the World Bank, in a broad plan aimed at accelerating its efforts to combat climate change, vowed to ramp up by 50% its funding of clean energy projects by 2020. The "World Bank Group Climate Change Action Plan" calls for increasing from $2.3 to $3.5 bn its direct funding of clean energy projects over the next five years. It pledged to use a variety of financial instruments to "de-risk" renewable energy investments to double the gigawatts of renewable energy generation associated with its projects, with an eye toward producing a total of 20 GW of renewable energy over five years [16].

However, in many developing countries new generation projects show high average budget and time overruns. Usually, the larger and more complex the project, the greater the overruns. The private sector has performed slightly better than the public sector in all areas other than hydro projects. Many international developers already include a premium or contingency of 20–30% for planning timelines (for example, productivity rates) and budgets on African projects compared with similar projects in a developed economy. But even those contingency amounts often fail to account for the total project expenses. Currently, a true "African cost premium" would be about 60% [8]. As more projects are built across these regions, and skills develop within to execute them, the cost premium should drop significantly.

No detailed information on the availability of jobs in the low carbon energy sectors in Africa and Latin America are available. Only IRENA is starting to coalesce this information for some countries in the area of renewable energies [17]. Worldwide, renewable energy jobs reached an estimated 8.1 million in 2015, excluding large hydropower, increasing 5% from the estimates reported in 2015 and the regional shift toward Asia continued, especially in manufacturing. In Latin America, Brazil (918,316) is the dominant player, followed by Colombia (106,420), Argentina (30,000) and Mexico (45,595), Uruguay (15,150) and Chile (8374). Brazil is the world's second largest renewable energy employment country after China, mainly because of employment in liquid biofuels (820,900). On the contrary, renewable energy-related data available shows that employment remains low in Africa: South Africa (27,634), Algeria (6880), Egypt (4170), Kenya (3008), Tunisia (2378), Morocco (2131) and Tanzania (2035).

3.4.3 Energy Subsidies

There are energy subsidies in many countries, mostly for supporting fossil fuels purchase but also electricity. The fossil fuel subsidies can be easily detected by comparing the price of gasoline and diesel fuel at the pump station in Africa (Fig. 3.15) and Latin America (Fig. 3.16) [18]. These graphs show that the most relevant subsidies are located in the oil producing countries but many other countries also subsidize fossil fuels. Moreover, information about diesel engine promotion policies can be obtained by comparing gasoline and gasoil prices in each country (see Chap. 11).

Subsidies have wide-ranging economic consequences as they aggravate fiscal imbalances, depress private investment, encourage excessive energy consumption, artificially promote energy-intensive industries, reduce incentives for investment in low carbon energy technologies, accelerate the depletion of natural resources, and can also produce shortage of supply. Fuel smuggling in Venezuela is considered to produce a deficit in fuel supply [20]. Most subsidy benefits are captured by higher-income households, reinforcing inequality. On a pre-tax basis, subsidies for petroleum products, electricity, natural gas, and coal (Table 3.4) reached $480 billion in 2011 (0.7% of global GDP or 2% of total government revenues). The cost of subsidies is especially acute in oil exporting countries, which account for about two-thirds of the total. On a post-tax basis, which also factors the negative

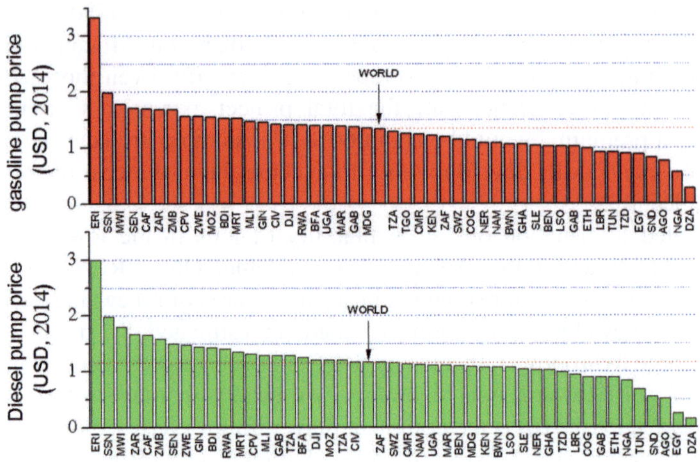

Fig. 3.15 2014 average gasoline and diesel fuel price at the pump station in African countries [18, 19]

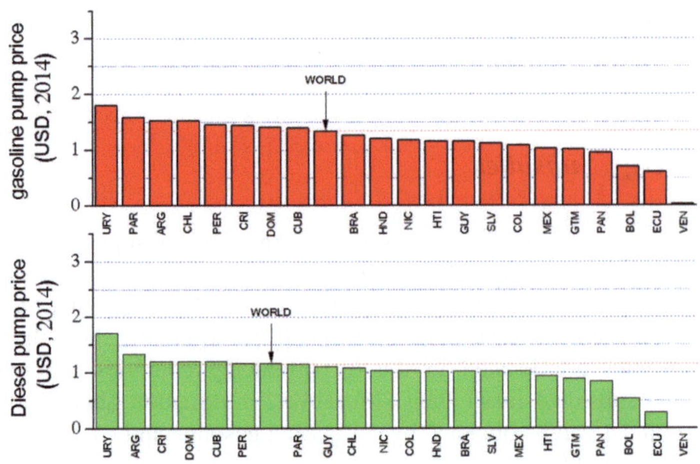

Fig. 3.16 2014 average gasoline and diesel fuel price at the pump station in Latin American countries [18, 19]

externalities of energy consumption, subsidies are much higher, reaching $1.9 trillion (2½% of global GDP or 8% of total government revenues) [21].

The IMF suggests six key elements for subsidy reform [21]: (i) a comprehensive energy sector reform plan entailing clear long-term objectives, analysis of the impact of reforms, and consultation with stakeholders; (ii) an extensive communications strategy, supported by improvements in transparency, such as the dissemination of information on the magnitude of subsidies and the recording of

Table 3.4 Known pre-tax subsidies (% GDP) for petroleum products, electricity, natural gas and coal (2011) [21]

Country	Petroleum	Electricity	NG	Coal
Argentina	0.00	1.03	0.77	0.00
Bolivia	2.40			
Ecuador	6.31	0.18	0.00	0.00
Panama	0.02			
Venezuela	5.58	1.02	0.59	
Algeria	4.30	1.08	5.36	0.00
Angola	1.30	0.27	0.00	0.00
Benin	0.00	1.78		
Burkina Faso	0.00	0.78		
Djibouti	0.00	0.45		
Libya	6.40	1.85	0.59	0.00
Mauritania	0.00	0.85	0.80	
Morocco	0.66			
Sudan	1.37			
Tunisia	0.77	2.23		
Cameroon	1.69	2.16		
DR Congo	0.00	1.57		
Congo	1.20	2.62		
Cote d'Ivore	0.00	2.72		
Equatorial Guinea	0.28			
Ethiopia	0.19	1.24		
Gabon	0.16			
Ghana	0.62	2.86		
Lesotho	0.00	0.85		
Madagascar	0.11	0.89		
Malawi	0.00	1.60		
Mozambique	0.00	4.93		
Namibia	0.00	0.52		
Nigeria	1.42	1.31		
Rwanda	0.00	0.29		
ST & Principe	0.33			
Senegal	0.00	2.26		
South Africa	0.01	0.55	0.00	0.00
Tanzania	0.00	2.10		
Uganda	0.00	1.32		
Zambia	0.00	4.85		
Zimbabwe		14.52		
World	0.30	0.22	0.16	0.01

Table 3.5 Summary of African and Latin American energy subsidy reform episodes [21]

Country	Product	Year	Successful	Impact
Brazil	Fuel	2010	Yes	0.8% GDP in mid-1990s
Chile	Fuel	1990s	Yes	n.a.
Peru	Fuel	2010	Partial	0.1% GDP
Brazil	Electricity	1993–2003	Yes	0.7% GDP
Mexico	Electricity	1999–2002	No	n.a.
Ghana	Fuel	2005	Partial	50% price increase in average
Kenya	Electricity	Mid 1990s	Yes	Subsidies declined from 1.5% GDP in 2001 to zero in 2008
Mauritania	Fuel	2008	No	n.a.
Namibia	Fuel	1997	Partial	0.1 GDP
Niger	Fuel	2011	Partial	0.9% GDP
Nigeria	Fuel	2011–2012	Partial	Subsidies declined from 4.7% GDP to 3.6% GDP in 2012
South Africa	Fuel	1950s	Yes	Successfully avoided subsidies and secures supply
Uganda	Electricity	1999	Yes	2.1% GDP

subsidies in the budget; (iii) appropriately phased price increases, which can be sequenced differently across energy products; (iv) improving the efficiency of state owned enterprises to reduce producer subsidies; (v) targeted measures to protect the poor; and (vi) institutional reforms that depoliticize energy pricing, such as the introduction of automatic pricing mechanisms.

Sub-Saharan Africa accounted for about 4% of global energy subsidies (1.5% of the regional GDP, and 5.5% of total government revenues), with electricity subsidies accounting for over 70%. Total subsidies exceeded 4% of GDP in three countries (Mozambique, Zambia, and Zimbabwe). The Latin America and Caribbean made up over 7% of global energy subsidies (approximately 0.5% of regional GDP or 2% of total government revenues), with petroleum subsidies accounting for nearly 65%. Energy subsidies exceeded 5% of GDP in two countries (Ecuador and Venezuela) [21].

Some countries have tried to implement reforms, sometimes with the support of the IMF, with different outcomes. A summary of these reforms is provided in Table 3.5. Many efforts have been made to reduce fuel subsidies but these changes are difficult because the private transport sector does not support change. Reducing electricity subsidies in these countries is also difficult as the power infrastructure is very costly and when these costs are passed on to the consumer, the price of electricity reaches values considerably above developed countries, as discussed in in Chap. 4.

3.5 Energy Security and Trading

The IEA defines energy security as “the uninterrupted availability of energy sources at an affordable price”. Energy security has many dimensions: long-term energy security mainly deals with timely investments to supply energy in line with economic developments and sustainable environmental needs. Short-term energy security focuses on the ability of the energy system to react promptly to sudden changes within the supply-demand balance [22]. The IEA is devoted to promote energy security between its members and the IAEA, an organization within the United Nations family, emphasizes energy security related to the safe, secure and peaceful use of nuclear technologies.

IEA membership requires countries to meet two key security obligations: to hold oil stocks equivalent to at least 90 days of net oil imports; and to maintain emergency response measures that can contribute to an IEA collective action in the event of a severe oil supply disruption. Response measures include stock draw, demand restraint, fuel switching and surge oil production. The IEA has the ability to respond rapidly to an oil supply disruption through real-time communication with member countries and major players outside the IEA, and it also has a framework for decision making which is tested and updated through regular simulation exercises. As in the electricity market, oil and oil products have already been largely replaced by natural gas for electricity production, half of the member countries of the IEA have developed specific stockholding measures related to gas that would provide some resilience in the event of a disruption [22]. No IEA members are located in Africa and Latin America, but Chile has presented its candidacy to the IEA and is making progress in this area (Fig. 3.17).

Appropriate emergency response measures can differ substantially between fossil fuels. For example, natural gas is capacity-bound to a highly capital-intensive transportation and distribution infrastructure, and there is little demand-side response in some large consumer sectors. Also, natural gas is far less fungible than oil, particularly with regard to transporting the fuel to end users. Downstream gas transport is mostly performed by fixed infrastructure (i.e., pipelines). While there are also many downstream oil distribution pipelines in use, a large-scale disruption to one pipeline can be isolated, and tanker trucks can be used to distribute the oil instead. Repairs to oil pipelines are also less costly than repairs to gas pipelines because of the elevated pressure of the gas system. Furthermore, gas transport is more difficult to scale up than oil transport as the available spare capacity, either physically or contractually, is sometimes limited in existing gas pipelines [22].

As it has been mentioned above, fossil fuel and uranium reserves are very irregularly distributed in Africa and Latin America and many countries have no reserves but consume fossil fuels and uranium, mostly in Latin America (Table 3.6). This irregular distribution of conventional resources leads to trading between countries but also to energy security concerns that can generate regional tensions.

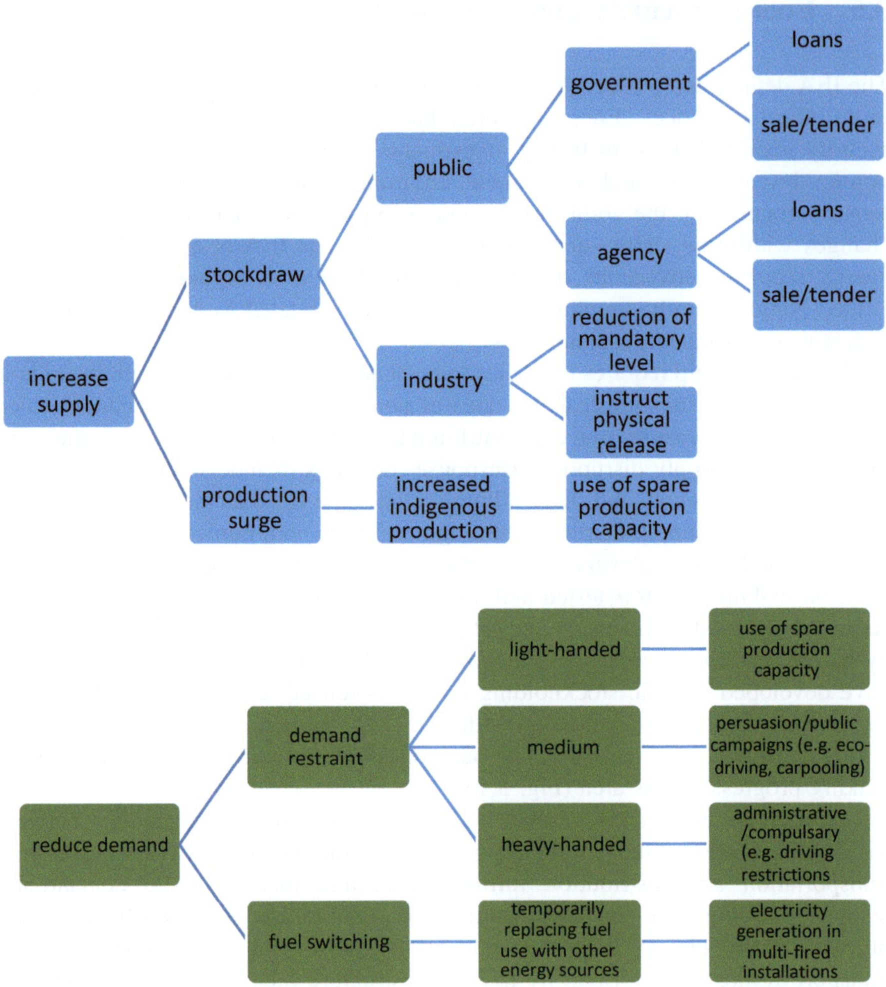

Fig. 3.17 IEA Emergency response system [22]

There are countries that consume fossil fuels and have their own reserves but are not producing (coal: Algeria and Egypt), and countries that do not produce nor consume, but have reserves (coal: Zimbabwe, Bolivia and Ecuador; natural gas: Benin, Ethiopia, Namibia and Sudan). Additional and specific security factors are discussed in chapters devoted to the technologies (e.g., electricity supply and losses in Chap. 4, deforestation in Chap. 5, uncertain inflows to hydropower plants in Chap. 8). Energy security concerns derived from periods of armed conflicts (e.g., before the 1989–2003 war, Liberia had an installed capacity of 191 MW, comprising a 64 MW large-hydro plant at Mount Coffee and heavy fuel oil and diesel generators, but by the end of the war this capacity was almost entirely destroyed

Table 3.6 Countries with no coal, oil, natural gas or uranium reserves but consuming these commodities [1]

Coal	Oil	Natural gas	Uranium
Costa Rica	Botswana	Dominican Republic	Mexico
Cuba	Costa Rica	Senegal	Argentina
Dominican Republic	Dominican Republic	Uruguay	
Guatemala	El Salvador		
Honduras	Eritrea		
Kenya	Ethiopia		
Namibia	Haiti		
Nicaragua	Honduras		
Senegal	Kenya		
Panama	Mozambique		
Uruguay	Namibia		
	Nicaragua		
	Panama		
	Paraguay		
	Senegal		
	Tanzania		
	Togo		
	Uruguay		
	Zambia		
	Zimbabwe		

and, currently, the integrated national utility has an installed capacity of only 27 MW, most of which is based on heavy fuel oil generation [9]) that is not considered in this book.

In a more or less defined strategy, countries should have defined plans and decision-making structures against security threats. However, a well defined emergency response plan against energy security threats is only detected in Chile. Chile has experienced several serious energy supply incidents including major droughts, a sustained gas supply cut from Argentina (in 2004 gas-fired power plants switched to diesel, increasing diesel demand for power generation from 4% in 2000 to almost 30% in 2008, but the construction of two liquefied natural gas terminals in 2010 attenuated the demand for gasoil), and a major earthquake in early 2010 which affected electricity networks and refineries and caused several black-outs. Chile does not currently hold strategic oil stocks but its oil industry participants have an obligation to hold stocks covering 25 days of their sales over the previous 6 months, or average imports over the same period if imports are for their own consumption. However, this requirement is currently not monitored or enforced. Chile has no specific response system for handling a oil or natural gas supply emergency [22].

Also a mention to cyber security should be included in this section. Securing a fuel or power grid requires a combination of common and advanced cyber security technologies. To maintain the stability of the whole system, most subsystems need to maintain functionality under all circumstances. Standard cyber security best practices, such as data encryption, antivirus provision, firewalls, intrusion prevention systems, network security design, defense in depth and system hardening. In addition to these common protection mechanisms, advanced cyber security is required to protect against more sophisticated attacks, including security information and event management systems, application whitelisting and security features embedded at the processor level, among others. Also, with the addition of more sensitive equipment the physical security of critical infrastructures is required. Greater participation from customers through demand response programmes can also support security of energy infrastructure, e.g., microgrids either connected to the larger grid or operating fully in isolation (in "island" mode) can provide grid recovery services during disturbances. Customer's data and privacy must also be secure under key principles as notice and awareness, choice and consent, etc. [23]. Currently no specific programmes concerning these issues have been detected in Africa and Latin America.

3.6 Energy Efficiency

The literature shows that higher socio-economic levels are acheived by those socities that have easy and reliable access to energy and energy products (mainly, electricity and fuels). Thus, if we consider the last data available from Africa and Latin America, it is evident that there is a linear relation between electricity consumption and GDP per capita (Fig. 3.18 and Table 3.7). However, growing concerns about the unsustainable increase in energy consumption worldwide are

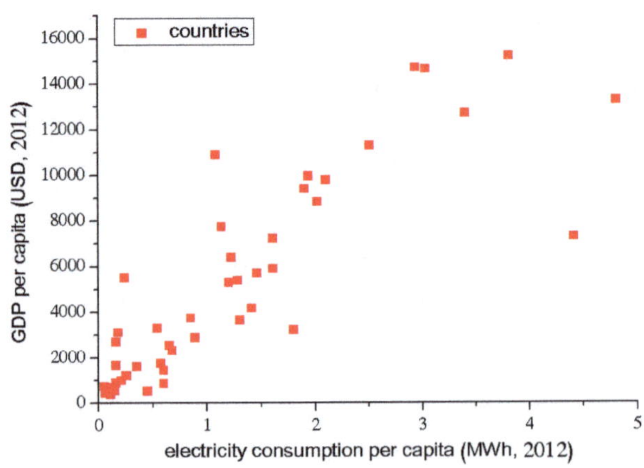

Fig. 3.18 Graph showing a linear dependency between electricity consumption and GDP per capita [2, 25]

Table 3.7 Annual electricity consumption ranking and GDP per capita in 2012 for African and Latin American Countries [25]

2012	MWh	GDP_{pc}($)	2012	MWh	GDP_{pc}($)
Libya	4.81	13303	Chile	3.81	15245
South Africa	4.41	7314	Venezuela, RB	3.40	12729
Mauritius	2.02	8862	Argentina	3.03	14680
Egypt, Arab Rep.	1.80	3256	Uruguay	2.93	14728
Botswana	1.61	7255	Brazil	2.51	11320
Namibia	1.61	5931	Mexico	2.10	9818
Tunisia	1.41	4197	Panama	1.94	9982
Algeria	1.20	5310	Costa Rica	1.90	9443
Gabon	1.08	10930	Dominican Republic	1.46	5733
Morocco	0.89	2902	Paraguay	1.30	3680
Zambia	0.60	1463	Ecuador	1.28	5425
Zimbabwe	0.60	909	Peru	1.22	6424
Mozambique	0.45	570	Colombia	1.13	7763
Ghana	0.35	1646	El Salvador	0.85	3782
Cameroon	0.26	1220	Honduras	0.68	2339
Cote d'Ivoire	0.25	1244	Bolivia	0.65	2576
Angola	0.24	5539	Nicaragua	0.57	1777
Senegal	0.21	1023	Guatemala	0.54	3341
Congo, Rep.	0.18	3154	Haiti	0.05	776
Kenya	0.16	933			
Nigeria	0.16	2742			
Sudan	0.16	1695			
Togo	0.15	589			
Congo, Dem. Rep.	0.11	418			
Tanzania	0.10	609			
Benin	0.09	751			
Eritrea	0.06	504			
Ethiopia	0.06	467			

producing new terms as "flexiwatts", shifting the timing of consumption, and "negawatts", using less electricity but with greater productively [24].

The initial threshold level of electricity consumption for rural households (i.e., 5 people and, 1 mobile phone, 1 fan and 2 compact fluorescent light bulbs for about 5 h per day) can be assumed to be 250 kWh per year, and for urban households (also including an efficient refrigerator, a second mobile phone and another appliance [e.g., small TV or computer]) it is 500 kWh per year [26]. Therefore, Table 3.7 can be also related to the level of urban development existing in African and Latin American countries.

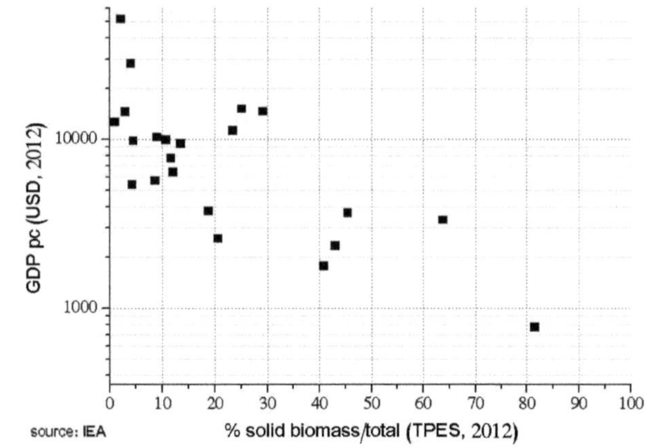

Fig. 3.19 GDP per capita versus percentage of solid biomass in the total primary energy supply for African and Latin American countries in 2012 [2]

As noted earlier, a logarithmic dependence exists between the percentage of solid biomass in the total primary energy supplied to a country and the GDP per capita (Fig. 3.19) [2]. Moreover, it is expected that countries with the highest percentage of solid biomass also show small industrial and service sectors.

Increasing energy efficiency is an urgent priority in Africa and Latin America, trying to decouple economic growth from carbon emissions [27]. The appliance ownership level is a good indicator of how energy demand will continue to grow, and in these regions it is generally much lower than in developed countries, and in rural communities it drops dramatically. With significant unmet energy demand in Africa and Latin America, energy efficiency can offer the opportunity to fundamentally alter the energy consumption growth and, simultaneously, avoid demand crisis by outstripping supply, improving air quality, better access to their electricity systems and investing savings in other more valuable economic activities.

In fact, there are African and Latin American countries promoting energy efficiency activities, mainly South Africa, Brazil and Mexico. Innovative efficiency programmes are making Mexico a leading country for energy efficiency investment in Latin America, acting on light and appliances (the Efficiency Lighting and Appliances Project has helped to replace over 1.6 million refrigerators, 0.2 million air-conditioning units and 45.8 million incandescent bulbs), equipment in the commercial sector, improve street lighting efficiency, energy-efficient buildings for low-income households (insulating ceilings and walls, reflective paint, efficient gas heaters, supplying new LED TVs, etc.). Brazil is also investing in energy efficiency, targeting a 10% electricity consumption savings in 2030 and alleviating energy poverty, and in the public sector, energy labelling appliances and equipment, and improving industrial energy efficiency, street lighting programs, energy efficiency in buildings, etc. South Africa is promoting end-use efficiency to improve system reliability. South Africa and also Mexico have made mass purchases of energy-efficient products such as CFL lighting, water heaters and refrigerators. Energy efficiency targets are being promoted in the Economic Community of West

African States and South Africa [27]. However, market forecast of smart meters penetration rate is close to zero in Africa and to 5% in Latin America [28]. In Africa, utilities in South Africa and Zimbabwe are among the leaders pursuing smart meters, and in Latin America utilities also remains moving slowly. Brazil, Mexico, Argentina and Chile are the only countries with modest deployments of smart meters [28].

Improving electricity system reliability is also a key driver of utility investment in energy efficiency in developing countries. Improved energy efforts designed to reduce technical losses can help expand energy access by enabling existing systems to provide electricity to more consumers, which can also drive energy efficiency investments of utilities [27]. In this respect, utilities can deliver end-use energy efficiency savings and consultancy services.

3.7 National and Regional Energy Plans

Most African and Latin American countries have introduced national energy plans (NEPs) to meet a growing energy demand with affordable, sustainable energy services that enable socio-economic development. Also regional plans are required and are in progress, mainly for cross-border hydropower projects, power pools and interconnections (e.g., Mali, Mauritania and Senegal share of the 200 MW Manantali hydropower plant on the Bafing river and a high voltage network of 1300 km transmission lines). NEPs are mainly approved to develop a strategy for diversification of the energy mix, ensuring energy security by using all endogenous resources, but also improving energy access (e.g., rural electrification, extension of the national grid and quality of the electricity supply), technical expertise, energy efficiency and savings, and an institutional framework to attract private investment. A brief description of most NEPs in African and Latin American countries is included in Table 3.8.

In conclusion, African and Latin American countries have adopted a wide variety of policy mechanisms to promote renewable energy deployment. The prevailing policies are fiscal incentives, including tax reductions (in sales, energy, CO_2, VAT, etc.), public investments, loans and grants, but also capital subsidies, grants or rebates. Tax reductions—the most widespread incentive—have many advantages, as they require no additional budget allocation, fewer administrative procedures and minimal regulatory supervision compared to other support policies. Also, as power sectors and institutional frameworks mature, regulatory policies such as auctions and net metering are becoming more prevalent. Several countries have adopted feed-in-tariffs (FiT), as exposed in Table 3.8. However, a current alternative that is gaining ground is based on auctions, in part to reduce support

Table 3.8 National energy targets and main policies in African and Latin American countries [9, 29, 30]

Algeria	Final energy target from renewables: 40% by 2030 Electricity generation from renewables: 27% by 2030 (2013: 0.5%) 22 GW renewable electricity by 2030 Feed-in policies (2002) and auctions
Angola	E10 biofuel blend mandate
Argentina	Electricity generation from renewables (non-hydro): 20% by 2020 (2014: 2.4%) 3 GW renewable electricity by 2016 Feed-in policies (2006) E10 and B10 biofuel blend mandate Auctions, VAT tax rebate and accelerated depreciation benefit 15% GHG emissions below BAU scenario in 2030
Benin	50% of rural electricity from renewable sources by 2025
Bolivia	160 MW renewable electricity added 2015–2025
Botswana	Final energy target from renewables: 1% by 2016 (electricity 2013: 0.1%) No specific incentives for clean energy
Brazil	19.3 GW biopower, 7.8 GW small hydro and 15.6 GW wind by 2021 Feed-in policies (2002) and tax incentives E27.5 and B7 biofuel blend mandate A fee discount for renewable energy transmission and distribution Net metering, but deployment has been slow
Burundi	Final energy target from renewables: 2.1% in 2020 4 MW biopower from solid biomass, 212 MW hydro, 40 MW PV, 10 MW wind
Cabo Verde	Electricity generation from renewables: 50% by 2020 Net metering and auctions
Cameroon	Energy plans to electrify 660 localities through grid extension and mini-grids Renewable energy projects can qualify for various tax reductions Pow er tariffs subsidized through direct budget transfers from the government to the utility
Chile	Electricity generation from renewables: 20% by 2025 (excl. large hydro) (2014: 13.1%) Renewable portfolio standards quota policies (2008) Auctions: generators bid to supply electricity during a certain time blocks (11 pm–8 am; 8 am–6 pm; 6 pm–11 pm) Net metering legislation approved (<100 kW) Renewable generators receive a reduction on transmission tariffs
Colombia	6.5% grid-connected and 30% off-grid renewable electricity by 2020 E10 and B10 biofuel blend mandate Income tax exemption for wind and biomass generators Import duties exemption on equipment for activities that generate carbon credits
Costa Rica	Electricity generation from renewables: 100% by 2021 (2013: 88.3%) E8 and B20 biofuel blend mandate Auctions; import, VAT and income tax exemptions for renewable energies A 5-year pilot net metering program for reaching 10 MW has 117 clients connecting renewable facilities to the grid during 15 years

(continued)

Table 3.8 (continued)

Côte d'Ivore	Final energy consumption target from renewables: 15% by 2020 (2014: 2.2%) Reduced VAT rates for solar panels
RD Congo	Investors are eligible for a range of tax reductions and import duty exemptions
Cuba	Electricity generation from renewables: 24% by 2030 (2013: 4.3%)
Djibuti	Electricity generation from renewables: 100% by 2020 30% of off-grid and rural electricity from PV by 2017
Dominican Republic	Electricity generation from renewables: 25% by 2020 (2013: 14.3%) 20% of distributed power in 2016 Feed-in policies (2007), clean energy fund and multiple tax credits Renewable generators dispatch priority and open access to the grid Net metering program with 59 customers connecting renewable facilities Conditional 25% GHG reduction in 2030 compared to 2010 levels
Ecuador	Electricity generation from renewables: 85% by 2017 (2013: 49.0%) Non-solar feed-in policies (2015) B5 biofuel blend mandate Tax exemptions
Egypt	Primary energy target from renewables: 14% by 2020 Electricity generation from renewables: 20% by 2020 (2013: 8.6%) 2.8 GW hydro by 2020, 700 MW PV by 2027, 2.8 GW CSP by 2017, and 12% of generation and 7.2 GW of wind power by 2020 Feed-in policies (2014), net metering and auctions
El Salvador	Generation open to private operators (around 13 IPPs* are active) Import duty exemption to clean energy equipment and machinery and income tax exemption to renewable generators First auction (2014) contracted 94 MW of solar PV at an average price of $116.2/MWh
Eritrea	Electricity generation from renewables: 50% (no date) (2013: 0.1%) 50% of wind power (no date)
Ethiopia	103.5 MW biopower from bagasse (no date) 1 GW geothermal (independent power producer) and 22 GW hydropower by 2030 770 MW wind by 2014 E10 biofuel blend mandate, and E20 upon commissioning of a new ethanol manufacturing plant Investors eligible for tax reductions and import duty exemptions
Gabon	Final energy target from renewables: 80% by 2020 Electricity generation from renewables: 70% by 2020 (2013: 37.9%)
Gambia	Electricity generation from renewables: 35% by 2020
Ghana	Electricity generation from renewables: 10% by 2020 (non-hydro 2014: 0.1%) Feed-in policies (2011), RPS*, net metering, TRE* certificates and auctions Investors are eligible for accelerated depreciation and import duty exemptions

(continued)

Table 3.8 (continued)

Guatemala	Primary energy target from renewables: 20% by 2020 Electricity generation from renewables: 80% by 2027 (2013: 67.2%) E5 biofuel blend mandate State and private players act in generation, transmission, energy trading and distribution segments (unbundled) Auctions since 2012, and tax exemptions
Guinea	6% PV and 2% wind in 2025
Guinea Bissau	Electricity generation from renewables: 2% by 2015 2% primary energy from PV
Haiti	Electricity generation from renewables: 50% by 2020 (2013:12.8%)
Honduras	Electricity generation from renewables: 80% by 2038 (2013: 37.8%) Feed-in tariff (2014): 10% price premium to clean energy projects for the first 15 years of operation One auction and tax incentives
Kenya	20 GW of power capacity in 2030, comprising 51% renewable 100% electrification in 2020 Solar water heating: 60% (buildings consuming > 100 l/day) (no date) 5 GW geothermal by 2030 423 MW PV, 794 MW hydro and 635 MW wind in 2016 Feed-in policies (2008), auctions and PPAs* Net-metering (expected in 2015)
Lesotho	260 MW renewable electricity by 2030 35% of rural electrification from renewables by 2020 Net metering and auctions
Liberia	Electricity generation from renewables: 30% by 2021 (2014: 15%) Biofuels in 5% of total transport by 2015 Increase of the electrification rate to 30% by 2030 (2014: 2%) Renewable projects are eligible for tax reductions and import duty exemptions
Libya	Primary energy target from renewables: 10% by 2020 Electricity generation from renewables: 10% by 2025 (2013: 0.0%) Solar water heating: 250 MW_{th} by 2020 844 MW PV, 375 CSP and 1 GW wind power by 2025
Madagascar	Final energy target from renewables: 54% by 2020 Electricity generation from renewables: 75% by 2020
Malawi	Primary energy target from solar and wind: 7% by 2020 346.5 MW hydropower by 2014 E10 since 2004, often not met due to short supply 2 PPAs* on course (hydro and coal), tax reductions and import duty exemptions
Mali	Primary energy target from renewables: 15% by 2020 Electricity generation from renewables: 25% by 2033 (excl. large hydro)
Mauritania	Primary energy target from renewables: 15% by 2015
Mauritius	Primary energy target from renewables: 35% by 2025 Electricity generation from renewables: 35% by 2025 (2013: 20%) Feed-in policies (2010)

(continued)

Table 3.8 (continued)

Mexico	Electricity generation from renewables: 25% by 2026 (2013: 13.3%) Accelerated depreciation for renewable energy projects and machinery Power generation is open to private developers (auctions and clean certificates) Net metering: surplus generation to the grid obtains billing credits 30% GHG reduction in 2020 with respect to 2000 Accelerated depreciation of investment in equipment State funds on rural electrification, efficiency and R&D
Morocco	Solar water heating: 1.2 GW_{th} by 2020 42% electricity renewable, 2 GW hydro, 2 GW PV-CSP, 2 GW wind by 2020 Auctions
Mozambique	100 MW onshore wind and 125 MW small-hydro by 2025 Solar off-grid electrification for 2.1 million people by 2021 Solar water and space heating: 100,000 systems in rural areas (no date) 1000 biodigesters for biogas and 3000 wind-water pumping (no date) 2 GW hydro, 2 GW PV, 2 GW wind, 82,000 solar home systems (no date) 5000 renewable energy systems (no date) E10 in 2012–2015; E15 in 2016–2020; E20 from 2021 Tax reductions and import duty exemptions for large infrastructure investors
Namibia	Electricity generation from renewables: 10% by 2020 (non-hydro 2013: 0.0%)
Nicaragua	Electricity generation renewables (voluntary): 91% by 2027 (2013: 52.3%) Feed-in policies (2015): USD 66/MWh - USD 80/MWh (lowest range) for wind projects up to USD 103/MWh - USD 118/MWh (highest range) for solar plants Full range of tax breaks Generators can compete freely in the market
Niger	Primary energy target from renewables: 10% by 2020 (electricity 2013: 0.9%)
Nigeria	Non-legally binding targets: Electricity generation from renewables: 10% by 2020 (excl. hydro > 30 MW) (2013: 0.0%) 400 MW biopower, 2 GW small-hydro, 500 MW PV (> 1 MW) by 2025 40 MW wind power and 5 MW CSP by 2025 15-year fixed feed-in policies for wind, solar, small-hydro and biomass (2012) E10 and B20 by 2020
Panama	Primary energy target from renewables: 18.3% by 2023 A 5% price premium to renewable projects up to 3 MW that sell electricity to La Empresa de Transmisión Eléctrica (ETESA) E10 biofuel blend mandate Auctions to contract renewable energy Import duty exemption for clean energy equipment; income tax credit and transmission and distribution tax exemption for generators
Paraguay	E24 and B5 biofuel blend mandate Little incentive to other renewable policies
Peru	Electricity generation from renewables: 60% by 2025 (2013: 53.0%) Auctions: 500,000 off-grid solar PV systems in operation in 15,000 remote locations (2019), 450 GWh from small hydro projects up to 20 MW, and 1300 GWh of biomass, wind and solar PV E7.8 and B5 biofuel blend mandate Accelerated depreciation of investment in renewable generation Grid operator guarantees priority dispatch for renewable sources

(continued)

Table 3.8 (continued)

Rwanda	300 MW biogas, 310 MW geothermal, 340 hydro, 5 MW off-grid by 2017 42 MW small-hydro by 2015 70% access to electricity by June 2018 Feed-in policies (2012): currently USD 67–166/MWh Tax incentives and accelerated depreciations
Senegal	Electricity generation from renewables: 20% by 2017 (2013: 10.0%) RPS*: 50 MW solar by 2015 Tax reductions and import duty exemptions in rural electrification concessions
Sierra Leone	Electricity generation from renewables: 36% by 2030 5% penetration of solar water heaters in hotels, guest houses and restaurants
South Africa	Electricity generation from renewables: 9% (2013: 1.8%), 17.8 GW (9.2 GW wind and 8.4 GW solar), 42% new capacity by 2030 Feed-in policies (2009, abandoned in 2011), auctions (7 GW, from 2011), net metering and RPS* E10 and B5 biofuel blend mandate from 2015 Accelerated depreciation, and tax deductions for energy efficiency measures
Sudan	Electricity generation from renewables (non-hydro): 11% by 2031 (2013: 0.0%) 54 MW biopower from solid biomass, 68 MW biopower from biogas, 63 MW hydropower, 667 MW PV, 50 MW CSP and 680 MW wind by 2031 E5 biofuel blend mandate
Swaziland	20% of all public buildings installed with solar water heaters
Tanzania	Electricity generation from renewables (non-hydro): 14% by 2015 (2013: 0.6%) 100 MW of solar and 200 MW each of geothermal and wind by 2025 A new feed-in tariff for projects up to 10 MW, superseding the standardized PPA program (<10 MW, 15 yr.) Accelerated depreciation (solar), and tax deductions
Togo	Electricity generation from renewables (non-hydro): 15% by 2020 (2013: 4.5%)
Tunisia	Electricity generation from renewables: 30%, 4.6 GW by 2030 (2013: 2.3%) 300 MW from solid biomass, 1.5 GW PV, 500 MW CSP, 1.7 GW wind by 2030 Net metering
Uganda	Electricity generation from renewables: 61% by 2017 (2014: 25%) 30 MW from organic MSW, 45 MW geothermal, 1.2 GW large-hydro, 85 MW mini- and micro-hydro, 700 kW PV, and 2200 million litres biofuels by 2017 Solar water heating: 21 MW_{th} by 2017 Competitively allocated 20-year tariffs for limited amounts of small hydro, biomass, biogas, geothermal and wind. Auctions for on-grid PV

(continued)

Table 3.8 (continued)

Uruguay	Primary energy target from renewables: 50% by 2015 (2013: 47.5%) Electricity generation from renewables: 92% by 2015 (2013: 80.9%) 200 MW biopower and 1.3 GW wind power by 2015 Auctions for wind and solar energy E5 and B5 biofuel blend mandate Income tax reduction for renewable generators and VAT exemption for wind equipment
Venezuela	613 MW renewable new capacity (500 MW wind) installed 2013–2019
Zambia	Small hydro and solar developers are eligible for tax holidays and import duty exemptions
Zimbabwe	10% of transport final energy demand by 2015 (2013: 0.0%) Universal electricity access by 2040 (2014: 31%) E5 biofuel blend mandate raised to E10 and E15 (no date) Energy investors are eligible for ten-year income tax holidays

*TRE *tradable renewable energy*, *RPS* renewable purchase standards, *PPA* power purchase agreements, *IPP* independent power producer

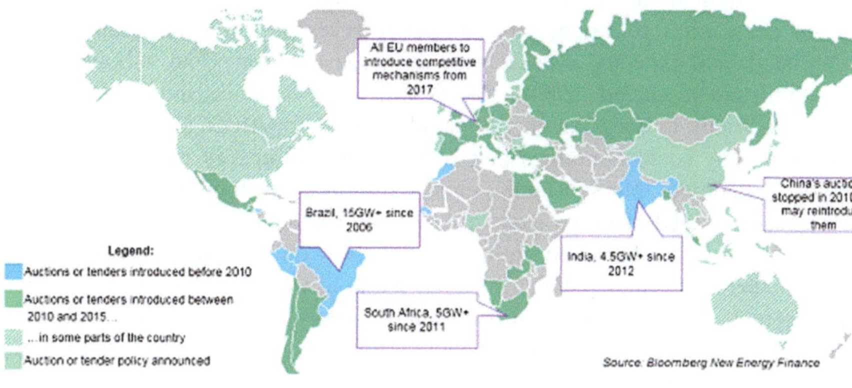

Fig. 3.20 Rapid increase in popularity of renewable energy auctions worldwide [31]

costs compared to FiT, since auctions can be effective in bringing out real prices (Fig. 3.20). Also power purchase agreements (PPAs), tradable renewable energy (TRE) auctions and renewable purchase standards (RPS) are used [30].

Thus, Brazil and South Africa have emerged as clear champions of the auction process, with more than 12.3 GW of onshore wind and 2.3 GW of solar PV procured through auctions since 2006 in Brazil alone [31].

3.8 Conclusions and Future Perspectives

Africa and Latin America possess tremendous potential for significant growth of renewable energy resources to support more than 23% of the 2015 global population. Much of this potential is derived from access to large land areas with more than 37% of the combined global surface area, the availability of abundant renewable resources (i.e., wind, water and solar), commitments to cut global emissions by 2030 and a growing recognition by some nations that low carbon development can enhance economic prosperity while reducing potential impacts on climate change. Also, Africa and Latin America produce more than 23% of the global renewable energy, with much of this energy provided by traditional biofuels and waste (36% of total global production) to support daily sustenance needs for heating and cooking rather than electricity production characteristic of nations with mature economies.

Moreover, Africa and Latin America have diverse energy reserves that vary between nations but seldom achieve capacity levels necessary for the long-term sustainment, or more importantly growth, of an individual national or regional economy. Current demand for energy in many nations is low (i.e., energy consumption per capita) so existing fossil fuel reserves are often adequate to meet existing needs and expected average growth in consumption. Coal, oil, natural gas and uranium are produced in both Africa and Latin America and nuclear power provides electricity to both. An energy paradox exists for Africa and Latin America driven largely by socioeconomics that suggests current energy resources are adequate for sustainment, but substantial energy resource development is necessary to improve regional GDP. Comparisons of GDP and CO_2 emissions per capita for the different countries in Africa and Latin America show only a few nations have been capable of increasing their GDP per capita while at the same time reducing their CO_2 emissions. Both regions have substantial opportunity to increase their GDP based on the expanded penetration of low carbon indigenous energy technologies into their infrastructure.

In conclusion, the future prospects for low carbon energy technologies in Africa and Latin America are significant, with energy resources and low carbon energy production increasing in general. Many nations have established national targets, commitments to establish and enlarge interconnections and have approved incentives at different levels that are helping to consolidate the low carbon energy future in these regions.

References

1. Energy Information Administration (2015) US Department of Energy. http://www.eia.gov/cfapps/ipdbproject/IEDIndex3.cfm?tid=5&pid=57&aid=6. Accessed 30 Nov 2015
2. International Energy Agency database (2015) http://data.iea.org/. Accessed 30 Nov 2015

3. World Nuclear Association (2015) http://www.world-nuclear.org/info/Nuclear-Fuel-Cycle/Uranium-Resources/Supply-of-Uranium/. Accessed 30 Nov 2015
4. Countries of the World (2015) http://www.theodora.com/pipelines/africa_oil_gas_and_products_pipelines_map.html. Accessed 30 Nov 2015
5. COMELEC (2010) Statistiques Annuelles http://comelec-net.org/IMG/pdf/Depliant_COMELEC_2009.pdf. Accessed 30 Nov 2015
6. IRENA (2013) http://www.irena.org/DocumentDownloads/events/2013/July/Afriac%20CEC%20session%202_EAPP_Gebrehiwot_220613.pdf. Accessed 30 Nov 2015
7. Southern African Power Pool (2015) Annual Report 2014 http://www.sapp.co.zw/docs/Annual%20report-2014.pdf. Accessed 30 Nov 2015
8. McKinsey & Company (2015) Brighter Africa. Electric Power & Natural Gas
9. Climatescope (2015) The clean energy country competitiveness index. The Multilateral Investment Fund, UK Department of International Development, Power Africa and Bloomberg New Energy Finance
10. SIEPAC (2014) Informe General, Antecedentes, estado actual y perspectivas del Sistema de Interconexión Eléctrica para los Países de América Central
11. Sistema de Interconexión Eléctrica Andina. www.sinea.com
12. Banco de Desarrollo de América Latina (2012) Nuevas Oportunidades de Interconexión Eléctrica en América Latina. CAF, 2nd edn
13. IEA/IRENA (2015) Joint policies and measures database. http://www.iea.org/policiesandmeasures/renewableenergy/. Accessed 30 Nov 2015
14. Global Climatescope (2015) http://global-climatescope.org/. Accessed 30 Nov 2015
15. BNEF (2016) Week in review. Available on 5 Apr 2016
16. BNEF (2016) Week in review. Available on 12 Apr 2016
17. IRENA (2016) Renewable energy and jobs. Annual Review 2016
18. World Bank Data (2015) http://data.worldbank.org/indicator/EP.PMP.SGAS.CD. Accessed 30 Nov 2015
19. World Bank Data (2015) http://data.worldbank.org/indicator/EP.PMP.DESL.CD/countries?display=map. Accessed 30 Nov 2015
20. Vidoza JA, Gallo WR (2016) Projection of fossil fuels consumption in the Venezuelan electricity generation industry. Energy 104:237–249
21. IMF (2013) Energy subsidy reform: lessons and implications
22. IEA (2014) Energy supply security. Emergency response of IEA countries 2014
23. IEA (2015) How 2 guide. Smart grids in distribution networks. Roadmap Development and Implementation
24. Dyson M, Mandel J, et al (2015) The economics of demand flexibility: how "flexiwatts" create quantifiable value for customers and the grid. Rocky Mountain Institute
25. World Bank Data (2015) http://databank.worldbank.org/. Accessed 30 Nov 2015
26. IEA (2014) Africa energy outlook
27. IEA (2015) Energy efficiency market report 2015. Market trends and medium-term prospects
28. Navigant Research (2014) Smart meters. Executive summary research report. https://www.navigantresearch.com/wp-assets/brochures/AMI-14-Executive-Summary.pdf. Accessed 30 Nov 2015
29. REN21 (2015) Renewables 2015. Global status report
30. IRENA (2015) Africa 2030: roadmap for a renewable energy future
31. Bloomberg New Energy Finance (2016) New energy outlook 2016. Powering a changing world

References

Chapter 4
Power Grids

Ricardo Guerrero-Lemus and Les E. Shephard

Abstract This chapter shows that the electricity mix in Africa and Latin America is highly varied but is typically tied directly to endogenous resources controlled by individual nations. While much of the electricity is generated using antiquated fossil fuel technology, it should be recognized that Latin America is the most decarbonized region in the world because of the abundance of hydropower, a dispatchable source of renewable energy. Non-dispatchable renewable energy (i.e., wind and solar) is prevalent in only a few selected nations. Sub-Saharan Africa contains one of the least electrified regions in the world with an overall electrification rate below 50% and with 17 nations that have electrification rates less than 20%. This contrasts significantly with North Africa and the majority of urban Latin America (Haiti is an exception), where more than 99% of the urban population has access to electricity. Electricity consumption in both regions remain well under the global average (21 and 60% of the world electricity output per capita, respectively), and rural electrification remains challenging for both regions. On the other hand, electric grid reliability remains problematic in many nations where proper maintenance is limited, power theft is prevalent and power outages frequent, limiting the availability of power with the concomitant impact on local economies. Moreover, in many countries the traditional grid extension mode requires a multidimensional approach covering regulation, finance, economic development and social dimensions to analyze the business case of each rural electrification project. The additional costs of decentralized electricity supply is driven by the cost of alternative power generation, and it is related to scale and to variations in daily power demand. However, increasing electrification rates, grid extension, reliability and trading, and diminishing electricity price will raise regional GDP and living standards in these regions.

Keywords Energy · Africa · Latin America · Electricity · Power grid · Rural electrification · Net metering · Energy storage · Power storage · Renewable energy · Transmission network · Offgrid power · Batteries · Transformers

R. Guerrero-Lemus and L.E. Shephard, *Low-Carbon Energy in Africa and Latin America*, Lecture Notes in Energy 38,
DOI 10.1007/978-3-319-52311-8_4

4.1 Introduction

In an ever more electrified world, power grids play a central role connecting power plants and consumers. However, worldwide 1.3 billion people continue to live without access to electricity, mainly in rural areas [1], and rapid population growth can continue to outpace the rate of electrification in many countries. Africa is one of the least electrified regions in the world but provides a striking contrast between the North, where more than 99% of the population has access to electricity, and the Sub-Saharan regions with an overall electrification rate below 50%. Most countries in Latin America have successfully reached near universal electrification, based on grid extension and strong state support [2]. Countries with electrification rates of less than 80% of the population consistently suffer from reduced GDP per capita except those with significant wealth in natural resources (e.g., Angola, Botswana and Gabon), but even they fall well short of economic prosperity [3].

Power plants provide electricity to the power grids, and can be classified according to fuel/principle: thermal (engine, conventional steam, nuclear, gas turbine, combined cycle, and combined heat and power), mechanical (hydraulic and wind), electrochemical (fuel cells and batteries) and photovoltaic. Also power plants can be classified according to the load supply: base, intermediate or peak load. The power plants are responsible for the power regulation of the grid and, consequently, for its frequency regulation but also, to a certain extent, for its voltage regulation. Also the grid can regulate voltage through the use of alternative components like transformers, compensators (capacitors and inductors) and FACTS (flexible AC transmission systems).

As summarized in Chap. 2, some technological advances in the energy sector can help to increase efficiency and decarbonize the energy sector through the replacement of coal fired power plants by natural gas combined cycle power plants (Sect. 2.2), nuclear power plants and/or renewable technologies (Sect. 2.3), and carbon capture and storage techniques (Sect. 2.5). There are also technologies directly related to the improvement of the power grid that can also contribute to the decarbonisation of the energy supply. They are based on (i) smart grid management devices; (ii) storage systems; and (iii) distributed generation. There are also alternatives to the placement of low carbon energy technologies in suboptimal locations by extending the transmission grid to optimal locations. This chapter emphasizes the key components of a reliable, decarbonised power grid.

4.2 Technology State of the Art

Power grids utilize AC power transmission lines to avoid losses ($\propto I^2R$) across long distances or/and submarine connections or HVDC (High Voltage Direct Current) connections. The power lines can be aerial or underground cables but the later are

Table 4.1 Voltage ranges and risks

IEC voltage range	AC (V_{rms})	DC (V)	Defining risk
High voltage (HV)	>1000	>1500	Electrical arcing
Low voltage (LV)	50–1000	120–1500	Electrical shock
Extra-low voltage	<50	<120	Low risk

much more costly and difficult to repair. In the aerial configuration, the cables are suspended between towers or poles, separated by insulators (the length of the insulator increases with the voltage) with the specific characteristics mainly dependent on the terrain, climatic risks, environmental aspects, security and mechanical loads. The power grid can be divided in two categories depending on the voltage range: (i) transmission grid (220, 400 kV) for the transportation of power over long distances; and (ii) distribution grid (10 kV, 20 kV, 400 V) connecting the end users. Also the power lines are classified in terms of high or low voltage range and by risks (Table 4.1). The conductor is a stranded cable that in aerial high voltage networks usually consists of an aluminium conductor steel reinforced (ACSR) cable. In low voltage grids and underground lines the wire is normally insulated to protect against direct lightning strikes into the phase.

Power transformers are used to connect power networks of different voltage, but also power and current transformers are used to connect measuring or protection equipment to the power grid. The transformers can be immersed in oil, in a synthetic liquid (e.g., silicone liquid) or can be dry (core and winding are resin-encapsulated).

Substations (with transformers) or switchgear installations (without transformers) are used to connect the different parts of a network. They are the "nodes" (points where the lines meet) of the network. Substations can be found in power plants, transmission networks, distribution networks and industrial networks, and the design and construction depends on the voltage level. The main criteria for differentiating the substations are: (i) indoor or outdoor; (ii) metal-enclosed, gas-insulated (GIS) or outdoor air-insulated (AIS); and (iii) type of insulator (e.g., solid materials, air or SF_6). The SF_6 GIS has a smaller size and no ambient influences compared to AIS, but it is more expensive, not easily accessible and maintenance and expansion are more complicated.

The switchgears are used to switch on and off lines, transformers, generators, motors, compensation elements (capacitors and inductances), among others. The most important distinction between switchgears is their ability to withstand certain currents while switching off and on equipment. There are disconnectors to provide isolating distance in the open position, circuit-breakers for making or breaking all short-circuit and service currents occurring at the operational site, and fuses that melt and interrupt the current when a certain value is reached.

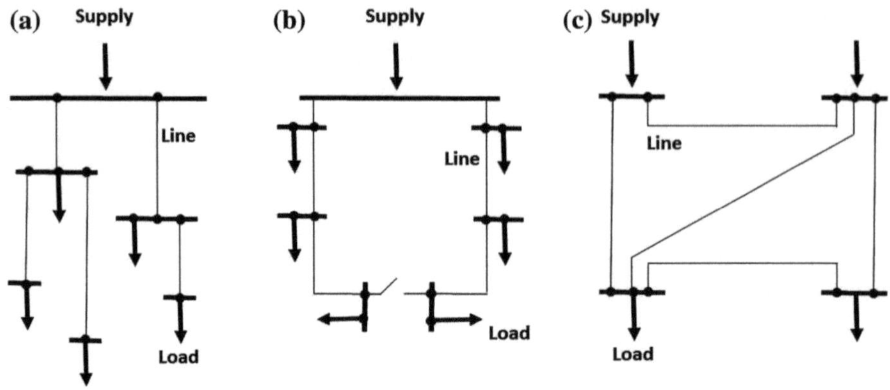

Fig. 4.1 Scheme of a (**a**) radial network; (**b**) ring or loop network; and (**c**) mesh network

A substation contains the switchgear and many other components that may be divided into different categories:

(i) Primary installations (all equipment connected to the voltage network): switchgears, fuses, bus bars, voltage and current transformers, surge or lightning arresters, power transformers and reactive power compensators (capacitors and inductances).
(ii) Secondary installations: protection relays and equipment for control and communications.
(iii) Auxiliary systems: power supply, heating and air-conditioning, and light.

There are three main possible network topologies (Fig. 4.1):

(i) Radial network: simple (simple protection, low costs, simple operation), low reliability because there is no redundancy, high voltage drop, high loses, and application in LV distribution.
(ii) Ring or loop network: can be operated open and closed, simple structure, simple operation, simple protection, higher reliability and higher costs compared to the radial grid, and application to high voltage distribution.
(iii) Meshed network: high reliability, low voltage drop, low losses, more complex protection, higher costs, complex start-up after blackout, more complex planning and operation, and application in transmission networks (sometimes also in distribution networks).

The frequency of the AC can be 50 or 60 Hz depending on the country (see Chap. 3) and some small frequency deviations are admitted (e.g., ±0.15 and ±0.25 Hz for intervals under 5 min). Also, voltage deviations should not exceed ±10%. There is also a N − 1 security rule in the grid that means in a system with N elements, if a single (1) element fails, the system will continue to function properly: (i) no element is overloaded; (ii) all voltages are within the specific limits; and (iii) stability of the HV-grid is given. The probability of two elements failing independently at the same time is usually very low, but there are cases where two elements fail at the same time due to a common cause/event.

Finally, many times it is possible to retrofit old thermal power stations, extending their lifetimes or improving their operation performance. Retrofitting can improve the economic performance of old stations substantially in power systems with increasing shares of non-dispatchable energy technologies, as discussed below. Retrofitting can involve a wide variety of measures ranging from a rapid and not very costly exchange of valves in the boiler system to a complete replacement of the steam-turbine system. Conventional power units are usually not standardized but individually designed for specific requirements and locations (e.g., local coal quality). Then, the retrofit measure implemented may be adapted on the individual component to be exchanged.

4.2.1 Power Transmission Grids

The traditional choice for the transport of electricity is high voltage alternating current (HVAC) combined with the assistance of transformers to increase or lower the tension, since it is the less costly option for connections across distances under 100 km, compared with the HVDC technology, where also DC/AC converters are required. Tensions up to 800 kV in HVAC are used in some electricity systems.

HVDC technology has been used since 1954, when the island of Gotland was connected to Sweden by a submarine cable of 96 km, 20 MW and 100 kV. There are two types of connections: (i) HVDC systems with line commuted converters (LCC), where the conversion efficiency AC/DC/AC is about 97–98%, and (ii) the most recent voltage source converter (VSC), based on insulated gate bipolar transistor (IGBT) technology.

HVDC cables experience low losses particularly when they are placed on the seabed, where transmission losses are mainly attributed to capacitance effects (reactive power generation) (Table 4.2). These cables can also be placed at shorter distances to objects sensitive to electromagnetic radiation, because it avoids the problems of variable electromagnetic fields associated with HVAC lines and, therefore, facilitates the planning of new grid lines in urban areas. However, HVAC lines can transport electricity very simply, in addition to the fact that AC motors and generators are simpler and easier to maintain than DC motors.

Table 4.2 General characteristics of HVAC and HVDC technologies [4]

	Unit	HVAC	HVAC	HVDC	HVDC
Operation voltage	kV	760	1160	±600	±800
Overhead line losses	%/1000 km	8	6	3	2.5
Sea cable losses	%/100 km	60	50	0.33	0.25
Terminal losses	%/station	0.2	0.2	0.7	0.6

Table 4.3 Average costs of different HVAC and HVDC technologies [5]

	Unit	HVAC	HVAC	HVDC	HVDC
Operation voltage	kV	760	1160	±600	±800
Overhead line cost	MUSD/1000 km	623–1167	1556	623–700	389–467
Sea cable cost	MUSD/1000 km	4980	9182	3891	2801
Terminals cost	MUSD/station	125	125	389–545	389–545

The costs of transmission grids depend on available technologies and characteristics of the grid in which the grid components are introduced. However, average costs of transmission lines in different configurations are exposed. Thus, as shown in Table 4.3, the average costs of transmission lines are well defined, with variations based on market prices of raw materials, mainly aluminium and steel. We can observe in Table 4.3 that the costs are lower for HVDC than HVAC transmission lines. Consequently, the former is more economical for long-distance connections while the latter is more economical for short and medium distances. "Supergrids" is a recent term applied to grids that connect electric stations over very large distances (often on the seabed) with HVDC technologies that reduce the losses normally attributed to long distances.

In terms of costs per unit of energy transmitted, the values depend on the country to be analysed, although it ranges from USD 5.8 to 8.4/MWh [5], equivalent to a 5–10% of total cost in electricity supplied to end users. Superconducting cable prices have decreased about 90% from 1990's levels and even though the price remains ten times higher than that of an ordinary copper cable, at around USD 15–25 per kA per metre, serves as the industry's preferred unit [6].

Case Example 4.1

Mobile Phone Call Data as an electricity proxy indicator in Côte d'Ivoire

The recent publishing of anonymized call detail records from mobile phones open the door for temporal and spatial human dynamic studies. These studies could be also used for creating new power proxy indicators that will not rely only on local or state institutions. A recent study uses the frequency of calls at different times of the day, in different small regional units (sub-prefectures) in Côte d'Ivoire to analyze power grid data. The spatial-variance of calling frequency identifies rural regions (north) lacking electricity. In these areas the morning calling frequency increases to an average of 10% of the average calling frequency at an early time compared to cities and regions to the south. This is attributed to the fact that the people in those areas live more according to the daylight and do the most arduous tasks before temperatures increase. The results also show that the north suffers from shortages in power generation and distribution. The results show a high correlation between mobile phone activity and power demand, and can be useful for the planification of grid extensions and the expansion of other services [7].

4.2.2 Power Distribution Grids

The distributed grid is usually radial and is designed to handle maximum possible load flows. A hierarchical tree structure of a distribution network is illustrated in Fig. 4.2, giving a typical example of a rural area [8]. Usually, there are several local network stations with MV/LV transformers in a village. Depending on their size, several villages are connected to one MV feeder. In the HV/MV substation, the MV network is interconnected with the HV grid. To estimate the load on the different grid elements the patterns and corresponding maximum loads of the single consumers and producers (e.g., distributed PV power) and the aggregated impact of multiple systems have to be analysed. Considering only consumption, the maximum load on the grid is not the sum of the maximum loads for single households. Empirical results and operating experience show a significant decrease in the maximum occurring load per house connection (HC) and also for sizing ratios in different sections of the grid (Fig. 4.2).

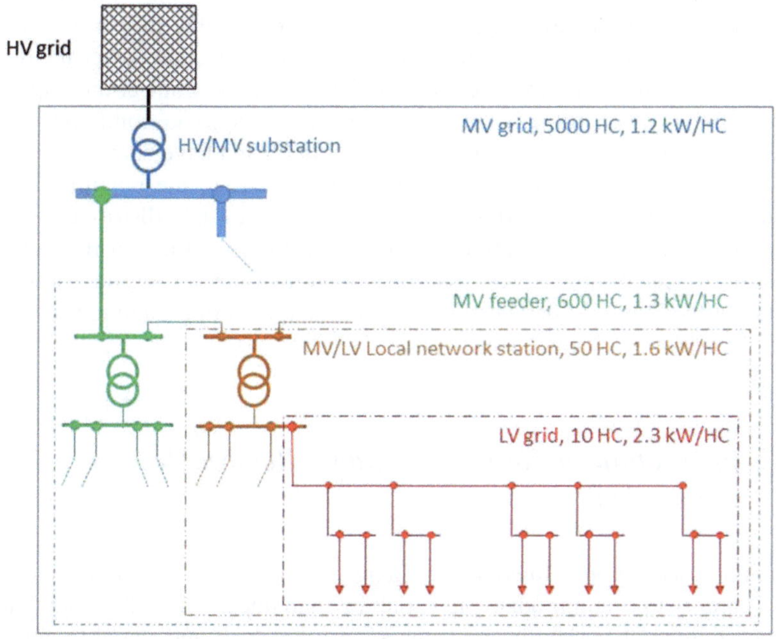

Fig. 4.2 Typical network structure with different sections and their maximum loads normalized to the number of house connections (HC). Additionally, the different spatial scales of the used datasets and analyses are shown [8]

Case Example 4.2

Electrification planning tool applied to Ghana

The study uses Network Planner, a decision support tool for exploring costs of different electrification technologies for providing electricity to 2600 non-electrified communities in Ghana within a 10-year planning period. The web-accessible model is written in Python, and developed by a team from Modi Research Group at the Earth Institute in Columbia University (New York, USA). The model requires electricity demand, geospatial, socio-economic, demographic and cost data, incorporates Geographical Information System (GIS) tools to perform spatial processing and analysis, and generates a comprehensive, cost optimized electricity plan, including a map of the projected grid extension. The results show that the cost-optimized option for most communities is grid connection (85%). This can be attributed to the extensive pre-existing grid network coverage over the country, which reduces the distances and thus the costs, to connect remaining communities. In the initial stages of grid development Ghana chose to connect the grid to all regional and district capitals, establishing a basic grid that makes it easier to connect communities later. Sensitivity analyses show that lowering diesel costs or lowering household electricity demand increase mini-grid compatible communities. Even though the model recommends some communities for off-grid electrification, previous experiences in Ghana have shown that communities have often preferred grid connectivity due to the potential use of high-wattage appliances and a perspective that renewable distributed power systems (e.g., PV) are inferior. Some prefer to discontinue using off-grid systems to put political pressure on governments for local grid extension, particularly if the grid has been extended to neighbouring communities [9].

4.2.3 Integration of Non-dispatchable Renewable Generation

Integration costs can be defined as the change in production costs associated with a system's ability to accommodate the variability and uncertainty of the net load [10]. The two sources of variability and uncertainty are load and variable generation. There are four components of production costs affected by variable net load: cycling costs, non-cycling variable O&M costs, fuel costs, and reserves provisioning costs.

To understand the technical and economic challenges presented by the integration of non-dispatchable renewable generation (mainly solar and wind energy) in

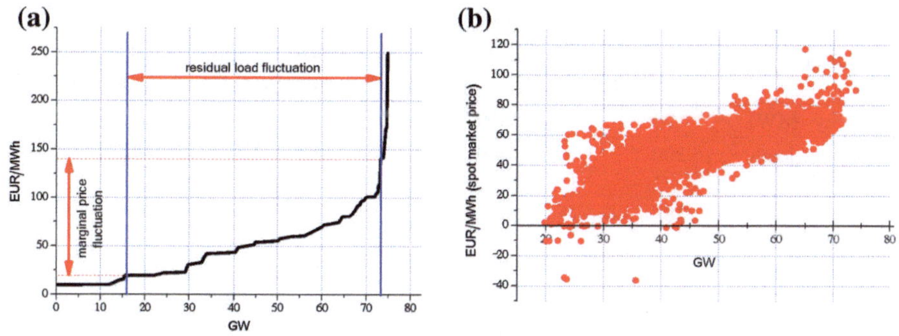

Fig. 4.3 **a** Example of the determination of power price in a market model; and **b** relationship between residual load and spot market power prices for a power system of a developed country [11]

a power grid, it is important to understand how conventional power plants operate. The usual market model for power purchase determines the least-cost generation schedule of the power plant fleet to meet residual load in every hour of a year (Fig. 4.3a). The hourly power prices thus correspond to the variable generation costs of the last unit needed to satisfy demand in this hour. The price is determined by the intersection of a merit-order curve with a (price-inelastic) residual load curve. Although in reality there clearly exists a strong positive correlation between prices and residual load, this relationship is not perfect but varies because of multiple circumstances (Fig. 4.3b) [11].

There should be some technical and economic restrictions considered for determining the variable generation cost of any conventional power unit (Table 4.4): (i) start-up cost, start-up duration and power output trajectory during start-up; (ii) shut-down power output trajectory and minimum downtime; (iii) efficiency loss when operating in part-load and minimum load level; and (iv) ramping gradients. Starting-up a power plant is often costly due to the extra fuel use for pre-heating the steam cycle and additional wear-and-tear associated with start-up processes. Fuel (mainly gas) order errors can also increase costs. Operation below full-load usually comes at a cost due to a part-load efficiency loss. During minimum load operation the plant usually incurs operational losses as the revenues in these hours do not cover all variable costs, but the power unit remains connected to the grid and can thus ramp up rapidly when prices increase, resulting in additional revenue opportunities. Then, stopping and starting a fast-start unit can be less expensive than leaving it running at minimum generation levels during low-load time periods.

Increasing shares of non-dispatchable energy resources impact residual load, backed up by the conventional power plant fleet, and therefore change power prices, operational patterns and operational income of thermal power plants. Thus, with increasing non-dispatchable energy, residual load is lower and more volatile. Volatility means that the bandwidth of residual load fluctuation increases

Table 4.4 Key technical and economic parameters of benchmark power plants [11]

Parameters	Hard coal (old) (1981–1985)	Hard coal (new) (state-of-the-art)	CCGT (old) (1976–1980)	CCGT (new) (state-of-the-art)
Capacity (MW)	400	800	350	400
Minimum load (MW)	200	200	210	160
Máximum efficiency (%)	38	46	48	59
Part-load efficiency (%)	31	40	37	48
Emissions (tCO_2/MWh_{th})	0.339	0.339	0.202	0.202
Start-up cost (EUR)	30,000	40,000	9000	7000
Fixed O&M costs (EUR/kW)	28	28	23	23
Ramping gradient (% capacity/min)	1.5	3.0	2.0	4.0
Duration of start-up (h)	3	3	2	2
Minimum down time (h)	3	3	3	2

substantially. This fluctuation can reach negative values, implying the need for storage or curtailment of non-dispatchable energy. Hourly residual load swings also become more pronounced, emphasising the need for fast ramping capabilities in the power plant fleet when non-dispatchable capacity is being expanded.

On the contrary, increasing shares of non-dispatchable energy depress prices in most hours of the year [11]. In many power grids the peak loads and associated price spikes occur in the evenings, when PV output is diminishing and wind energy is very stochastic. Consequently, the highest price hours are hardly affected while the prices of many other hours are depressed, resulting in a steeper price duration curve. Also, there are important variations in wind capacity factors from one year to the following. Of course, increases in wind and PV capacities depress on average power prices and, consequently the profitability of the thermal plants will decrease. Cycling costs in a more flexible scenario can be higher (increasing the cost of gas turbines), but the overall generation costs can decrease. However, the magnitude of the effect depends on the characteristics of the power system.

Full load hours of all dispatchable power plants decrease with increasing shares of non-dispatchable power. The state-of-the-art coal and gas plants slip from baseload to mid-merit operation and the number of power plant start-ups increase. For old plants, full load, part-load and number of start-up hours drop, confirming the difficulty old stations have in a market with increasing non-dispatchable energy supply. This highlights the increasing need for flexibility in a system with increasing shares of non-dispatchable energy sources [11]. While renewable energy technologies variable annual yield has a direct effect on profitability, the profitability of thermal units depends on the stochastic nature of non-dispatchable renewable generation. Thus , if wind and solar capacities are the only augmented parameters in the analysis, the direction of the effect is unambiguous: power prices will be depressed on average and start-ups will increase. Consequently, profitability of thermal plants will decrease, but the magnitude of this effect is still unclear [11].

Conventional power plants can be retrofitted to increase ramping capabilities and decrease minimum load and start-up costs. For example, coal power plants ramping capabilities have been increased by enhancing the control of coal conveyor belts that serve the boiler and introducing new coal pulverizing mills to decrease the particle size of pulverized coal and, consequently, improve combustion and efficiency. Also the replacement of old inefficient open-cycle gas turbines by state-of-the-art aircraft gas turbines for easy ramp up and ramp down can add competitiveness to these units.

4.2.4 Rural Electrification

In developing countries the traditional grid extension mode of electrification has been supplemented, if not challenged, by an alternative, decentralised mode of supply. The combination of poor and unreliable energy supply with the emergence of modular renewable energy technologies and the support for low-carbon energy solutions in response to climate change concerns have given further credence to the alternative options recognized as technically feasible [12]. A multidimensional approach covering regulation, finance, economic development and social dimensions is required to analyse the business case of each rural electrification project, from simple centralized grid extension-connection systems to off-grid systems based on renewables and/or diesel generation, supplemented with battery-operated systems (Fig. 4.4).

Decentralised and off-grid power generation technologies for rural electrification can be broadly classified into two categories: direct current (DC) producer and alternating current (AC) producer. The DC can either be used directly or converted to AC using inverters. The later is the preferred option as most appliances available in the market are currently AC driven (maybe in the future more low power appliances will be DC driven). The AC comprises the prime mover-generator (steam turbines, gas turbines, water turbines, wind turbines and a variety of reciprocation engines fuelled by diesel, petrol, gas, biofuel, etc.) but many of them require skilled manpower which will be difficult to obtain in rural areas. The most common generating technologies used are based on diesel generation, photovoltaics and mini/micro hydro systems (vulnerable to seasonal variations in sun and water supply) combined with rechargeable batteries for energy storage. The consuming technologies are mainly high efficiency lamps (either compact fluorescent or LED) and an outlet for televisions or other low power consuming appliances [14].

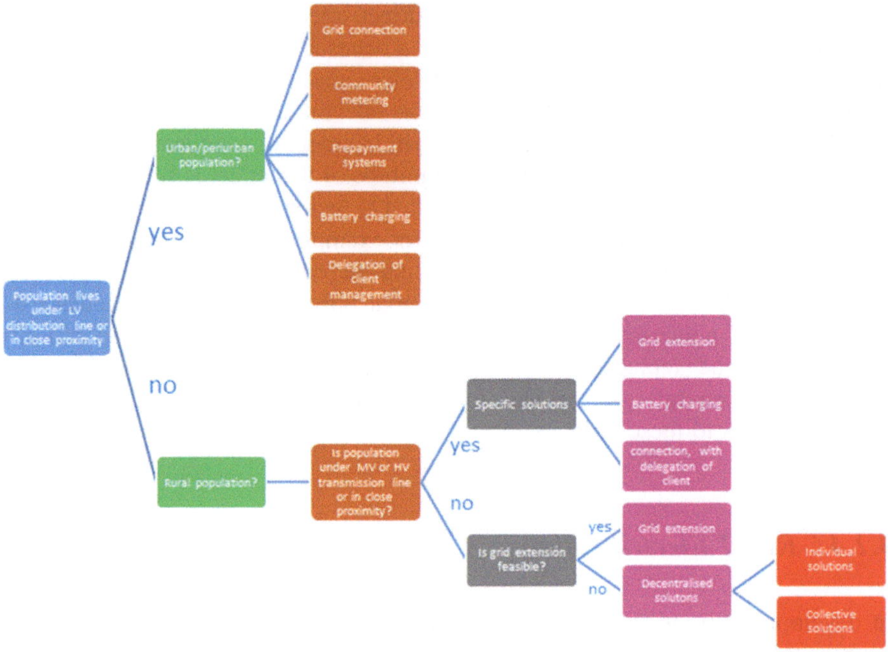

Fig. 4.4 Decentralised solution tree [13]

The extra-cost of decentralised and off-grid electricity supply is not only based on the cost of alternative power generation, but also to scale (achieving energy conversion efficiencies to the extent allowed by thermodynamics at reasonable cost) and skewed power demand (essentially lighting load during 3–4 h after sunset and occasional demand for electric motors). The base load tends to be zero (or very low) for decentralized generation until the local economy develops to some level, thereby further reducing the financial viability (and also technical viability in many cases) of the system [15].

Many financial instruments for off-grid electrification can be used (Fig. 4.5) [16]. Financial instruments are typically divided into project-level financing and end-use level financing both relying on finance from donor agencies, budgetary support from the state and private investors. The World Bank suggests that to become sustainable an off-grid project has to be beneficial to all main stakeholders (consumers, service providers, financiers and government) and that designing an off-grid system is not an exact science because of a combination of complex factors (e.g., high cost, poor consumers and new technologies) [17].

Where off-grid solutions are being offered, they appear as a temporary solution, until grid extension becomes feasible. Thus, promoting off-grid solutions appears for customers as inferior, creating concern and reducing attractiveness due to a sense of discrimination and isolation, adversely affecting the success of programmes for access to electricity. Further, the introduction of modern technologies in rural areas with

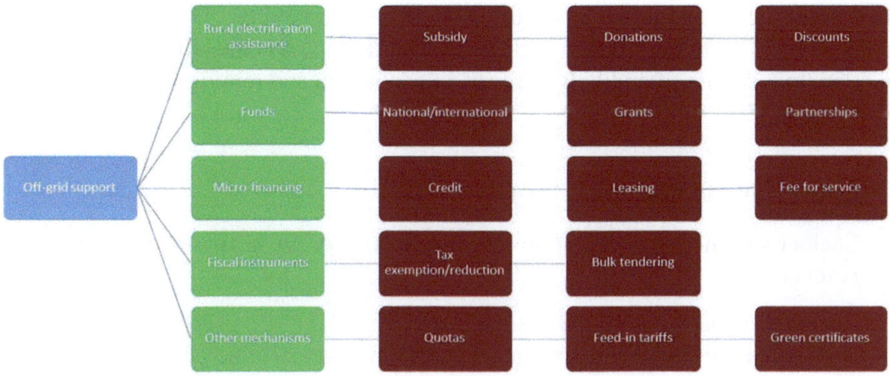

Fig. 4.5 Financial instruments for off-grid electrification [16]

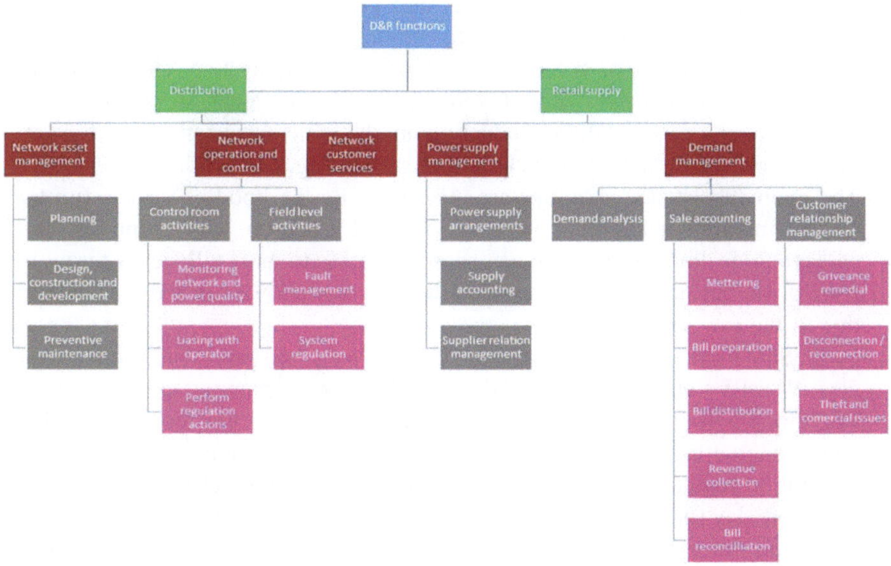

Fig. 4.6 Typical distribution and retail supply activities of an electricity supplier [19]

limited support networks creates the challenge of sustaining such options over a long period. Poor quality of components, using local solutions for cost reduction, poor workmanship and lack of technically skilled maintenance staff are common challenges faced in most cases [18]. Another challenge is creating an institutional framework that can provide an efficient long-term service (Fig. 4.6) [19] on an inclusive social basis, while ensuring local accountability. This challenge may sometimes involve the need to create new institutions, whether for delivery or regulation, but is more often likely to mean an adaptation and reinvigoration of existing institutions [20].

Case Example 4.3

Socioeconomic impacts of access to electricity in a Brazilian Amazon reserve

The Mamirauá Reserve has a total extension of 1124 ha, 265 villages with approximately 11,000 inhabitants. Two Amazonian riverine populations were analyzed: one was partially electrified by solar home systems and a communitarian diesel generator (São Francisco de Aiucá: 38 wood-made homes) and the other (Punã: 105 homes) almost totally electrified by grid extension in 2011 (before that, it had a diesel generator operating for 5–6 h per day and had problems related to fuel supplies and generator stoppages). Houses in Punã have electricity meters and receive a monthly electricity bill. Also Punã has public illumination, a community water pumping system, electrification in some public centers (school and churches) and local shops. Fishing, agriculture and cassava flour production are the main subsistence activities at the Reserve. Other activities include timber, hunting, craftwork, and services. The main subsistence activities are influenced by the flooding season (up to 10 m): mainly fishing and planting in the dry season, and cropping and cassava flour production in the wet season. Access to most villages is through river boats but navigation to some of them is not allowed in the dry season due to the low water levels.

In São Francisco de Aiucá there are 23 solar systems and each has a 200 W_p PV plant, two 150 Ah batteries, one inverter and they pay R$20/month for battery replacement. When this study was conducted, the diesel engine had not been in operation for several months and 3 solar systems were not working (6.5 years after installation). Electricity is used for 3–4 h per day for illumination, TV, radio, small fans, and cell phones (there is a demand for refrigerators, but they can't be used due to power limitations). The mean annual electricity consumption per person is 16 kWh. Three houses have diesel generators (2.0–2.5 kVA) used to extend hours of light and conserve food in refrigerators. Lanterns are also used, and batteries produce local pollution. In Punã almost 55% of houses (6–7 persons) consume less than 100 kWh per month and almost all the village has a consumption lower than 200 kWh. In Punã refrigerators, freezers, washing machines and blenders are also used.

Impacts of both types of electricity supply were positive mainly due to a reduction in complementary energy consumption (and associated cost).

People with electricity supplied by grid extension described a greater impact in terms of their productive activities (expanding illumination hours to perform various tasks: manufacturing gillnets, fish and meat processing, food preparation, handicraft production, etc.), and also improvements in education (extended operating hours), health (availability of fresh water), and access to internet [21].

4.2.5 *Smart Grids*

A smart grid can be defined as an electricity network system that uses digital technology to monitor and manage the transport of electricity from all generation sources to meet the varying electricity demands of end users, coordinating the needs and capabilities of all generators, grid operators, end users and electricity market stakeholders in such a way that they can optimise asset utilisation and operation and, in the process, minimise both costs and environmental impacts while maintaining system reliability, resilience and stability [22]. Traditionally, the supply of electricity within grids has been managed by a supervisory control and data acquisition (SCADA) system, which monitors and controls this process from generation to substation to detect generation needs or respond to system instability. The term "smart grid" covers the next step in the supply and distribution of electricity, using information technology to make more visible and controllable the conventional grid itself and contributing with new grid elements, especially designed to manage the demand response and control small generation and storage systems (Fig. 4.7). The smart grid can be considered an electricity network that uses digital and other advanced technologies to monitor and manage the transport and distribution of electricity from all generation sources to meet the varying electricity demands of end-users in African and Latin American urban and rural areas.

The control of smart grids is provided by high-speed, two-way communications, effective sensors and real-time coordination of all grid elements. In addition, measurement and control of grid elements can be carried out below the substation level and behind the meter, from controlling only the transport grid to also control the distribution grid. Thus, the smart grid will act to:

1. Guarantee the active participation of consumers.
2. Accommodate all generation and storage options.
3. Enable the development of new products, services and markets in the electricity sector.
4. Optimize the operation of the grid elements.
5. Anticipate and respond to system disturbances.
6. Resist attacks and natural disasters.

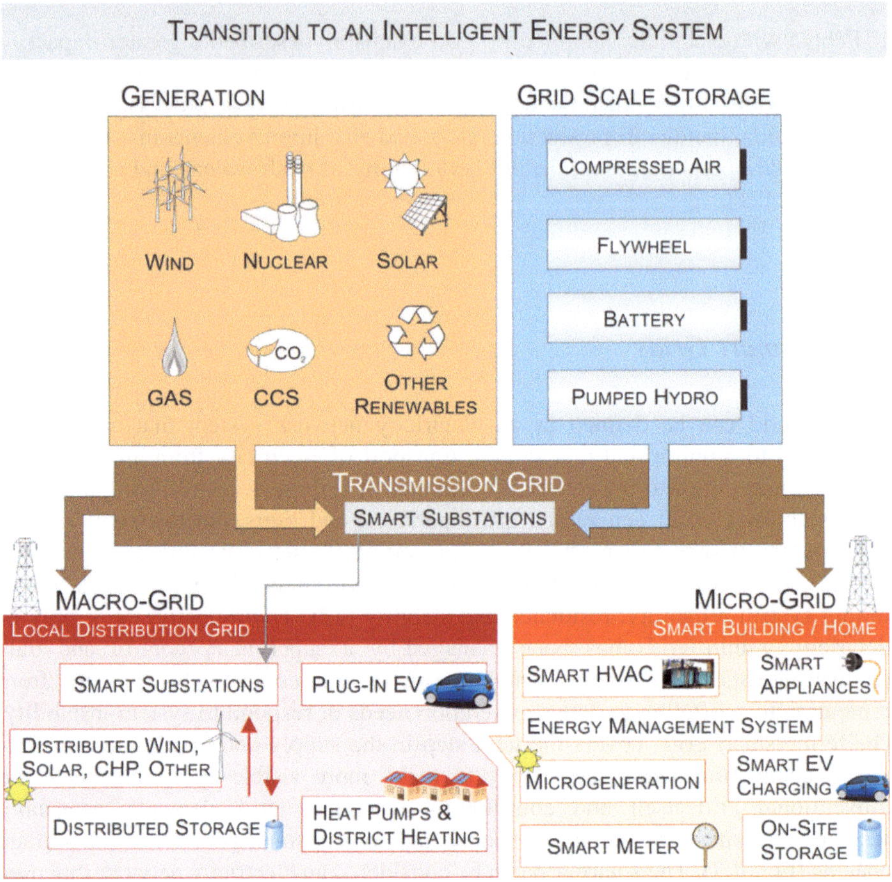

Fig. 4.7 Scheme of the transition to an intelligent energy systems connected by an smart grid

Case Example 4.4

Human behaviour in household energy efficiency in a town in Nigeria

The study area is Bauchi town, the capital of Bauchi State. The target population included all heads/representatives of household units (male or female) resident in Bauchi town, living in formal housing typology of flats/apartments and not in traditional settings. Twelve interviews and 365 questionnaires for households of three geographical districts (2 clusters per district) were made, mostly in academic and government environments. Results show a very low level of energy efficiency practiced by the occupants in the households due largely to lack of awareness and knowledge on energy efficiency, a lack of adequate policies and programmes in place and a lack of incentives and motivation. The study is expected to be useful for seeking ways of improving the energy use situation in the country [23].

These capabilities can automatize and reduce the complexity of many activities required for an electricity supplier (Fig. 4.6) and are useful for making decentralised and off-grid projects more competitive in terms of cost and readiness for African and Latin American customers compared to grid extensions.

It is usually argued that the use of smart grids can contribute to the decarbonisation of electricity systems in the future. Thus, some scenarios of an IEA report [4] considers that CO_2 emissions from electricity generation can decrease from 40% of the total current global emissions to 21% in 2050, representing a reduction of 20.2 Gt CO_2/year, and smart grids would contribute with annual savings of 0.9–2.2 CO_2/year Gt. This reduction in CO_2 emissions is not only attributed to a higher penetration of renewable energies, but also to a lower demand in grid infrastructure (and the correlated reduction in power losses) and a more efficient use of fuels supplied to distributed power plants. New smart grids placed in Africa and Latin America can contribute to this decarbonisation goal.

4.2.6 Storage Systems

Electricity storage technologies have emerged in response to the need to synchronize supply and demand for electricity that would otherwise have to be produced and consumed instantaneously. For non-dispatchable renewable resources (mainly wind and solar energy) and power demand this instantaneous exigency cannot be always fulfilled. The technologies involved in electricity storage are diverse, both in terms of storage procedure and system (chemical energy, gravitational, electrical, thermal, etc.) and in terms of storage system characteristics (power capacity, energy capacity, weight, size, efficiency, charging rate, discharging half-life, etc.). The usual storage technologies, in different phases of development, are pumped hydro, batteries (lead acid, metal-air, sodium sulfur, redox flow, Li-ion, ZnBr and NiMH), compressed air energy storage (CAES), flywheels, superconductors, and electrochemical capacitors (supercapacitors). The types of batteries discussed in this book are secondary (rechargeable) batteries.

The storage systems can, in general, be classified as large or small, and stationary or mobile systems. In this book the stationary systems will be considered, since they are the best adapted for the introduction of low carbon energy technologies in Africa and Latin America. Electricity storage systems can also be classified in relation to the application considered (Table 4.5).

Table 4.5 Application category specifications [24]

Application category	Discharge power range	Discharge time range	Stored energy range	Representative applications	Cost effective technology
Bulk energy storage	10–1000 MW	1–8 h	10–8000 MWh	Load leveling spinning reserve	Pumped hydro
Distributed generation	100–2000 kW	0.5–4 h	0.05–8 MWh	Peak shaving, transmission deferral	Batteries
Power quality	0.1–2 MW	0–30 s	0.028–16.67 kWh	End-use power quality and reliability	Batteries, flywheels and super condensers

1. Bulk energy storage: used to decouple generation and power consumption. A typical application is the leveling of the demand, which involves charging the storage system when the cost of energy is low so that it can be discharged when the cost is high.
2. Distributed generation systems: they operate at intervals of seconds to minutes to ensure continuity of service when the source of energy generation is replaced.
3. Systems to improve power quality: only operate in second and sub-second intervals to ensure the stability of the power supplied.

Another interesting classification is provided in terms of the services the storage system provides [25]:

1. ISO/RTO services: energy arbitrage, frequency regulation, spin/non-spin reserves, voltage support and black start.
2. Utility services: resource adequacy, distribution deferral, transmission congestion relief and transmission deferral.
3. Customer services: time-to-use bill management, increased PV self-consumption, demand charge reduction and backup power.

While some systems can operate in all functional categories, costs impose some restrictions to this versatility. Consequently, different storage systems provide different applications within the electrical system in terms of power required and the time interval in which the storage unit can respond. Thus, systems with little storage capacity and power are used to improve power quality in the network, while larger power and storage systems are used to ensure the stability of the electrical transmission system. With increased storage capacity and readiness response, storage systems can allow larger penetrations of non-manageable renewable energy sources in the grid and better control grid frequency and voltage. The selection of the storage system has implications in terms of capital cost per energy and power units and energy density. Orientative cost values are shown in Figs. 4.8 and 4.9 [26], but a deeper analysis should mainly consider capacity and service provided.

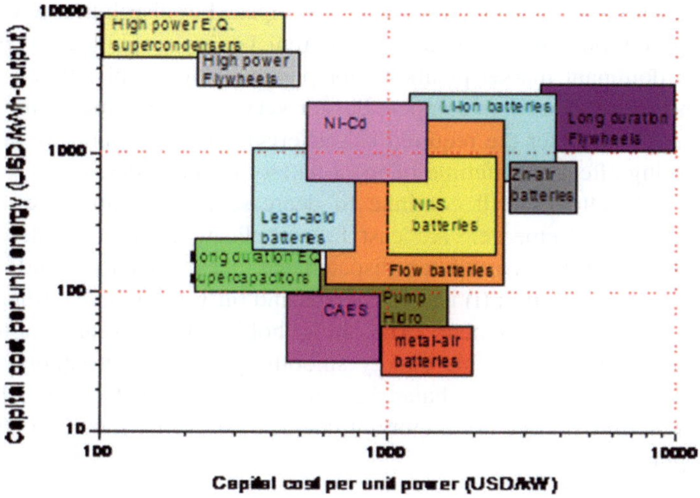

Fig. 4.8 Scheme of costs per unit of energy and power for different storage technologies [26]

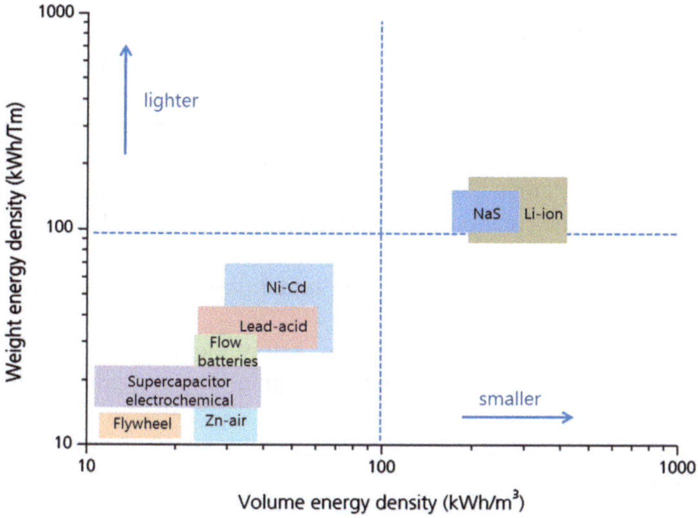

Fig. 4.9 Energy density per unit mass and volume for different energy storage technologies [26]

A survey of the different electricity storage technologies in terms of different characteristics can be found elsewhere [27].

In stationary systems devoted to electricity storage, water pumping systems account for nearly 100% of the total, because the technology is mature and there is a large amount of hydropower available in many locations. Hydropower generation is likely to increase in the future (see Chap. 9) in emerging countries where hydro resources have not been thoroughly developed. For distributed generation the main storage options are batteries, but different battery technologies compete to achieve the dominant market position. For power quality applications, batteries, flywheels and supercapacitors compete but for very short regulation only flywheels and supercapacitors offer the required characteristics because short-cycle charging and discharging affect the lifetime of most battery technologies.

The cost of batteries will continue to decrease in the future and many battery applications will emerge. At least four applications can be identified and related to increasing the share of non-dispatchable renewable energy and improving electricity supply reliability: (i) island systems and off-grid/rural electrification with renewable energy deployment; (ii) households with solar photovoltaics; (iii) non-dispatchable renewable energy smoothing and energy supply shift; and (iv) fast, short-term electricity balancing in ancillary markets [28]. Also, when batteries are selected, there are important characteristics that should be considered for any application:

(i) capacity: number of hours a specific current (A) can be delivered by a fully charged battery before it is completely discharged.
(ii) lifetime: this is a conditional statement, as the battery should be kept in a specific temperature band and float voltage. If these guidelines are not followed, the life of the battery will be reduced. A fully charged battery that can only deliver 60–80% of its original capacity may be considered at the end of its cycle life.
(iii) temperature of operation: normally at 25–27 °C, the capacity of the battery increases with temperature, whereas the life of the battery decreases. High ambient temperatures cause internal reactions that produce capacity losses, corrosion and the creation of gases requiring ventilation. The electrolyte may also freeze.
(iv) self discharge: in terms of % per month and depending on the battery technology.
(v) other characteristics: energy density, safety elements, company warranty, O&M costs, efficiency, charging/discharging time and grid requirements.

Case Example 4.5

Flow battery for a telecommunications base transceiver site (TBS) in Dominican Republic

Two zinc-bromide, free-maintenance flow batteries were installed and connected to the −48 V_{DC} telco bus to replace a bank of 24 outdated lead-acid batteries already on site, which had been significantly degraded over a short period of time due to the site's constant grid outages and no active cooling. The system is located in a remote location, accessible via rocky road and difficult terrain. With no other sources of power, the site experiences an average of 2–3 outages per day, often lasting about 6 h but sometimes longer than 20 h. The motivation was to avoid the effects of an intermittent grid (44.6% site down-time without batteries) in the service for the customers of the mobile network operator. Site up-time with the new batteries reached 96.4% on average. Site average temperature was 29.4 °C and maximum of 42.2 °C. Also, when tropical storm Erika hit the island in August 29, 2015, the site was in operation for 12 h after the grid outage (lasting 21.5 h), enabling emergency communications. No tampering or theft was recorded over the 3 month trial [29].

If batteries will be used to participate in providing ancillary services the location of the batteries becomes an important and critical decision. There are studies that suggest the best location for batteries is in the nodes of the distribution grid to offer the lowest costs and reliability [30] while others suggest that the best location should be as far downstream in the electricity system as possible (behind-the-meter) to provide the largest number of services to the electricity grid, even if storage deployed behind the meter is not always the best option [25].

In terms of technology, the market has shifted from sodium-sulphur battery deployment to other types, namely lithium-ion batteries (deep discharge cycle life, large energy and power density, and other attributes). These have proven favourable on a cost and performance basis compared to other battery types. Batteries can be used in multiple ways, so different battery types offer their own relative advantages. This means a whole range of batteries will continue to be active in the market. A battery installation includes battery cells (for cell-based batteries), a power conversion system, materials in the module, a battery management system and other components. While individual cell costs may be a good economic indicator for comparison purposes, they only represent around 20% of all relevant costs. The cost of a complete system can be best represented by the calculation of the LCOE [28]. Lithium-ion batteries utilised for distributed energy storage applications of various capacities and power outputs ranged from about USD 0.5–2/kWh and upwards [31]. Cost statistics should be viewed with care, due to the current lack of defined standards and approaches by which companies provide cost and performance data.

Fuel cells can also provide reliable energy and power to the grid particularly when combined with hydrogen or other low stoichiometry hydrocarbons as fuels. Fuel cells are devices that use hydrogen (or a hydrogen-rich fuel) in an electrochemical process to produce electricity and heat. Generally, the fuel cell consists of a system of two electrodes (anode and cathode) immersed in an electrolyte, in which hydrogen is supplied to the anode and oxygen to the cathode. A catalyst ionizes the hydrogen atoms into electrons and protons, taking each particle an opposite path towards the cathode. Thus, the electrons move through an external circuit (supplying an output power), while the protons move through the cell electrolyte. At the cathode the electrons and protons react with oxygen to produce water and heat. Different fuel cell technologies are available, mainly based on proton exchange membrane (PEMFC) and solid oxide (SOFC), but also based on molten carbonate (MCFC), phosphoric acid (PAFC) and alkaline media. Several fuel cells are generally connected in series and/or parallel to achieve the desired output power and voltage. The main advantages of fuel cells are their high efficiency for electricity production (about 60%), and near zero emissions if hydrogen is introduced as fuel, in contrast to the very polluting internal combustion fuel engines (efficiency about 20% or lower). However, fuel cell technology is expensive, and requires additional R&D activity to become more attractive in markets where proper operations and maintenance will be limited.

Thermal storage for electricity production, mainly confined to concentrated solar power (CSP) plants could also be applicable in power grids in the near future. A brief description of the different thermal storage technologies is included in Chap. 7.

4.2.7 Distributed Generation

Distribution grids usually have a radial configuration where electricity flows from the center to the end nodes accompanied by a voltage drop between the center and the end nodes. However, households and commercial buildings with non-dispatchable distributed generation (mostly PV systems) connected to the grid have fundamentally different characteristics as the distributed power production can exceed consumption during certain periods of time and subsequently dominate the load flow and increase the voltage of the grid. This leads to new patterns in the grid, including power reversal to overlying grid levels. Even though the increase in the distributed generation can be gradual, their locations and rating are often random if proper planning and adaptation of the grid configuration to the new generators is not taking place simultaneously.

Such single-phase bidirectional power flow caused by the new distributed generation can have an adverse effect on the voltage imbalance of a three-phase distribution network: (i) overheating and de-rating of all induction motor based loads, (ii) mal-operation of protection relays and voltage regulation equipment, (iii) non-characteristic harmonics, etc. These adverse effects are mainly due to unequal system impedances, unequal distribution of single-phase loads or large number of single-phase transformers. These voltage imbalances can be severe at far end nodes and might increase to more than the standard limit.

In Africa and Latin America it is difficult to properly manage the distribution grid because of these concerns and the lack of adequate workmanship and electricity suppliers. To avoid the inconveniences that can result in power shutdown periods of time and/or distributed generation curtailment, new distribution grid configurations are being modelled and tested. The solutions proposed are mainly based on the application of series and parallel custom power devices as distribution static compensators and dynamic voltage restorers. These devices may be required to improve voltage imbalance problems in these feeders [32], but additional modelling and testing is required specifically in the low voltage distribution grid as this network is the less supervised by the system operator.

Case Example 4.6

Uganda's liberalized energy market and energy poverty

Uganda is one of the few sub-Saharan African countries to have liberalized its energy market, as its generation, transmission and supply segments were unbundled in 2001. The generation and transmission companies are both state-owned, while the supply company is privately owned. Independent power producers currently account for 58% of generation capacity, though UEGCL, the state-owned utility, intends to triple its capacity developing a 600 MW hydropower project [33]. Uganda is also an early adopter of new business models to reduce energy poverty, as pre-paid electricity meters are used to charge mobile phones and offer appliance packages (e.g., a few LED lights, a mobile phone charger and a radio, all powered by a solar panel and a battery, and paid by mobile phone). The price of the packages is spread out enough to bring the monthly cost below that of buying kerosene for lamps, and default rates are negligible as the service can be disconnected remotely by disabling the box that links the panel to the appliances. The trick, executives say, is to convince clients that they are buying appliances rather than electricity and to make the appliances more energy efficient so that they can operate with rooftop PV panels. The next step is to provide low-energy refrigerators that could help customers open small restaurants or groceries and to educate potential customers about the social and economic benefits of small but life-changing amounts of power [34].

4.2.8 Lighting

Lighting in low-income households of Africa and Latin America is usually provided by candles or kerosene/diesel lanterns. Candles and low-efficiency lanterns emit smoke and increase health pathologies, as it will be discussed for rudimental biomass cooking in Chap. 5. Kerosene lamps produce better light, but they are uncomfortably hot in tropical climates and can be difficult to light. Use of kerosene also imposes health risks, through fires and small children drinking fuel stored in soft drink bottles, and there is emerging evidence of links to tuberculosis and cancer. Switching to electrical lighting eliminates most of these risks and increases luminous efficacy (Table 4.6). For example, a paraffin wax candle has 1 lm (lumen) intensity and 0.01 lm/W [35], while a 15 W compact fluorescent lamp has an intensity of 1150 lm and 40 lm/W [36].

Table 4.6 Different types of light sources, luminous efficacy and average operating hours [35, 36]

Product type	Luminous efficacy (lm/W)	Operating hours
LED (warm white)	67–93	30,000–50,000
High intensity discharge system (high watt)	115	9000–30,000
Linear fluorescent system	108	30,000
High intensity discharge system (low watt)	104	9000–30,000
Compact fluorescent lamp (CFL)	70	8000–10,000
Halogen	20	2000–6000
Incandescent	15	1000

Fig. 4.10 Shop in Tanzania offering solar lighting and other portable lighting devices (*Source* Simone Tuchella and Engineers without borders)

There has been much success in the dissemination of compact fluorescent light bulbs (CFLs) in many developing countries [37]. Thus, increasing electrification of lighting in Africa and Latin America improves comfort and can reduce costs and health risks. Also, small lighting systems where energy is supplied by small photovoltaic cells integrated in the lighting system (Fig. 4.10) are an alternative that is gaining market share in many rural areas.

Case Example 4.7

Battery selection for a off-grid school lighting in Angola

The project consists of a 200 W/1300 Wh lithium-nickel-cobalt-aluminium battery coupled with a PV system, adding lighting to a school in Angola, enlarging activities and providing more services to the community. The estimated cost to light four classrooms for up to 8 h is USD 2 per day. This battery type was selected because it is more resilient to high temperatures than lead-acid, which is very important in this hot environment. Given lithium's tendency to overheat, the batteries are protected by a thermal management material and placed in the shade. No other voltage or temperature control was employed. The most important parameters in this case are temperature and long charge and discharge cycles. The remote location also imposes unique maintenance requirements. As hands-on maintenance is not possible because of the lack of expertise in this area, the battery technology has been selected to attract commercial activity on replacing batteries instead of providing maintenance [28].

4.2.9 Net Metering and Interconnections

Net metering is an option that utilities can offer to electricity auto-producers to obtain incomes or billing credits for injecting their surplus production in the power grid. This possibility is especially valuable for non-dispatchable renewable technologies, mainly solar and wind. Only a few countries in Latin America have in place such policies (Table 4.7). Most of the auto-producers in these countries get billing credits for the electricity surplus injected into the grid and these credits can be rolled over subsequent months to compensate excesses. Currently, only Panama pays excess production. Most countries limit the capacity of the auto-producer facilities and many of the regulations consider not only renewable technologies but also CHP [18].

The current design of retail electricity rates and the presence of net metering in many countries have elicited concerns that the possible under-recovery of fixed utility costs from renewable (mainly PV) system owners may lead to a feedback loop of increasing retail prices that accelerate renewable technologies adoption and further rate increases. However, a separate and opposing feedback loop could offset this effect as increased renewable deployment may lead to a shift in the timing of peak-period electricity prices that could reduce the bill savings received under net metering where time-varying retail electricity rates are used, thereby dampening

Table 4.7 Net metering main characteristics for the Latin American countries that have promoted this bidirectional relationship between utilities and customers

Country (starting date)	Characteristics
Brazil Apr 19 2012	Hydro, solar, biomass, wind and qualified cogeneration up to 1 MW Billing credits obtained and used within 36 months Consumers must pay the availability cost of the power grid
Chile Oct 22 2014	Renewable or CHP <100 kW Obtain credits for excess electricity that would be equivalent to the price/kWh charged by the utilities to their customers If the customer's credit exceed its own consumption, the billing credit would be rolled over the subsequent months
Costa Rica Oct 15 2010	Self-generation facilities Obtain credits for excess electricity Regulation about the quality of renewable generation If the customer's credit exceed its own consumption, the billing credit would be rolled over the subsequent months
Dominican Republic Jul 29 2011	Self-generation facilities Obtain credits for excess electricity Regulation about the quality of renewable generation If the customer's credit exceed its own consumption, the billing credit would be rolled over the subsequent months Every 31 January utility pays 75% of accumulated credits based on the previous month's retail tariff
Guatemala Sep 4 2014	Hydro, solar, biomass, wind and geothermal Billing credits obtained and used within subsequent months Consumers must pay the availability cost of the power grid
Mexico Mar 10 2010	Renewable and CHP: low voltage (<1 kV) residential (<10 kW), and commercial (<30 kW), and medium voltage (1–69 kV, <500 kW) Billing credits obtained and used within subsequent 12 months
Panama Jun 27 2012	Renewable <500 kW Paid for excess production injected to the grid: computed from January to December and paid by February of the following year; amount paid equivalent to the average price/kWh charged by the utilities to their consumers in the year that the excess was generated
Uruguay Jul 1 2010	Wind, solar, biomass and mini-hydro limited in terms of power injected Billing credits obtained at the appropriate rate and used within next month Exempt of income tax on their self generation

further renewable capacity adoption. The result of these two competing feedback dynamics can nearly offset one another depending on the retail rate structure [38].

Case Example 4.8

Adaptation of feed-in tariff for remote mini-grids in Tanzania

This work takes an integrated approach (geographical, technical, economic and institutional) to examine the feasibility of a mini-grid feed-in tariff (FiT) based on PV-diesel or mini-hydro for existing and new remote grids in

Tanzania. Only a 18% of Tanzanian rural population has access to electricity but, on the other hand, Tanzania has been ranked in 2013 among the top 10 countries in the world for establishing sustainable business models for renewable energy-based mini-grids. In 2009 Tanzania introduced a standardized power purchase (SPP) agreement to interconnect renewable energy generators and export excess power, and in 2010 adopted a standardized tariff methodology (STM) to calculate the SPP electricity prices. There are 26 isolated mini-grids owned by utilities and based on diesel generation, and many tens of PPAs approved totalling 40.1 MW for biomass, PV and hydropower. However, results show the current SPP tariff seems adequate to attract small hydropower investors, but insufficient to attract PV developers, creating differences between regions. These results suggest that specified tariffs by technology would be more suitable to support the supply of electricity in all geographical areas depending on the natural resources available [39].

Establishing uniform technical and procedural requirements for interconnection of generation on the electric grid is a key policy and technology challenge. Interconnection standards are defined for safety (people and property), grid reliability, performance, cost considerations and fairness [40]. In developed markets there are many interconnection standards mostly depending on size, participation in the market and location. More robust markets are expected also in Africa and Latin America with the implementation of net metering and interconnection standards, specifically to improve the penetration of PV systems [41].

4.3 Regional and National Perspectives on Technology

4.3.1 Electricity Output and Power Mix

It is very important to consider the power output and mix in Africa and Latin America when a transition to low carbon energy technologies is promoted. African electricity output in 2014 reached 765 TWh, 3.2% of the world output, and Latin America 1592 TWh, 6.7% of the world output (Table 4.8) [42]. Therefore the use of electricity in both regions is under the global average, but more so in Africa. This assumption is confirmed if we consider the electricity output per capita (Table 4.8) and observe that African and Latin American values only represents 21 and 78%, respectively, of the world average electricity output per capita. The effect of fossil fuels for the lower GDP per capita countries also deteriorates the power mix by the fact that the fleet of fossil fuelled power plants consists largely of technologies with the lowest efficiencies, often favoured because of their lower upfront capital costs. There is also captive generation by industry, predominantly

Table 4.8 Global, African and Latin American electricity output, kWh/inhabitant and power mix in 2014 [1, 42]

	Africa	Latin America	World
Electricity output (TWh)	765	1592	23,903
kWh/inhabitant	684	2548	3252
Coal (%)	33.7	6.4	40.6
Natural gas (%)	36.9	26.1	21.6
Oil (%)	9.3	12.6	4.3
Nuclear (%)	1.8	2.7	10.3
Hydro (%)	16.5	46.4	16.6
Geothermal (%)	0.2	0.3	0.3
PV (%)	0.1	0.07	0.7
Wind (%)	0.3	1.5	3.0
Biomass (%)	0.2	3.9	1.2
CSP (%)	-	0.01	0.04

owned by sugar (e.g., Malawi and Zambia) and mining (e.g., Sierra Leone) companies, that should be considered [33].

If we consider the power mix percentage in both regions (Table 4.8), we can conclude that Latin America is the world's most decarbonised region because of the strong share of hydropower. Also the share of oil in both regions is higher than world average, and it can be attributed to the role of low capacity generating units in many small countries and isolated grids, as we will show below. A higher share of natural gas exists in Africa and Latin America than the global average largely attributed to the existence of important natural gas fields in some countries in these regions. The small share of nuclear energy is due to the technological challenges and economies of scale associated with this resource. Relative to other renewable technologies, wind and solar technologies are below global averages in contrast to the enormous potential it has in many areas. This is attributed to the lack of specialised manpower and the associated risks of investment. Renewable energy shares however are improving in some African and Latin American countries, as described in the following chapters. Biomass plays an important role in the electricity mix in Latin America, mostly attributed to the role of agricultural waste and large farms in some countries (Brazil, Chile, Guatemala, Nicaragua and Uruguay).

The electricity mix in African countries, based on available data, shows an enormous variability (Fig. 4.11) driven mostly by a dependance on the indigenous energy resources each country own. Indeed, countries with significant amounts of natural gas such as Algeria and Nigeria rely largely on this fuel for producing electricity. However, there are still many African countries that do not use this low carbon emission fuel. The same is true for countries with large amounts of hydro resources such as Ethiopia, Democratic Republic of Congo, Mozambique, Namibia and Zambia. In the case of Botswana and South Africa, almost all of the electricity is produced with coal due to the large amount of coal reserves both countries own. This coal relation is also observed, in a lesser extent, in Morocco and Zimbabwe. Similarly, where oil is predominant in the electricity mix it is

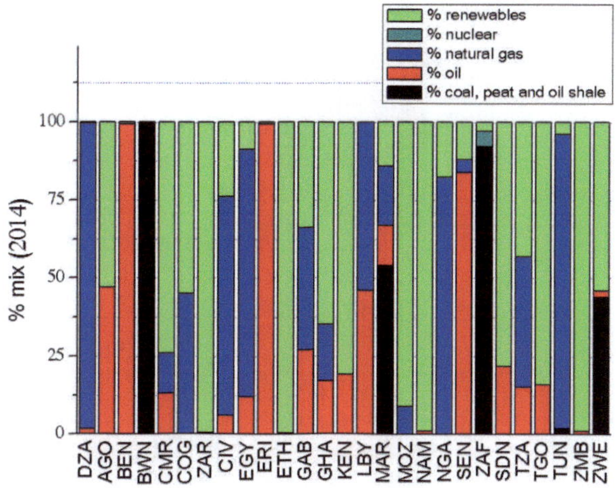

Fig. 4.11 Power mix in African countries [1]

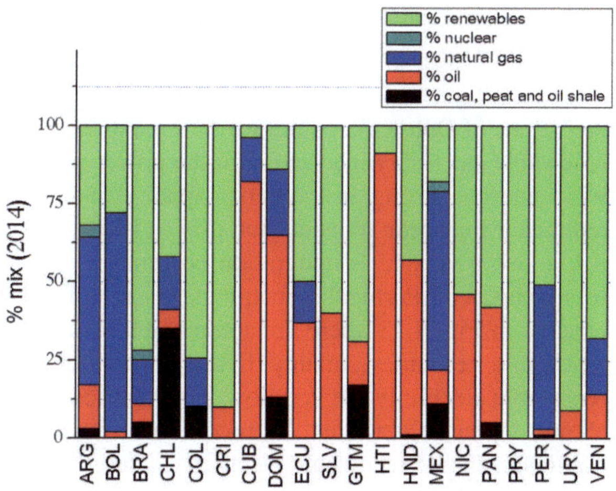

Fig. 4.12 Power mix in Latin American countries [1]

related to oil producing countries like Angola and Libya, and with small and isolated grids such as Eritrea.

For Latin America the same relationship between energy mix and indigeneous resources is observed. Hydropower contributes a large share of energy (Fig. 4.12) and natural gas plays a limited role in most countries. Surprisingly, oil plays a significant role in electricity production compared to Africa largely because of the relatively small grid capacites and the lack of grid interconnections until very recently.

Another important matter to be considered is the relation between the percentage of electricity output produced by renewable resources in each country and the share of non-dispatchable renewable electricity. For this comparison we have included

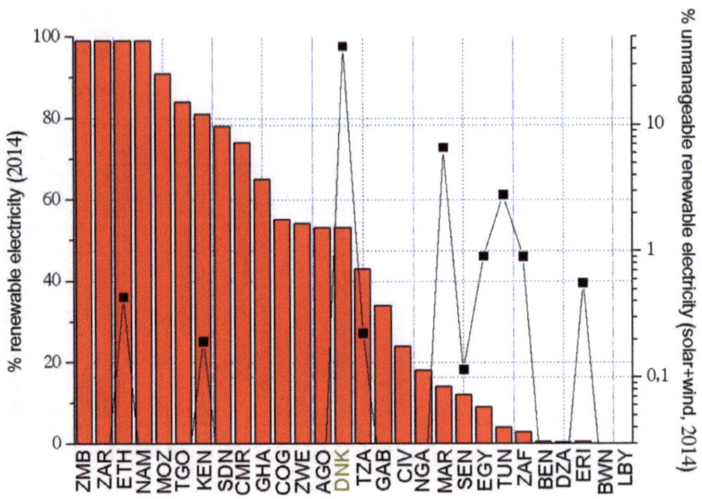

Fig. 4.13 Percentage of electricity output produced in 2012 by renewable resources in each African country and Denmark (bars) and the share of non-dispatchable renewable electricity (squares) [1]

the values of Denmark, as the world's most successful country in non-dispatchable renewable electricity integration in the power grid because of its leading position in wind energy, strong investment capacity and small contribution to a solid and redundant European transmission grid (it is also recommended to consider Spain, as it has also a very high integration percentage of non-dispatchable renewable electricity and, with Portugal, constitutes a near isolated Iberian power grid). As we can observe in Fig. 4.13, there are some countries reaching near 100% electricity from renewable resources, mainly hydropower, but the values for non-dispatchable renewable electricity are very low. Only Morocco, Tunisia, Egypt and South Africa shows appreciable values. Morocco and Tunisia could very easily increase their non-dispatchable renewable electricity share because of their grid interconnections to Spain and Algeria, respectively. The countries with the lowest percentage of non-dispatchable renewable energy in their mix are mostly countries near 100% based on hydropower, or countries with small electricity systems and/or with large oil and natural gas reserves.

On the contrary, in Latin America (Fig. 4.14) several countries have reached significant percentages of non-dispatchable renewable energy in their mix, mainly Nicaragua, Uruguay, Costa Rica and Honduras, and based on wind energy. Only Paraguay is fully dependent on hydropower for its electric supply. As expected, the countries with the lowest percentage of non-dispatchable renewable energy in their mix are islands, countries with small electricity systems or with large oil and natural gas reserves.

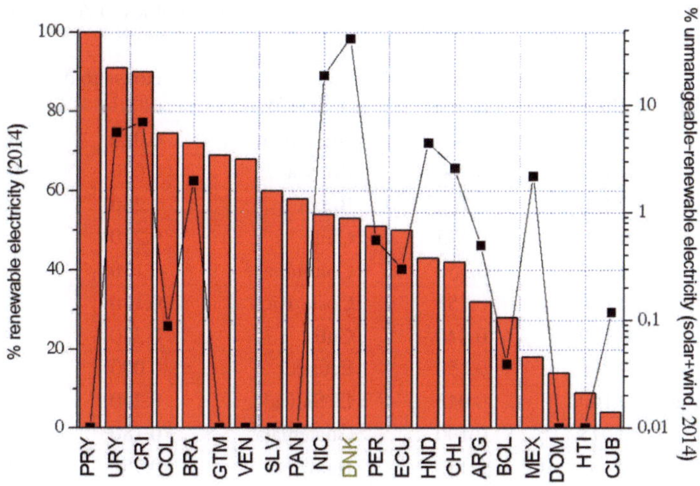

Fig. 4.14 Percentage of electricity output produced in 2012 by renewable resources in each Latin American country and Denmark (bars) and the share of non-dispatchable renewable electricity (squares) [1]

4.3.2 Electricity Price

Another important parameter is the price of electricity (industrial/household) but, unfortunately, no accredited energy agency supplies these data from most African and non-OECD Latin American countries. Data summarized from different sources and dates (Table 4.9) [43, 44] indicate that in many underdeveloped countries the electricity is very expensive and often influenced by internal policies. These high prices offer important opportunities to accelerate the deployment of low cost alternative energy systems.

4.3.3 Electricity Trade

Another important parameter to be analysed in this regional context is the percentage of net consumed electricity being imported or exported (2014). In Africa there are many countries that are not trading electricity with neighbours or the amount of electricity traded is very small (Fig. 4.15). Conversely, other countries are actively involved in trading and are nearly fully dependent on electricity supply from other countries as Benin, Niger and Togo (mostly supplied from Ghana and Cote d'Ivori), and Botswana, Namibia and Swaziland (mostly supplied from South Africa). This interdependence is a result of two mature African power pools, SAPP and WAPP. Mozambique exports 82% of its electricity and imports 61% compared to local consumpiton and serves as an exception to all of Africa. Electricity trading

Table 4.9 Grid el ectricity prices by end-use sector in selected countries in Africa [43] and Latin America [44]

Country (USD/kWh)	Household (2013)	Industrial (2013)	Country (USD/kWh)	Household (2011)	Industrial (2011)
Angola	0.040	0.045	Argentina	0.020	0.030
Cameroon	0.160	0.200	Bolivia	0.090	0.060
DR Congo	0.150	0.100	Brazil	0.260	0.190
Côte d'Ivore	0.200	0.175	Chile	0.210	0.150
Ethiopia	0.205	0.150	Colombia	0.190	0.200
Gabon	0.150	0.155	Costa Rica	0.140	0.130
Ghana	0.170	0.440	Cuba	0.230	0.100
Kenya	0.205	0.150	Ecuador	0.090	0.060
Mozambique	0.100	0.045	El Salvador	0.240	0.180
Nigeria	0.075	0.100	Guatemala	0.180	0.160
Rwanda	0.102	0.133	Haiti	0.350	0.400
South Africa	0.103	0.055	Honduras	0.140	0.200
Tanzania	0.100	0.070	Mexico	0.100	0.110
Zimbabwe	0.110	0.105	Nicaragua	0.220	0.220
			Panama	0.170	0.160
			Paraguay	0.080	0.050
			Peru	0.130	0.060
			Rep. Dominicana	0.180	0.200
			Uruguay	0.280	0.130
			Venezuela	0.020	0.010

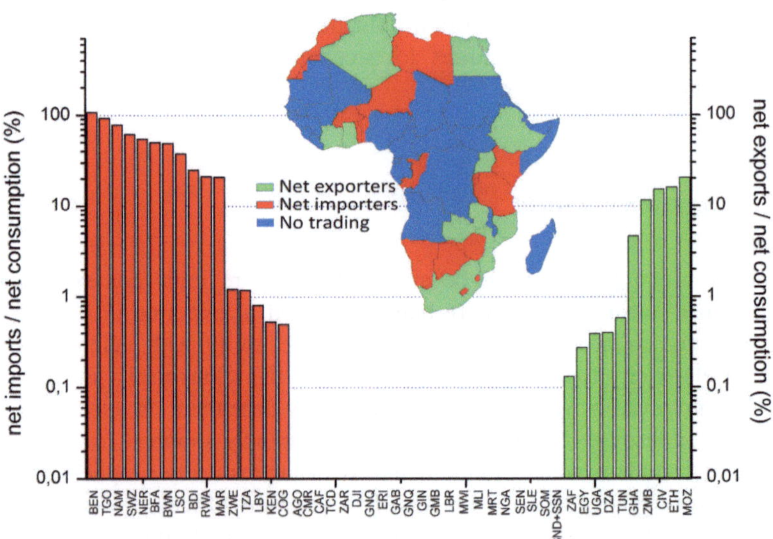

Fig. 4.15 Percentage of net consumption attributed to net imports and exports of electricity (2014) in African countries [1]

data supplied by the IEA [1] and the EIA [45] are not well balanced in general (maybe due to the difficulties to obtain real data from many of these countries) for a clear understanding of the electricity trading between the African countries.

Case Example 4.9

The economics of grid interconnection in Africa

Sometimes the optimal solution supplying cost competititve electricity to national economies is from distant energy resources using high voltage lines to transmit substantial renewable energy resources instead of relying on domestic resources. This solution can also contribute to the expansion of new transmission infrastructure to meet the growing demand of many areas in Africa and Latin America. Sanoh et al. [46] indicate that more than 11 million new customers per year through 2025 are expected in Africa and about two-thirds of the discounted system cost (8% of the continent's GDP approx.) are associated with new generation. The remaining one-third of the discounted cost is associated with the development of the transmission network. However, the expansion of interconnections can reduce the total system cost by 21% relative to the business as usual case. Some general conclusions of this work include: (i) the vast hydro potential of Central Africa can be shipped to any place on the continent at a moderate cost (i.e., Inga Hydro electricity cost is estimated about USD 0.13/kWh transmitted from Democratic Republic of Congo to Egypt and USD 0.09/kWh to South Africa); (ii) hydro resources from Central Africa are competitive in West Africa, but the availability of inexpensive natural gas in Nigeria makes the connection of these two regions less optimal in the long term; (iii) although high wind potential is available on the coasts of Somalia, Morocco, and Tanzania the transmission costs can make transnational projects non-competitive (e.g., wind energy can be produced at USD 0.085/kWh in southern Morocco but can rise to about USD 0.25/kWh transmitted to Egypt); (iv) transmission of solar energy from the desert and the Sahel areas to other parts of the continent becomes feasible only if solar investment costs decrease enough to compensate for the high transmission costs; and (v) in the short term it is more attractive to invest in transmission lines that ship hydro power from Central Africa to Southern Africa and from Eastern to North Africa.

In contrast, in Latin America there are two main trading systems: (i) Brazil, Argentina, Uruguay and Paraguay, where Paraguay plays the net supplier role; and (ii) the small Central American countries (Fig. 4.16). There is no electricity trading from or between islands and unfortunately none between Haiti and the Dominican Republic, largely attributed to political and economic discrepancies. Similiar challenges limit trading between Peru, Bolivia and Chile. It is important to mention

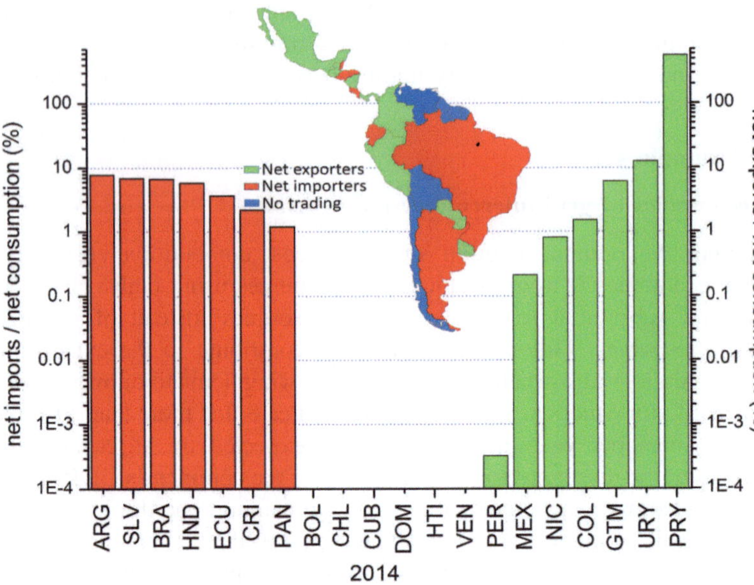

Fig. 4.16 Percentage of net consumption attributed to net imports and exports of electricity (2014) in Latin American countries [1]

the role of Paraguay as the South American power "battery" and hub, and that Uruguay has evolved from net importer to net exporter after 2012 mainly due to the huge increase in wind energy. Also, some countries evolve from net exporters to importers and vice versa depending on the rainy season, mainly in Central America.

Case Example 4.10

Simulating electricity market coupling between Colombia and Ecuador

Colombia and Ecuador have been trading electricity for more than 10 years (interconnection inaugurated in March 2003) and offer an interesting case example because of their complementarity in terms of hydropower supply in different months through the year and the high dependency on hydroelectricity for both countries (67% of the total installed capacity for Colombia and 44% for Ecuador). Also El Niño-Southern Oscillation has opposite effects in Ecuador and Colombia in terms of droughts and floods. There are 4 transmission lines with a total capacity of 500 MW (5% of the peak demand for Colombia and 14% for Ecuador) actively used for trading electricity between both countries. Both electricity markets have similar technologies, dispatch mechanisms and transmission charges. Ecuador is also planning new hydroelectric projects with the aim of becoming a net exporter of electricity during the rainy season. The model developed scenarios with/without capacity payments and isolation versus grid expansion and integration. By increasing integration, the price of electricity tends to converge, decreasing in Ecuador and increasing in Colombia that contributes to an asymmetric benefit. As a result price volatility increases in Ecuador and price peaks are shifted to coincide with those of Colombia (higher in the periods of flood and lower in the periods of drought in Ecuador) and the utilization factor of hydropower increases. Profitability increases at the expense of thermal units and incentives for investment can move from one to the other country. These asymmetries can be moderated or directly cancelled with capacity payments. The interconnector capacity also plays a key role in achieving mutual benefits. The study concludes that capacity and integration policies need to be coordinated [47].

4.3.4 Electrification Ratios

Unmet demand is not captured in electricity data and makes it difficult to measure electricity demand in a holistic sense. The last data available shows that the percentage of urban and rural electrification in Africa (Fig. 4.17) is near 100% in the Maghreb region and most state islands compared to sub-Saharan Africa because of very low investment [1]. A lack of proper maintenance also drives down

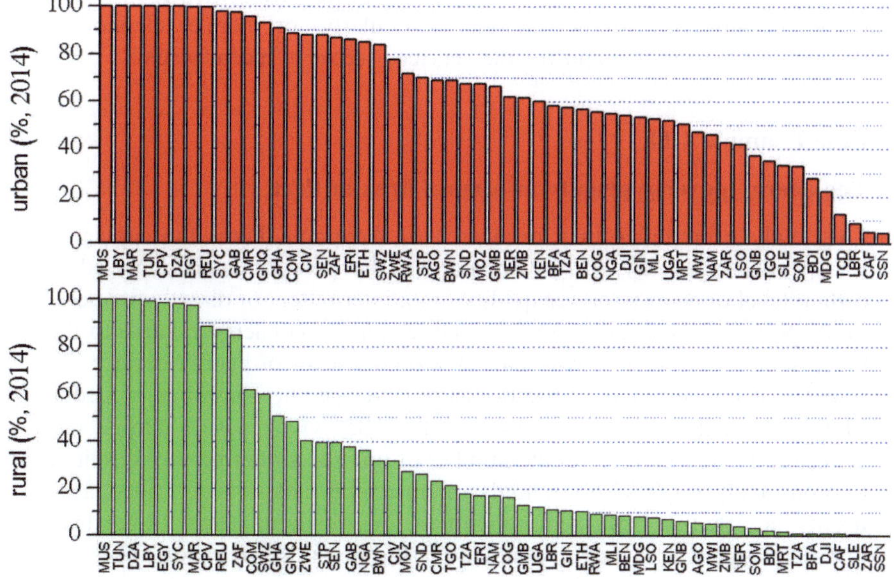

Fig. 4.17 Rural and urban electrification rate in Africa in 2012 [1]

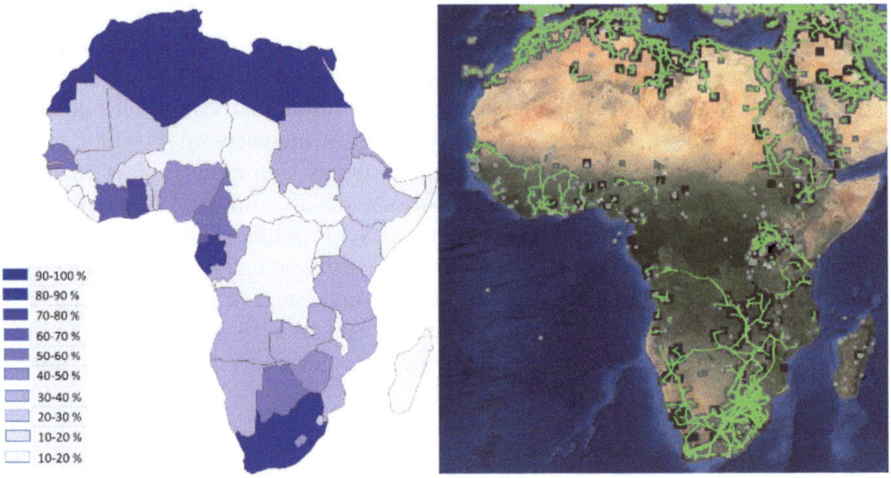

Fig. 4.18 National electrification rate in Africa in 2012 [1], and African power grid map in 2015 (*Source* IRENA Global Atlas for Renewable Energy—OpenStreetMap)

electrification rates (e.g., Zimbabwe plants suffer from reliability issues, and currently only around 50% of installed capacity is available, meeting 40% of peak connected demand, and imports from its neighbours have been curtailed due to non-payment [33]). An important gap exists between urban and rural electrification because of customer density and wealth, showing most of the Sub Saharan African rural electrification rates below 20%. There are also a several African countries with a very low rate of national electrification and power grid extension (Fig. 4.18). However, a boom in communications, particularly for mobile telephones, has helped rapidly expand demand for the services sector in recent years.

Case Example 4.11

The energy poverty penalty in a rural area in Peru

This study exemplifies in a rural area of Arequipa (Peru) the concept of *energy poverty penalty* (*EPP*), which argues that people that are deprived of a certain level of energy service quality (e.g., intermittent or directly not access to the power grid) spend more financial resources on energy relative to their total income than people who enjoys a higher level of energy service. The research team designed a questionnaire and specific sampling procedure. Mobile phone network coverage for a set of 3356 villages is used to generate data stratification in terms of the number of providers offering services, and a rigorous analysis of 342 households and micro-businesses is described. Study results indicate that being remote implies a multitude of structural handicaps that can aggravate a poverty situation. Further disaggregation of results reveals that the EPP is most prevalent in the lowest income classes. In conclusion, the authors propose to better target the policy auction in terms of energy inclusion measures at least to the lowest income segments, based not only on subsidies on energy technologies, but also investment in infrastructure facilitating the delivery of those [48].

In terms of national electrification rates (urban and rural areas) it is important to mention that there are many African countries (South Sudan 1%, Liberia 2%, Central African Republic 3%, Chad 4%, Sierra Leone 5%, Democratic Republic of Congo 9%, Malawi 9%, Burundi 10%, Guinea 12%, Niger 14%, Madagascar 15%, Somalia 15%, Uganda 15%, Rwanda 17%) with a rate of national electrification under 20% (Fig. 4.18) [1]. Sparse population, low GDP per capita and lack of national programs to promote electrification are the main factors that contribute to these low values.

Case Example 4.12

Local and national energy planning in Senegal:

A sensitivity analysis has been conducted in Senegal to capture connection cost and coverage (access) variations as a function of demand, fuel and policy uncertainties in two cases: a local area of 400 km^2 in northern Senegal, and the national case. The power sector in Senegal is dominated by the national utility, "Société National d'Éléctricité" (SENELEC). While SENELEC focuses on urban electrification, rural electrification has been the responsibility of the "Agence Sénégalaise d'Électrification Rurale" (ASER). The mandate of ASER is not only to increase access to electricity but also to the reduction of poverty, reaching 60% of the potential population in 2022. The leading program under implementation by ASER is the rural electrification priority program (PPER) which focuses on establishing 18 concessions (10–25 years, 5000–10,000 customers) via private sector participation available for competitive bidding.

The planning methodology took into account the needs and growth in demand from various sectors, explicitly modeling schools, health facilities, and productive capacity, making extensive use of geographical information systems. Precision about Precise population and location data is fundamental for low cost grid expansion. The two main questions underpinning the study were: (i) what electricity expansion planning approach will achieve the greatest number of customer connections at the lowest cost; and (ii) how do uncertainties in demand, prices, and policy choices affect the cost and length of the grid distribution network. The model considered two decentralized technology options, PV and PV-diesel generation, and grid extension. All cost data were obtained through discussion with experts from ASER and SENELEC. The study found at both the local level and the national level, that a high percentage of the currently non-electrified population lives in areas where grid costs are more favorable than solar PV and diesel minigrids if the cost structure considered is maintained [49]. However, PV costs assumed for this study are much higher than currently offered in the market.

In contrast, in Latin America, with the exception of Haiti, nearly 100% urban electrification has been reached and also high rates of rural electrification have been achieved in most countries (Fig. 4.19) [1]. As summarized above, the electrification of Latin America has been fairly recent and is mainly attributed to national programs approved and financed to enlarge the national grids.

In terms of national electrification rates (Fig. 4.20), in Latin America the most underdeveloped country is Haiti (28%). Other countries that have less than 100% penetration include: Nicaragua (74%), Guatemala (86%), Honduras (86%), Bolivia (88%) and Panama (89%) [1]. A rapid increase in electrification rates can be expected in most of these countries.

4.3.5 *Electricity Distribution Losses*

Finally, it is necessary to analyse the electric distribution losses in the different countries, as the values obtained shows the structural robustness of the power system (Figs. 4.21 and 4.22). It is important to distinguish between the percentage

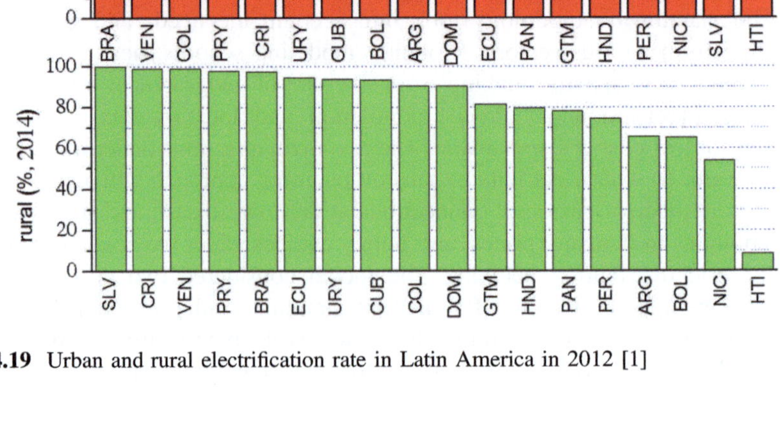

Fig. 4.19 Urban and rural electrification rate in Latin America in 2012 [1]

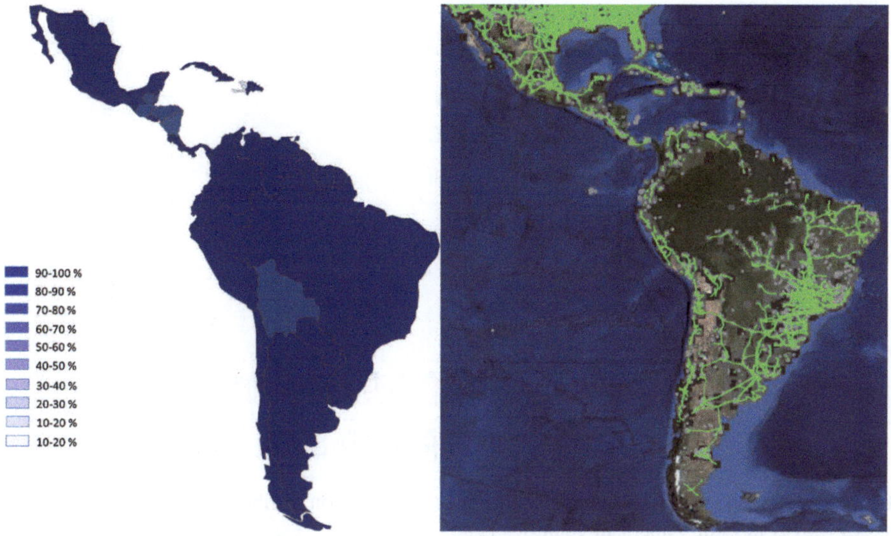

Fig. 4.20 National electrification rate in Latin America in 2012 [1], and Latin American power grid map in 2015 (*Source* IRENA Global Atlas for Renewable Energy—OpenStreetMap)

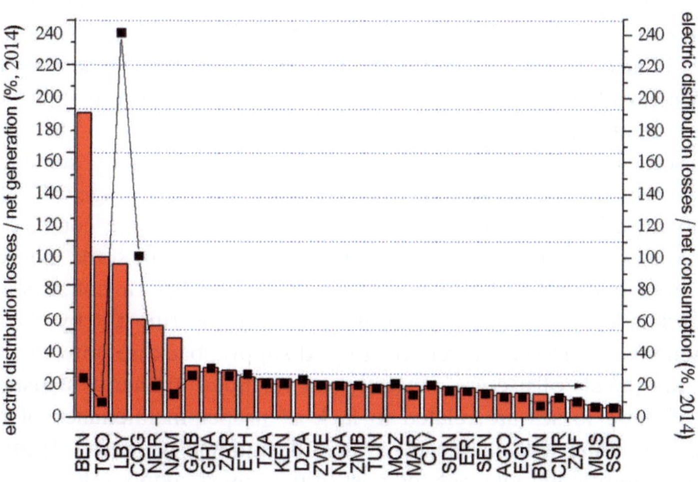

Fig. 4.21 Percentage of net generation and net consumption attributed to electric distribution losses (2014) in African countries [1]

of net generation attributed to electricity losses and the percentage of net consumption attributed to energy losses. For example, there are countries that import most of their electricity (e.g., Botswana imported 49% of the electricity consumed in 2014). Then, the percentage of net generation attributed to electricity losses could

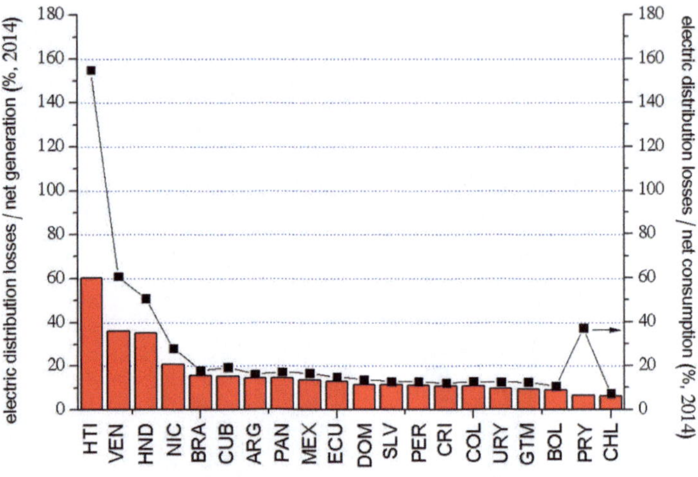

Fig. 4.22 Percentage of net generation and net consumption attributed to electric distribution losses (2014) in Latin American countries [1]

be overvalued compared to large net importers but not if the losses are compared to net consumption. On the other hand, if the country is a large net exporter of electricity (e.g., Paraguay exported 75% of the electricity produced in 2014) the percentage of net consumption attributed to electricity losses can be also overvalued. There is also the case of countries as Haiti and Congo, where no electricity trading exists with neighbour countries and, thus, it is clear that the electricity system is very inefficient.

Grid-based electricity supply is insufficient to meet electricity demand in many areas. It is reported to be unavailable for 540 h per year on average (6% of the year) in Sub-Saharan Africa, but this value is much higher in some countries, such as Nigeria, Guinea and the Central African Republic [44]. In some grid-connected areas, the high frequency of power outages (mostly in dry or hot seasons) means that demand is either unmet or met by other means (mostly by diesel-fuelled back-up generators). However, relative to grid supply, back-up power generation is expensive and levels of use are generally not recorded in energy statistics.

Much of these losses are related to lack of proper maintenance of the power plants and grid along with inefficient system design and operation. High loss rates reduce the reliability and increase the cost of the power supply (electricity prices in parts of sub-Saharan Africa are among the highest in the world [44]), which is already insufficient to meet demand in many countries, mostly in Africa. Electricity power theft is another concern in both developing and developed countries. Electricity power theft is the use of electrical power without a contract with a supplier and/or the total or partial bypassing of metering systems or interfering with the system to modify its measurements (e.g., in Venezuela over 99% of population has access to the power grid, however, only 74% is actually subscribed to a electricity supplier [50]). This activity increases the price of electricity for the subscribed customers and produces voltage drops in the distribution grid that also can lead to power shutdowns. The electricity supplier should detect such irregularities, correct them (e.g., use of boxes to seal customer's connections, construction of distribution networks less vulnerable to tampering and irregular connections and analysis of distribution feeders with the highest losses) and promote correct customer behaviour.

Other important causes of losses are inaccurate metering and billing (e.g., fraudulent practices and extortion), unmetered supplies and non-payments of electricity bills. The transmission and distribution line fault detection methods usually are very poor in developing countries, with no capability to know the specific location of a fault. The employees responsible for O&M are many times not well equipped to perform their jobs and very incapable of handling problems. Moreover, the voltage profiles at the distribution nodes can be out of safety margins due to power losses in the transmission and distribution lines and overloaded transformers (because there is no information at the control center on the loading and current status). In conclusion, improvements are needed in many areas to reduce electricity losses and other issues related to the quality of the power supply.

Case Example 4.13

Rural distribution meter failures in Colombia

A high failure rate of energy meters (electromechanical and electronic) installed in low voltage networks in rural areas with high keraunic level was detected in Caldas, Colombia. The case example is focused in the area covered by the Caldas hydropower plant. All the causes of meter removal between the years 2007 and 2012 were included, with the exception of the meters removed using administrative processes. Critical nodes are defined as those with more than 3 failed meters. The forensic analysis of the electromechanical meters showed that the main cause of failure was overvoltage (other causes were overload, moisture and blows and hits), and potential coils for electronic meters. These failures were attributed to lightning. Technical field visits also found poor engineering practices (no electrical ducts, protection systems, or protective boxes and poor connections), deficiencies of the grounding systems (in the down conductors, insulated electrodes and in the use of unapproved materials), and the installation of transformers [51].

4.4 Conclusions and Future Perspectives

Power grids are the central key element in increasing GDP in Africa and low income areas in Latin America. Renewable energy resources are irregularly distributed in these regions so regional integration, such as power pools, and promotion of renewable generation can result in significant savings relative to capital spending and considerable reductions of carbon emissions. Encouraging cooperation between countries is actively recommended in four different directions: (i) consistency in technical standards, meaning cooperation and coordination between system operators, but also development of major cross-border transmission lines; (ii) financial feasibility, as the cost of power in the exporting country must be lower than the price in the importing country; (iii) political acceptability, as exporters are supplying neighbours when they may have not achieved universal electrification themselves and have insufficient energy to meet current needs; and (iv) regional stability, as a large cross-border power project is a long-term investment that requires commitment and foresight of consecutive political leaders [3].

Sensitivity analysis should be conducted in the future to plan grid extension, decentralized or off-grid power supply in rural areas considering the cost of the technologies involved, the expertise of the technicians devoted to O&M activities, evolution of the demand, fuel cost and policy uncertainties. A percentage of the currently non-electrified population lives in areas where grid expansion could be more favorable than the decentralized or off-grid energy supply technologies, and expansion outcomes (costs and access) are very sensitive to demand levels and capital costs of medium voltage lines and transformers.

Electricity consumption in both regions remain well under the global average (21 and 60% of the world electricity output per capita, respectively). Sub-Saharan Africa contains one of the least electrified regions in the world with an overall electrification rate below 50% and where 30 nations have rural electrification rates less than 20%. This contrasts significantly with North Africa and the majority of urban Latin America (Haiti is an exception), where more than 99% of the population has access to electricity. Rural electrification remains challenging for both regions.

The electricity mix in Africa and Latin America is highly varied but is typically tied directly to endogenous resources controlled by individual nations. While much of the electricity is generated using antiquated fossil fuel technology, it should be recognized that Latin America is the most decarbonized region in the world because of the abundance of hydropower, a dispatchable source of renewable energy. Non-dispatchable renewable energy (i.e., wind and solar) is prevalent in only a few select nations. Moreover, those nations with the lowest amount of non-dispatchable renewables generally have large hydropower resources, small electricity systems and/or large oil and gas reserves.

Finally, it should be mentioned that electric grid reliability remains problematic in many nations where proper maintenance is limited, power theft is prevalent and power outages frequent, limiting the availability of power in many countries with the concomitant impact on local economies. In many countries the traditional grid extension mode requires a multidimensional approach covering regulation, finance, economic development and social dimensions to analyze the business case of each rural electrification project. The extra-cost of de-centralized and off-grid electricity supply is based on the cost of alternative power generation, and it is related to scale (achieving energy conversion efficiencies to the extent allowed by thermodynamics at reasonable cost) and to variations in daily power demand (e.g., lighting load 3–4 h after sunset). The base load for many of these nations tends to approach zero for decentralized generation, further reducing the financial viability of the system until the local economy develops to some minimal level that can sustain the base load. However, increasing electrification rates, reliability and trading, and diminishing electricity price will raise regional GDP and, subsequently, base load.

References

1. International Energy Agency database (2015) http://data.iea.org/. Accessed 30 Nov 2015
2. Bhattacharyya SC (2013) Rural electrification experience from south-East Asia and South America (chapter 7). In: rural electrification through decentralised off-grid systems in developing countries, Springer Green Energy and Technology Series. ISBN 978-1-4471-4672-8
3. Castellano A, Kendall A, Nikomarov M, Swemmer T (2015) Brighter Africa. The growth potential of the sub-Saharan electricity sector. Electric Power & Natural Gas. McKinsey & Company
4. IEA (2010) Energy technology perspectives 2010
5. IEA (2008) Energy technology perspectives 2008
6. The Economist (2011) Resistance is futile. Available 3 Dec 2011
7. Šćepanović S, Mishkovski I, Hui P, Nurminen JK, Yä-Jääski A (2015) Mobile phone call data as a regional socio-economic proxy indicator. PLoS ONE 10:4
8. Wirth G, Lorenz E, Spring A, Becker G, Pardatscher R, Witzmann R (2015) Modeling the maximum power output of a distributed PV fleet. Prog Photovoltaics Res Appl 23:1164–1181
9. Kemausuor F, Adkins E, Adu-Poku I, Brew-Hammond A, Modi V (2014) Electrification planning using network planner tool: the case of Ghana. Energy Sustain Dev 9:92–101
10. Stark GB (2015) A systematic approach to better understanding integration costs (NREL/CP-5D00-64930). Natl Rev Energy Lab
11. Trüby J. (2014) Thermal power plant economics and variable renewable energies. A model-based case study for Germany. Insight Series. IEA
12. Bhattacharyya SC (2013) Introduction (chapter 1). In: rural electrification through decentralised off-grid systems in developing countries. Springer Green Energy and Technology Series. ISBN 978-1-4471-4672-8
13. World Bank (2001) Best practice manual: promoting decentralized electrification investment. Report No. 248/01. Energy Sector management assistance program. Washington, DC
14. Palit D, Chaurey A (2013) Off-grid rural electrification experiences from South Asia (Chapter 4). In: rural electrification through decentralised off-grid systems in Developing countries. Springer Green Energy and Technology Series. ISBN 978-1-4471-4672-8
15. Kishore VVN, Jagu D, Gopal EN (2013) Technology choices for off-grid electrification (chapter 3). In: rural electrification through decentralised off-grid systems in developing countries, Springer Green Energy and Technology Series. ISBN 978-1-4471-4672-8
16. Ortiz B, Vetter M, Golz S, Bopp G (2007) Concepts for hybrid PV village grids to make their operation a local business, 22nd European photovoltaic solar energy conference and exhibition, 3–7 September, Milan

17. World Bank (2008) Designing sustainable off-grid rural electrification projects: principles and practices. Operational Guidance for World Bank Group Staff. Washington, DC
18. Bhattacharyya SC. Electrification experiences from sub-Saharan Africa (chapter 6). In: rural electrification through decentralised off-grid systems in developing countries, Springer Green Energy and Technology Series. ISBN 978-1-4471-4672-8
19. Bhattacharyya SC, Srivastava L (2009) Emerging regulatory challenges facing the Indian rural electrification programme. Energy Policy 37:68–79
20. Bhattacharyya SC, Dow S (2013) Regulatory issues related to off-grid electricity access (chapter 11). In: rural electrification through decentralised off-grid systems in developing countries, Springer Green Energy and Technology Series. ISBN 978-1-4471-4672-8
21. L.R. Valer LR, Mocelin A, Zilles R, Moura E, Nascimento ACS (2014) Assessment of socioeconomic impacts of access to electricity in Brazilian Amazon: case study in two communities in Mamirauá Reserve. Energy Sustain Dev 20:58–65
22. IEA (2015) How 2 guide. Smart grids in distribution networks. Roadmap development and implementation. International Low-Carbon Energy Technology Platform, Paris
23. Hussaini IU, Majid NHA (2014) Human behaviour in household energy use and the implications of energy efficiency delivery. A case of Bauchi, Nigeria. Int J Energy Sect Manage 8:230–239
24. SAND (2007) Long vs short term energy storage: sensitivity analysis. Sand Report (SAND2007-4253). Sandia National Laboratories
25. Fitzgerald G, Mandel J, Morris J, Touati H (2015) The economics of battery energy storage: How multi-use, customer-sited batteries deliver the most services and value to customers and the grid. Rocky Mountain Institute, USA
26. Energy Storage Association (2015) www.electricitystorage.org. Accessed 30 Nov 2015
27. Akhil AA, Huff G Currier AB, Kaun BC Rastler DM, Chen SB, Cotter AL, Bradshaw DT, Gauntlett WD (2015) DOE/EPRI Electricity storage handbook in collaboration with NRECA. Sandia Report SAND2015-1002
28. Kempener R, Borden E (2015) Battery storage for renewables: market status and technology outlook. IRENA, Abu Dhabi
29. Redflow Tower Power. www.redflow.com. Accessed 30 Nov 2015
30. Carigiet F, Niedrist M, Scheuermann C, Baumgartner F (2016) Case study of a low-voltage distribution grid with high PV penetration in Germany and simulation analyses of cost-effective measures. Prog Photovoltaics Res Appl 24:1523–1532
31. DOE/EPRI (2013) Electricity storage handbook in collaboration with NRECA, EPRI, Palo Alto, CA and DOE, Washington, DC. National Rural Electric Cooperative Association (NRECA), Arlington, VC. http://www.sandia.gov/ess/publications/SAND2013-5131.pdf. Accessed 30 Nov 2015
32. Shahnia F, Ghosh A (2014) High penetration of rooftop photovoltaic cells in low voltage distribution networks: voltage imbalance and improvement (chapter 4). In: Hossain J, Mahmud A (eds) Renewable energy integration. Challenges and solutions. Springer Green Energy and Technology Series. ISBN 978-981-4585-26-2
33. Climatescope (2015) The clean energy country competitiveness index. The multilateral investment fund, UK Department of International Development, Power Africa and Bloomberg New Energy Finance. http://globalclimatescope.org/en/. Accessed 30 June 2016
34. The Economist (2016) Power to the powerless. Ending energy poverty. Available 27 Feb 2016
35. Yadoo A (2010) Rural lighting. ITDG technical brief, practical action, Bourton on Dunsmore
36. Office of Energy Efficiency & Renewable Energy (2015) US department of energy. http://energy.gov/eere/ssl/led-basics#how_efficient. Accessed 30 Nov 2015
37. OECD/IEA (2010) Energy poverty. How to make modern energy access universal? Special early excerpt of the World Energy Outlook 2010 for the UN General Assembly of the Millennium Development Goals

38. Darghouth NR, Wiser R, Barbose G, Mills A (2015) Net Metering and market feedback loops: exploring the impact of retail rate design on distributed PV deployment. Ernest Orlando Lawrence Berkeley National Laboratory. LBNL-183185
39. Moner-Girona M, Ghanadan R, Solano-Peralta M, Kougias I, Bódis K, Huld T, Szabó S (2016) Adaptation of feed-in tariff for remote mini-grids: Tanzania as an illustrative case. Renew Sustain Energy Rev 53:306–318
40. Ellis A (2009) Interconnection standards for PV systems. Where we are? Where we are going? Sandia National Laboratories. http://www.uwig.org/pvwork/4-ellis-interconnectionstandards.pdf. Accessed 3 February 2017
41. Steward D, Doris E (2014) The effect of state policy suites on the development of solar markets. NREL/TP-7A40-62506 (2014)
42. IEA (2015) 2014 Key world energy statistics
43. IEA (2014) Africa energy outlook
44. Organización Latinoamericana de Energía (2013) Sistema de Información Económica-Energética
45. Energy Information Administration International Energy Statistics (2015) US department of energy. http://www.eia.gov/beta/international/data/browser/#/?ord=CR&cy=2013&v=H&vo=0&so=0&io=0&start=1980&end=2013&vs=INTL.44-1-AFRC-QBTU.A ~ INTL.44-1-ASOC-QBTU.A ~ INTL.44-1-CSAM-QBTU.A ~ INTL.44-1-EURA-QBTU.A ~ INTL.44-1-EURO-QBTU.A ~ INTL.44-1-MIDE-QBTU.A ~ INTL.44-1-NOAM-QBTU.A ~ INTL.44-1-WORL-QBTU.A. Accessed 30 November 2015
46. Sanoh A, Kocaman AS, Kocal S, Sherpa S, Modi V (2014) The economics of clean energy resource development and grid interconnection in Africa. Renew Energy 62:598–609
47. Ochoa C, van Ackere A (2015) Does size matter? Simulating electricity market coupling between Colombia and Ecuador. Renew Sustain Energy Rev 50:1108–1124
48. Groh S (2014) The role of energy in development processes—the energy poverty penalty: case study of Arequipa (Peru). Energy Sustain Dev 18:83–99
49. Sanoh A, Parshall L, Sarr OF, Kum S, Modi V (2012) Local and national electricity planning in Senegal: scenarios and policies. Energy Sustain Dev 16:13–25
50. Vidoza JA, Gallo WR (2016) Projection of fossil fuels consumption in the Venezuelan electricity generation industry. Energy 104:237–249
51. Marin OJS, Plata EAC, Farfan AJU, Younes C (2014) IEEE industry application society annual meeting. Vancouver. 5–9 Oct 2014

Chapter 5
Biomass for Heating and Power Production

Ricardo Guerrero-Lemus and Les E. Shephard

Abstract This chapter shows how biomass is the predominant fuel for many less developed African and Latin American countries. Nearly 38% of the world population (2014) relies on biomass for cooking, largely in rural areas, but in Africa this value increases to 67% with much of this contribution from sub-Saharan Africa. In Latin America, 15% of the population relies on biomass as it represents an affordable energy source for many households. The supply of biomass in Africa and Latin America has grown considerably since 1990 in both regions (greater than 81% and 42%, respectively) with the majority of the biomass being derived from solids. Moreover, biomass utilization in Africa varies significantly between Maghreb countries with almost no reliance on traditional biomass for energy consumption and Sub-Saharan countries where most countries have a strong dependence on traditional biomass. These variations are attributed to lower GDP per capita, the dominance of rural areas, ease of access to forest and agricultural waste and the limited availability of fossil fuels in sub-Saharan countries. In Latin America where GDP per capita is higher, the reliance on traditional biomass is much lower than in Africa. However, the use of biomass has contributed to more than 577,000 premature deaths in Africa and 74,000 in Latin Americas in 2012, mostly in low and middle income countries. Exposure is particularly high among women and young children with many premature deaths attributable to ischaemic heart diseases, strokes and chronic obstructive pulmonary diseases in adults and acute lower respiratory infections in children under 5 years. Then, modernising biomass technology is not only beneficial for the economy. It will also improve human health, create social benefits and reduce environmental damage.

Keywords Energy · Africa · Latin America · Electricity · Biomass · Cook stoves · Energy crops · Combustion · Forest wood · Carbon emissions · Efficiency · Arable land

R. Guerrero-Lemus and L.E. Shephard, *Low-Carbon Energy in Africa and Latin America*, Lecture Notes in Energy 38,
DOI 10.1007/978-3-319-52311-8_5

5.1 Introduction

Biomass is almost the only type of fuel used by mankind since ancient times up to the industrial revolution, with a current large use in wide areas of the less developed African and Latin American countries. Thus, supplies of biomass are abundant in many countries of these regions, although local scarcity exists. Therefore, a 38% of the world population still relies on the traditional use of biomass for cooking (2014), but in Africa this value raises to a 69% (mostly in sub-Saharan Africa) and in Latin America to a 14% [1] as biomass is the only affordable energy source for many households. Heavy dependence on biomass is concentrated in, but not confined to, rural areas, and the basic consumption of dry wood (15% moisture) varies between 1 and 8 kg/person·day (1 kg dry wood = 15 MJ = 4.17 kWh) depending on weather conditions [2].

When wealth is increasing, modern forms of energy are usually applied sparingly at first and for particular services (such as electricity for mobile phones, radio and television, or LPG for making tea and coffee) rather than completely supplanting an existing form of energy that already supplies a service adequately. The most energy-consuming activities in the household—cooking and heating—are the last to switch. Also, the use of multiple fuels provides a sense of energy security, since complete dependence on a single fuel or technology leaves households vulnerable to price variations and unreliable service. Some reluctance to discontinue cooking with fuelwood may also be due to taste preferences and the familiarity of cooking with traditional technologies [3].

However, each year 4.3 million premature deaths, of which 577,100 are in Africa and 74,400 in America's low- and middle-income countries (2012, last data available), can be attributed to household (indoor) air pollution resulting from the traditional use of solid fuels, such as fuelwood, crop wastes and charcoal [4]. In poorly ventilated dwellings, smoke in and around the home can exceed acceptable levels for fine particles up to 100-fold. Exposure is particularly high among women and young children, who spend the most time near the domestic hearth [5]. This situation can be improved if some affordable technologies are introduced, as we will expose in this chapter.

On the other hand, biomass can play a central role in the use on endogenous resources and the introduction of low carbon energy technologies in Africa and Latin America, as combustion, gasification and anaerobic digestion can be considered as almost carbon neutral if they do not produce significant transformation of the environment as deforestation. However, there are some concerns because the world forest area has been reduced in the period 1990–2015 from 42.6 to 41.3 million km^2 (−3.1%), in Africa from 7.2 to 6.3 million km^2 (−12.5%) and in Latin America from 10.4 to 9.4 million km^2 (−9.6%) [6] (Fig. 5.1). Fuelwood brings some localised deforestation, but depletion of forest cover on a large scale has not been found to be attributable to demand for fuelwood as it is more often gathered from the roadside and trees outside forests, rather than from natural forests. Clearing of land for agricultural development and timber are the main causes of

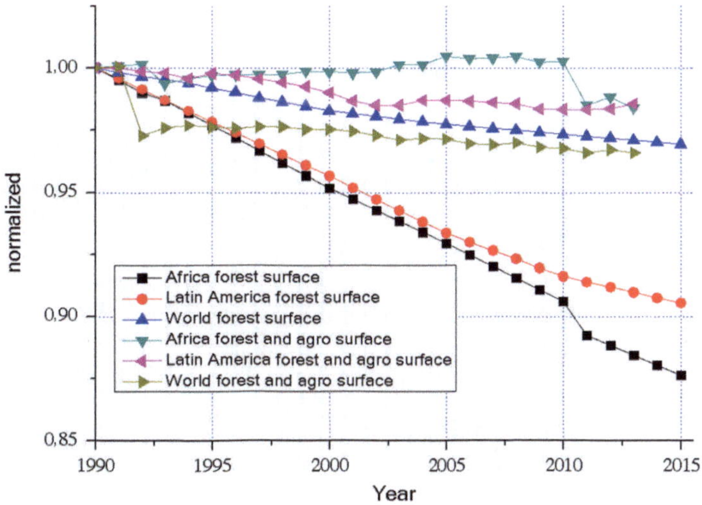

Fig. 5.1 Evolution of the world forest and agricultural surface area in Sub-Saharan Africa, Latin America and the world in the period 1990–2013 [6]

deforestation in developing countries (studies at the regional level indicate that as much as two thirds of fuelwood for cooking worldwide comes from non-forest sources such as agricultural land and roadsides) [3]. When agricultural surface is added to the forest surface, the values are decreasing very softly in the period 1990–2013 (Fig. 5.1). This evolution can be interpreted as the change in use of the soil from forest to agriculture as the main reason in deforestation. When animal dung and agricultural residues are used for fuel rather than left in the fields or ploughed back into fields, soil fertility is reduced and propensity to soil erosion is increased.

According to statistics from the International Energy Agency [7], the supply of biomass (primary solid biomass, biogas, liquid biofuels and waste) has grown steadily from 194 and 92 Mtoe in 1990 to 368 Mtoe (+89.8%) and 136 Mtoe (+48.1%) in 2014 in Africa and Latin America, respectively (last available data) [7]. In Africa, almost 100% of biomass supplied was solid (0.01% of liquid biofuels, and 0.001% of biogases) in 2014. In contrast, in Latin America a 82.3% of the biomass supplied was solid, a 17.4% was liquid biofuels, and a 0.3% biogases. No energy supply from renewable municipal waste has been reported in both regions, as it will be discussed in Chap. 12, and no production of heat from any renewable resource [7]. Also, it has been impossible to show data about the biomass potential in these regions due to fundamental discrepancies in terms of methodology and results shown in recent literature [8].

Only 1827 GWh of electricity in Africa were supplied in 2014 by biomass (1.3% of the total renewable electricity), mostly from solid biomass (Fig. 5.2) and a 0.02% from biogas. In Latin America 45,377 GWh were supplied from biomass (3.8% of the total electricity), also mostly from solid biomass and a 0.9% from biogas [7].

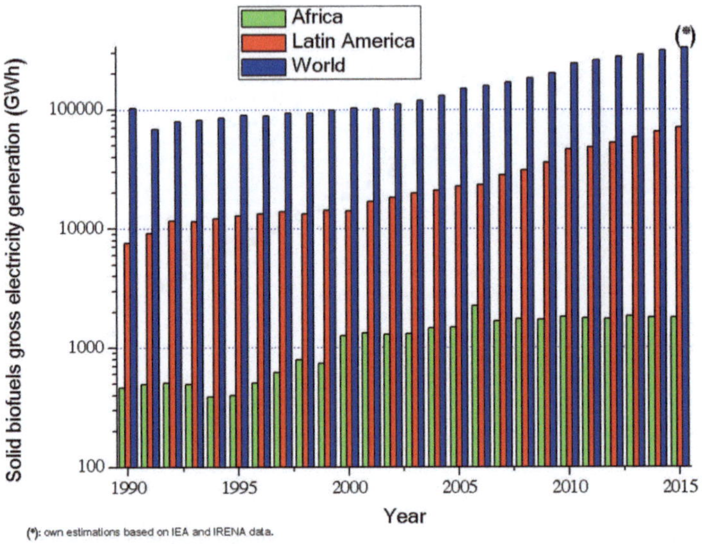

Fig. 5.2 Evolution of the world, Africa and Latin America gross electricity generation from solid biofuels in the period 1990–2015 [7, 9]

5.2 Technology State of the Art

Biomass is any living mass, both animal and vegetable, existing on earth, including its waste, and can be converted into heat or other energy carriers such as electricity when it reacts with oxygen in the combustion process. Biomass has a low energy density compared to fossil fuels. The energy stored in biomass varies between 8 GJ/t for green wood and 15 GJ/t for dry wood, which are very low compared to 55 GJ/t of natural gas (Table 5.1). However, its energy density can be increased substantially by conversion to biofuels through different mechanical, chemical or biological procedures. Biomass can also be considered as solar energy stored in chemical bonds of carbon-hydrogen as a result of metabolic activity or

Table 5.1 Energy content per unit weight and volume for various types of biomass [10, 11]

Feedstock	GJ/t	GJ/m^3
Dry wood	15	10
Paper	17	9
Dry manure	16	4
Baled straw	14	1.4
Dry sugar cane	14	10
Household waste	9	1.5
Crude	42	34
Coal	18	50
Natural gas	55	0.04

photosynthesis of different organisms (about a 0.05% of the solar power that reaches the Earth is considered to be absorbed to produce photosynthesis [10, 11]).

Unlike other technologies for exploiting renewable sources, biomass is extremely varied in its nature and, consequently, requires specific technologies in each case, both in the management of the particular crops, and in the energy use, like heat, electricity, etc.

5.2.1 Energy Crops

To produce bioenergy from agricultural crops, the most widely used plants are sugarcane and corn, as discussed in more detail in the chapter devoted to biofuels (Chap. 11). These plants are used for the production of liquid biofuels. Other plants such as sunflowers and soybeans are used for biodiesel production. Moreover, the yield per hectare of energy crops has grown significantly since 1960 until 2005 (e.g., 70% increase for sugarcane and 400% for corn) [12]. Table 5.2 shows both the biomass yield (t/ha) and the energy equivalent produced per hectare and year for different energy crops [11]. It is assumed that the values shown are the maxima obtained in practice and the biomass is almost dry.

For further comparisons with other energy technologies, the energy efficiency of photosynthesis is defined as the energy content (heat from glucose combustion to CO_2 and liquid H_2O at STP) of the biomass that can be harvested annually divided by the annual solar irradiance over the same area. Then, solar energy conversion efficiencies for crop plants in both temperate and tropical zones typically do not exceed 1%; for comparison, efficiencies of 3% are obtained for microalgae grown in bioreactors and about 5–7% for microalgae in bubbled bioreactors. On the other hand, a theoretical limit of about 12% can be considered for the efficiency of photosynthetic glucose production from CO_2 and water [13].

The yield of energy crops depends on many factors such as climate, soil type, available water resources and technology. The cultivation techniques vary very substantially between countries in terms of cost of labour and access to automation process systems. In general, the productivity of the cultivation process increases with the level of development of the country concerned. Consequently, it is estimated that

Table 5.2 Maximum performance energy harvests from dry biomass [1]

Feedstock	Yield (t/ha)	Energy (GJ/ha)
Forest wood	7	130
Tropical wood	11	200
Sorghum (3 harvest/year)	50	850
Sudan grass (6 harvest/year)	40	600
Corn	25	77
Wheat	22	
Sugar cane	30	150

global productivity can be substantially improved as new cultivation techniques are adopted by the most underdeveloped countries in Africa and Latin America.

Obviously, energy crops, after having been grown, should be harvested, dried, cut, and transported. The objective is to obtain an homogeneous product of appropriate size, high energy density and with minimal moisture. The harvest of forest residues also arises from the cleaning and maintenance of forests. These residues are rarely used but burned. Alternatively, these residues can be pressed up to 1 t/m^3 for easy transport [10], reaching a significant energy density of about 15 GJ/t [6].

To increase the energy densification of biomass, pellets are usually fabricated by applying pressure on biomass waste (mainly sawdust). Pellets are already subject to quality standards for a growing process of commercial activity (6–12 mm in diameter, 10–30 mm in length, moisture content below 10%, ~650 kg/m^3 and ~17 GJ/t) [14]. An additional option is to subject the biomass to torrefaction processes (200–300 °C) by which it is converted into a dry, hydrophobic and higher energy density (19–23 GJ/t) product [14], holding up 92% of the energy of the raw material. Torrefacted biomass can also be converted into pellets, process that can lower its cost per energy unit. Moreover, torrefaction facilitates biomass transport and storage activities.

Another procedure to improve biomass energy density is through pyrolysis. In this process the raw material is heated in the absence of oxygen, producing charcoal, bio-oils and biogas. The fraction of each product depends on the temperature and residence time of vapours in the process. The application of moderate temperatures (~500 °C) for 1 s is used to maximize the liquid fraction, in a process called fast pyrolysis. By contrast, the slow pyrolysis in traditional earth kilns with efficiencies between 10 and 20% is commonly used for obtaining charcoal. Improved metal, brick and retort kilns offer efficiencies between 25 and 40% [15]. The biogas produced is often used as energy input in the pyrolysis process itself. The bio-oils obtained have a calorific value of about 17.5 GJ/t, but the energy density per unit volume (20–30 GJ/m^3) doubles to quadruples that of the pellets and the product of torrefaction, although this is only yet about half the biodiesel energy density [14].

Case Example 5.1

Briquettes as an alternative to firewood and charcoal:

Briquettes have been successfully marketed as an alternative to firewood and charcoal in countries such as Egypt, Sudan, Rwanda, Namibia and Kenya. Their greater energy density means reduced transport costs, a longer burning time and, depending on the type of biomass and processing method, less emissions. However, many briquetting projects have failed in the past due to poor project planning, marketing, low quality products and lack of availability of appropriate stoves. For example, it has been found that in some cases it was essential to locate points of supply and points of sale in proximity to each other, as small-scale producers and buyers may not have access to transport [15].

5.2.2 *Cook-stoves*

In Africa and Latin America, biomass is mostly burned in cook-stoves. In order to evaluate stove performance of cook-stoves the most recognized testing procedure is the water boiling test (WBT) [16]. It is composed by three phases: cold-start, hot-start and simmer test; the amount of fuels used and emissions generated for the three stages define the WBT indicators [17]. Cook-stoves can be fed by a wide range of fuels (mainly dry wood, residues, dung, charcoal, kerosene, alcohols, butane, gases derived from petroleum or natural gas), solar energy (described in Chap. 7) or electricity.

Wood stoves are the most used for cooking in developed countries and countless models exist. The most traditional one is the three-stone fire (Fig. 5.3), where a fire flame originated on wood combustion is partially surrounded by stones, bricks or a mud base to support a boiler. The space in the base is used for fuel recharging. This technology presents very low performances, both in terms of fuel consumption and of pollutant emissions, and it is considered as the baseline for the evaluation of other stove performances (Table 5.3) [18].

An advance with respect to the three-stone fire is the traditional stove, having a metal or ceramic combustion chamber of a cylindrical shape. It achieves higher performances than the three-stone fire except for particulate matter (PM) because insufficient air draft or cold zones in the combustion chamber caused by an improper design [18]. The improved cook-stoves are mainly based on the rocket concept (Fig. 5.4), where the stove has a combustion chamber made up of two orthogonal parts: an insulated upright chimney (having a height of two or three times the diameter) and an horizontal zone where wood sticks are placed, greatly helping combustion efficiency [16], reducing fuel use by 33%, CO emissions by

Fig. 5.3 Image of the three-stone stove (first plane) and a rocket stove (*Source* www.solsolidari.org)

Table 5.3 Different stove models and associated energy and pollutant indicators [18]

Stoves	Energy use (MJ)	PM emissions (mg)	CO emissions (g)	Time to boil (min)
Three-stone fire	21.8	1674	65	38.1
Stoves without combustion chamber	11.6–20.4	585–3,509	16–50	14.0–40.5
Stoves with rocket combustion chamber	10.4–23.5	515–1,806	9–21	12.2–58.0
Charcoal gasifier	13.6	70	8	27.3
Forced air stoves	9.4–15.3	27–293	4–9	13.2–33.7
Charcoal stove	12.6–20.8	44–260	41–113	18.9–30.0
Propane	6.7	5	1	23.0
Ethanol	6.8	4	5	31.6
Kerosene	9.7	10	8	42.5

Fig. 5.4 Small manufactured rocket cooking stoves in use in Njewa Malawi (*Source* khym54 [https://www.flickr.com/photos/khym54/1804279896])

75%, and PM emissions by 46% on average in comparison to the three-stone fire [19]. Thus, heat transfer efficiency seems to be predominantly dependent on the geometry of the stove and how closely the hot gases pass around the bottom and sides of the pot. Also, CO can be burned up more completely with higher temperatures, and PM seems to be reduced by thorough mixing of air, gas and flame in a hot space, burning only as much fuel as can be combusted completely [19].

The presence of a chimney can greatly reduce harmful emissions but the heat reaching the pod diminishes, increasing fuel consumption. Consequently, stoves with a chimney have much better results if sunken pots or skirts are used (a pot skirt can reduce fuel use and emissions by 25–30% [19]). Then, stoves with chimney are not optimal for boiling or simmering food, but they are highly performing for frying [16]. In the same way, fan-assisted stoves allow for the control of fuel burning thanks to forced air convention, that permits reducing fuel use by an average 40%, and CO and PM emission levels over 90% in comparison to three-stone fires, but they need to be fed with electricity, which is not always available and reliable in developing countries. If charcoal is used instead of wood, the combustion chamber can be smaller because of the higher heating value. On the other hand, the effective available energy in charcoal is only 30–50% of the original energy available in the wood biomass used for producing the charcoal, therefore, energy use in charcoal stoves and emissions should be more than doubled if the 2-stages combustion process is computed [18].

Case Example 5.2

The flexible role of charcoal production in smallholders in Mozambique:

In Mozambique the percentage of population participating in the production, trade and sale of charcoal could be as high as 15%, three-quarters of them are based rurally, and up to 50% of revenues are retained in rural areas. A community of around 30,000 inhabitants in central Mozambique is studied during 9 months (the first author of ref. [37] resided in the study area) in 2014, most of them farmers relying in rain-fed 1–2 ha agriculture (mostly maize). Most families have woodland at the edges of their farms and do not have to travel far for firewood. About half of the households made charcoal within the past 12 months. Charcoal is regulated under the Mozambican forest law, which normally requires licensing and the pay of taxes (USD 0.22–1.32 by sack) for producing, but most producers in this study area operate without license. Charcoal is predominantly made on each household's own smallholding. A 2 m high and 5 m long kiln can yield about 25 sacks of approximately 40 kg each. Traders pay USD 1.37–1.83 per sack in pick-up points or taken on public transport (USD 0.69 per sack). About half of the producers make charcoal as part of the preparation for a new field, a third of them as a key livelihood, and a fifth when the household needs cash [20].

Another option are gasifiers, working in a 4-stage combustion process (drying, pyrolysis, char-gasification and gas-combustion) and permitting a large improvement in combustion efficiency (Table 5.3) also because gases burn above the generating zone and can directly reach the pot. When properly operated, gasifier stoves can reduce particulate matter a 90% in average compared to the three-stone fire [19]. This technology will be exposed below (Chap. 13) because it is also useful for producing

Fig. 5.5 Image of a gasifier combined with a thermoelectric generator (*Source* BioLite)

electricity in a turbine. Anaerobic digesters, useful for producing biogas from waste, are exposed in Chap. 12. Cook-stoves can be also provided with thermoelectric generator (TEG) modules for producing small amounts of electricity exploiting the Seebeck effect but increases the expensiveness of the system (Fig. 5.5) [21]. Finally, liquid fuel (not kerosene) stoves generally have high thermal efficiency and low polluting emissions (emitting high levels of PM than some improved wood stoves when not operating properly [19]) but in developing countries there is unavailability and high costs of many fuels. This drawback can be avoided if the fuels are locally produced from agriculture residues.

Some basic recommendations can be also provided for improving the performance and safeness of the stove: (i) burn only dry, seasoned wood and maintain a hot fire, (ii) have a certified technician for installing and annually service the appliance, and (iii) upgrade to an efficient certified appliance [22]. Also, it is important to mention that the success of a stove program for improving unhealthy cooking technologies in African and Latin American countries should be evaluated by the number of new stoves being used as intended in communities, rather than simply the number of stoves that are disseminated and built [23].

Case Example 5.3

Household fuel mix versus income in transition economies, Botswana:

In theory, a regional shift from biomass to "modern" fuels has long been expected to occur in tandem with urban growth. However, trends observed in a large cited paper [24] after 78 semi-structured household interviews conducted in urban and peri-urban Maun, Botswana, show that the population has often not followed such patterns and fuelwood continues to be important in most areas. This is attributed to the fact that the model of household fuel switching largely dismisses the importance of active (and strategic) decision making by urban consumers and their responsiveness to structural factors such as relative fuel prices. Multiple energy sources are employed in complex ways, each for specific purposes, such as modern fuel uptake largely complements fuelwood rather than leading to its abandonment. Thus, different fuels are not considered to be completely inter-substitutable and, consequently, fuelwood remains strategically important. Also, the importance of fuel prices to energy-use decision-making reveals that although income is clearly relevant, it is far from the sole driver of fuel choice patterns. Other factors as household characteristics, lifestyle, reliability of fuel distribution networks, policies and market trends also influence. Modern fuel penetration data are somewhat illusory as incomplete fuel switching is relatively common. While commercial fuel use has risen, the proportional contribution of wood to national energy consumption has declined slightly, but increasing in absolute terms [24].

5.2.3 Technologies for Producing Electricity

The direct combustion of biomass is the oldest technique used for energy recovery, but it is not very efficient on producing electricity (5–30% efficiency in developing countries, but 85–90% in CHP systems [14]). The efficiency for electricity production ranges between 10% for steam boilers with powers below 1 MW, and 40% for >50 MW steam turbines combined with the latest combustion technology in fluidised beds [14]. Normally, the combustion of biomass requires a first step consisting in a simple physical treatment: sorting, crushing, compressing, air drying, etc. The low conversion efficiency for green biomass is partly due to the energy necessary to evaporate all water contained in the organic matter. Furthermore, biomass contains a high proportion of volatile materials that are emitted as vapours or tars (causing sparking during combustion) and requires additional energy, reducing the efficiency of the process.

A technology improvement compared to the above process is the fluidized bed combustion (FBC), which allows a larger heat transfer during combustion, and also an enhanced temperature control (the flame temperature to ignite biofuel is lower than in conventional combustion) and reduction of gas emissions (NO_x and CO). In this technique, the fuel material is cut in small pieces and placed on a limestone bed. In this way, an enhanced mixture between biomass and air is obtained during combustion, thus increasing the performance, and allowing the use of a less homogeneous biomass. The fluidized bed may be stationary or circulating.

As an alternative to conventional combustion, organic Rankine cycles are being considered. This cycle uses an organic fluid with high molecular mass, instead of steam, to recover heat from low temperature sources. The plants work with powers of 0.5–2.0 MW, achieving efficiencies of 17% (slightly higher than with steam boilers) and lower working temperatures. However, the net efficiency, after deducting auto-consumption, is lower. Moreover, in a smaller range of power capacities are the Stirling engines (10–100 kW_e), whose technology is promising for domestic cogeneration. However, Stirling engines using biomass as fuel require further tests before they are introduced in the market. Electricity conversion efficiencies in these systems are in a range of 12–20% [14].

As mentioned above, among all plants generating electricity from biomass, only 10–40% of the fuel energy is exploited, since the rest is evacuated as heat to the atmosphere or to the cooling circuits. For this reason, it is much more efficient a simultaneous cogeneration of heat and power, and in this case the efficiency can reach 80–90%. In CHP plants, heat can be used to heat water for district heating or industrial processes. In industrial applications the heat is usually transferred through steam at a temperature of 130–200 °C and pressures of 3–16 bar [12]. Obtaining heat causes the electricity production efficiency to be reduced by several percentage points [14].

In the gasification technique, biomass is partially oxidized by means of water vapour and air (or oxygen) at high temperatures (800–1000 °C) to form syngas (mixture of CO, CH_4 and H_2), which can also be used for secondary purposes other than energy production. The gas can be used in steam boilers, gas turbines, or co-firing. The process is not very efficient, since the total energy of the gasification products is lower than that of the initial biomass. However, the fuel gas produced is more versatile and less polluting. If the gasification is carried out in air, the gaseous biofuel has a calorific value of 3–7 MJ/Nm^3, and if it is carried out in oxygen, it can reach 7–15 MJ/Nm^3 [12], although the production of oxygen requires also a considerable amount of electricity. In addition, up to 20 MJ/Nm^3 can be obtained in indirect combustion processes [12], which represent 10–45% of the heat capacity of natural gas.

The syngas can also be cleaned of particles and condensable hydrocarbons and burned in an internal gas combustion engine, providing an electricity efficiency of 22–35%. This efficiency is slightly higher than that achieved with the steam turbines used for biomass combustion processes [12]. This efficiency can be increased up to 40% if the syngas is burned in gas turbines [12].

Finally, anaerobic digestion, as the biological degradation of biomass in oxygen-free environments, produces biogas, rich in CH_4 and CO_2. The anaerobic digestion process takes place in two phases: (i) hydrolysis and acetogenesis, which converts biodegradable material into glucose and amino acids, subsequently converted into fatty acids; and (ii) methanogenesis from acetic acid, also producing CO_2. The two steps can be performed in a single reactor (single step anaerobic digestion) or in two steps (two-step anaerobic digestion). The second process is more efficient but more complex and expensive.

The biogas obtained from anaerobic digestion can be directly burned in CHP systems [12] or be upgraded to meet the standards of natural gas and injected directly into pipelines. Anaerobic digestion can operate with any type of biomass that can be digested by animals (that is, except biomass from wood). The process is particularly efficient with wet biomass, sludge from sewage treatment plants, as well as in landfill cells for municipal waste storage. If the biomass used as raw material is not contaminated, the solid digestate can be also used as fertilizer, adding value to the process.

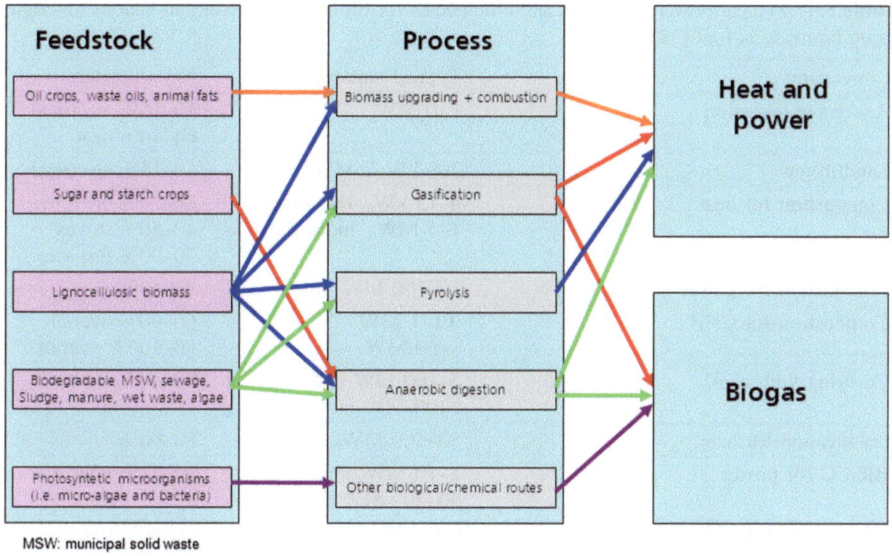

Fig. 5.6 Scheme of the different routes for the production of electricity, heat and gaseous fuels from solid biomass

Considering the different feedstock and processes involved, Fig. 5.6 shows a schematic overview of the pathways followed to convert solid biomass into electricity, heat and gaseous fuels.

Also, Table 5.4 shows a summary of typical power capacities and electrical and heat efficiencies for the different technologies involved for the production of energy from biomass.

5.2.4 *Costs*

Unlike other types of renewable technologies with freely available resources (sun, wind, etc.), the cost of biomass typically represents between 50–90% of bioenergy production (with the exception of waste). In developed countries the forest biomass costs are well established, typically ranging from USD 2.3 to 6.5/GJ for biomass delivered to an energy recovery plant, and an upper limit cost of USD 3.0–4.0/GJ if strong production growth is expected [14]. These costs vary significantly depending on the country, and are affected by the specific conditions of the plantation, distance to the energy recovery plant, techniques for collecting and processing biomass, etc. On the other hand, the use of biomass residues for other purposes such as animal feeding, fertilizers, materials, etc., can add value and, consequently, reduce the cost of energy production. It should also be noted that the structure of costs largely depends upon the available infrastructure, and the technology used to harvest.

Table 5.4 Typical power capacities and efficiencies for different electrical and heat technologies using biomass as fuel [12]

Conversion	Typical capacity	Net efficiency
Anaerobic digestion	<10 MW	10–15% electrical 60–70% heat
Landfill gas	200 kW–2 MW	10–15% electrical
Combustion for heat	5–15 kW_{th} residential 1–5 MW_{th} industrial	10–20% open fire 40–50% stoves 70–90% furnaces
Combustion for power	10–100 MW	20–40%
Combustion for CHP	0.1–1 MW 1–50 MW	60–90% overall 80–100% overall
Co-firing with coal	5–100 MW existing >100 MW new plants	30–40%
Gasification for heat	50–500 MW_{th}	80–90%
BIGCC for power	5–10 MW demo 30–200 MW future	40–50% plus
Gasification for CHP using gas engines	0.1–1 MW	60–80% overall
Pyrolysis for bio-oil	10 t/h demo	60–70%

Moreover, the lack of transparency in some markets, especially in relation to forest resources, obstructs the resource pricing in the market.

On the other hand, a detailed cost analysis of the densification processes can show wood pellet costs higher than natural gas in recent years [25]. This scenario is expected to continue as natural gas reserves are increasing worldwide mainly due to new fracking techniques to extract it. Moreover, there is no apparent link between the price of natural gas and pellets. Pellets costs are mainly related to the raw material (43%), but also to drying (35%), being other significant costs associated to pelletisation (7%), personnel (6%) and storage (3%) [2]. The costs for the production of heat in combustion boilers using pellets as fuel are USD 11–142/GJ, with an average value of USD 37/GJ, which makes it competitive with fossil fuels [14]. It is expected a cost reduction of only 4–6% in 2030 (at constant prices) as a consequence of increasing system lifetimes and efficiencies [14].

The prices of efficient cook-stoves range from USD 5 to USD 10 [15], but there is still a barrier to the uptake of efficient stoves because for poor households these low prices are still beyond the means of many. But costs can be further reduced by design and manufacture changes, or through subsidies and micro-financing. Some other issues that can result in the failure of cook-stove projects include a failure of the stove to meet the cooking requirements for particular dishes, the non-availability of suitable fuels, and religious beliefs in which the traditional cook-stove plays an important role in religious lore and practices [26].

The range of investment costs shown in Fig. 5.7 is indicative of the importance of economies of scale in energy production from biomass and the maturity of each

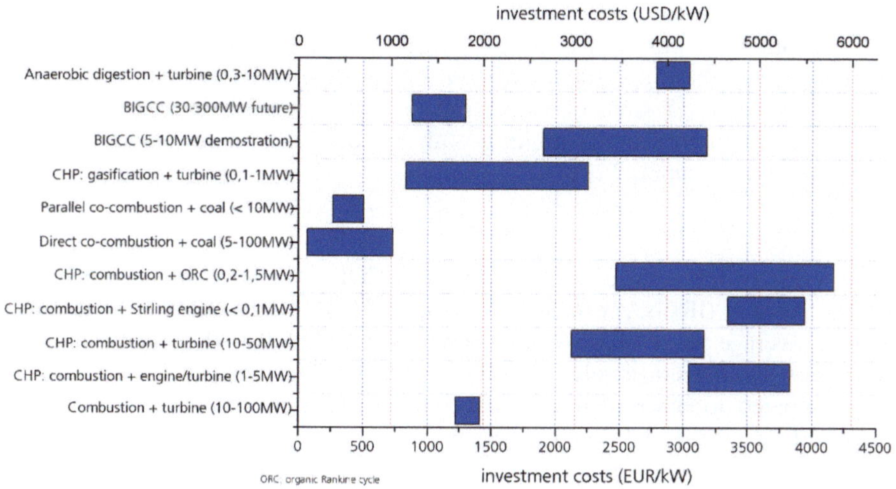

Fig. 5.7 Investment costs for different technologies for energy production from biomass [4]

technology (being Stirling engine, BIGCC and organic Rankine cycle the most immature). As it can be observed, the lowest costs are associated to the co-firing processes, as it capitalizes thermal plants fuelled by coal. However, this process cannot be considered a low carbon technology.

Case Example 5.4

Corn food processing and the gasification of cobs in Cameroon:

Researchers have studied a solution for sustainable energy production integrating corn food processing with the gasification of the cobs in Fongo Tongo, a 20,000 inhabitants municipality in West Cameroon not provided with electricity. The corn is already cultivated in West Cameroon and there are preexisting collecting points where the corn is shelled and milled. The byproducts of the process are the corn cobs that did not found a valuable application. The choices of the gasification derives from the high energy conversion efficiency, together with the chance to exploit existing skills on engines to maintain the plant. The thermo-conversion process was modeled and the results were compared with the energy required for corn processing, therefore demonstrating the sustainability and attractiveness of the solution. The higher heating value of the corn cobs was measured (15.9 MJ/kg), total moisture (9.4%), density (1.36 g/cm^3) and the composition. The main results of the corn cobs gasification model considered 20 kg/h of biomass consumption, specific volumetric tar amount of 1.32 g/Nm^3, volumetric syngas flow of 47.0 Nm^3/h, a syngas higher heating value of 4.64 MJ/Nm^3 and a cold gas efficiency of 76.7%. Then, the electrical energy produced in the gasification of corn cobs is almost 15 times the energy consumed for processing. Summing up, 100 kg of corn ears can be converted into 89 kg of corn and about 7 kWh of electricity. A commercial 20 kW_e gasifier was selected for supplying electricity to different public buildings assuring continuous power supply [27].

In terms of levelized cost of electricity (LCOE), the International Renewable Energy Agency (IRENA) establishes a range of USD 0.029–0.146/kWh in Africa with an average value of USD 0.060/kWh, and USD 0.030–0.131/kWh in South America with an average value of USD 0.030/kWh (Fig. 5.8) [9]. Thus, in average and compared to other regions, production of electricity from solid biomass is favourable in Africa and South America.

5.2.5 CO_2 Emissions

Most CO_2 produced by biomass energy technologies could be considered neutral from an accounting standpoint, since the employed biomass has previously captured CO_2 from the atmosphere. However, emissions from construction and decommissioning phases, as well as from biomass harvesting, treating and transporting must be also considered.

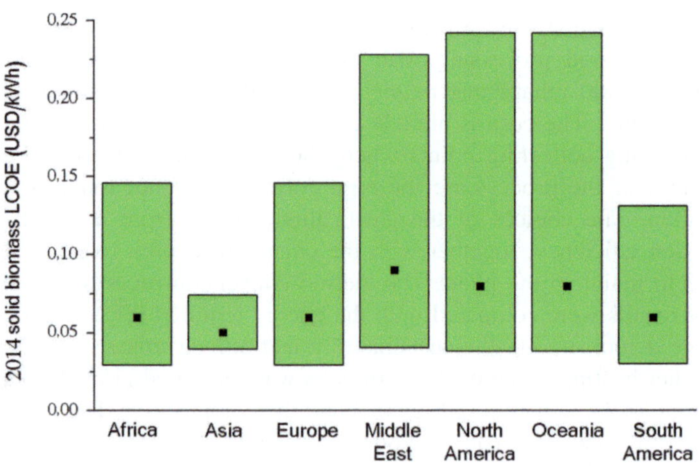

Fig. 5.8 Levelized cost of electricity from solid biomass in the different global regions (2014) [9]

In this regard, the literature shows [28] that biomass from wood and crops burned in CHP plants shows that CO_2 emissions are mainly originated on the biomass itself. Also, the amount of CO_2 emitted in the operation process is substantial, and attributed mainly to the use of fertilizers and to transportation. In contrast, the emissions attributed to plant construction and decommissioning are under 10% of the global emissions. Moreover, due to the production of fertilizers, large amounts of CO_2 are emitted but accompanied by ammonia and nitrogen oxide emissions. It should be noted that studies considers that straw, as it is a residue, does not occupy soil nor requires fertilizer for its production.

In a recent LCA study [29], different biomass plants for producing heat and/or power have been evaluated and compared with reference fossil energy systems likely to be displaced by the bioenergy system. In the cited case studies, bioenergy systems reduce GHG emissions by between 18% and 128% compared to their counterpart fossil reference systems, but it is difficult to generalize (GHG emission reductions above 100% mean that the involved process acts as a GHG sink). Also, it is concluded that GHG mitigation is larger where biomass is used for heat and electricity applications rather than for liquid transport fuels. Moreover, biomass energy production associated with the CCS technology (Chapter 2) can act as a negative CO_2 emitter, as biomass captures CO_2 from the atmosphere and is then stored.

Case Example 5.5

Gabon land-based strategy for contributing to the UN Framework Convention on Climate Change:

Gabon estimates that its forest captures four times the carbon dioxide that emits, and these emissions are due (60%) to changes in the affection of the lands (deforestation). However, Gabon plans to reduce a 50% its carbon emissions in 2025 compared to the base scenario. This plan is based in (i) a Law to protect the forest, expanding rotations from 15 to 25 years: (ii) the creation of 13 new national parks: (iii) a national Land Allocation Plan for optimally allocate areas to different uses, excluding intact forests. Also Gabon plans to base the electricity sector in a power mix 80% on hydropower and 20% on natural gas in 2025, up to 700 MW installed. Additional actions are planned to be deployed for the transport, waste, and water sectors [30]. However, we consider that the plan is not well designed, as the document also considers a population growth of 2.5%/year. Then, substantial changes in the affection of land (deforestation) due to urbanisation and agriculture activities are expected to continue in future.

5.3 Regional and National Perspectives on Technology

5.3.1 Biomass and Health

As it has been previously mentioned in the introduction to this chapter, household air pollution has been responsible of 577,100 premature deaths in Africa and 74,400 in Latin America in 2012 (last data available) [4]. Those premature deaths are attributable to ischaemic heart diseases, strokes and chronic obstructive pulmonary diseases in adults, and acute lower respiratory infections in children under 5 years (Fig. 5.9). As it is shown below, the difference in deaths between both regions can be attributed to the higher percentage of population relying on traditional biomass Africa compared to Latin America (Figs. 5.10 and 5.11).

In Africa there exists a huge contrast between Maghreb countries, where almost no population is relying on traditional biomass for energy consumption, and Sub-Saharan countries, where most countries have a strong dependence on traditional biomass. The main factors that help to explain different levels of fuelwood consumption are: (i) climate/seasonality; (ii) household size; (iii) ease of access; (iv) population density, as increasing urban population drives to alternative fuel consumption; (v) availability and price of alternative fuels and stoves; (vi) alternative uses (e.g., brick-making and tobacco curing); and (vii) cultural factors and

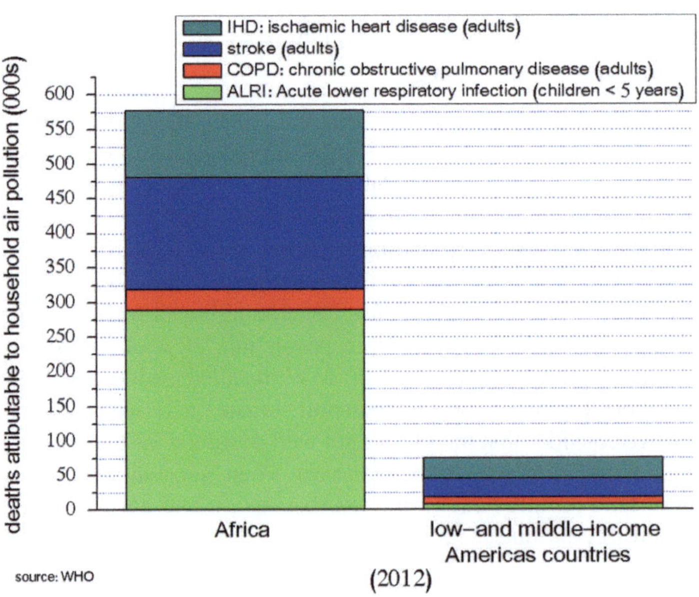

Fig. 5.9 Number of deaths attributable to household pollution in Africa and Latin America in 2012 [4]

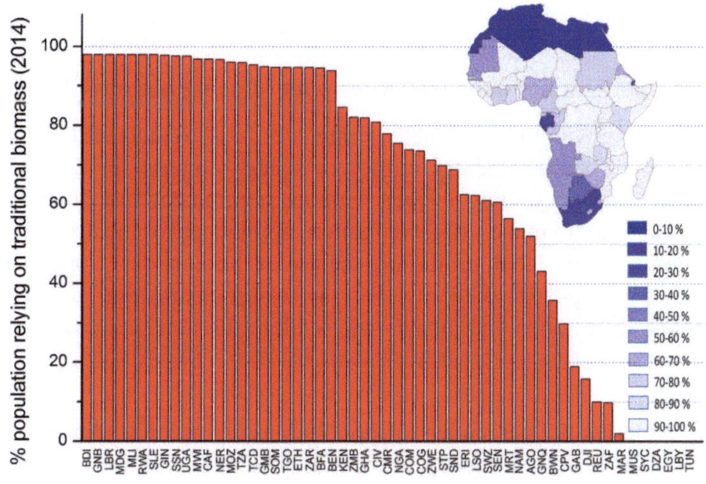

Fig. 5.10 Percentage of population relying on traditional biomass in Africa in 2014 [1]

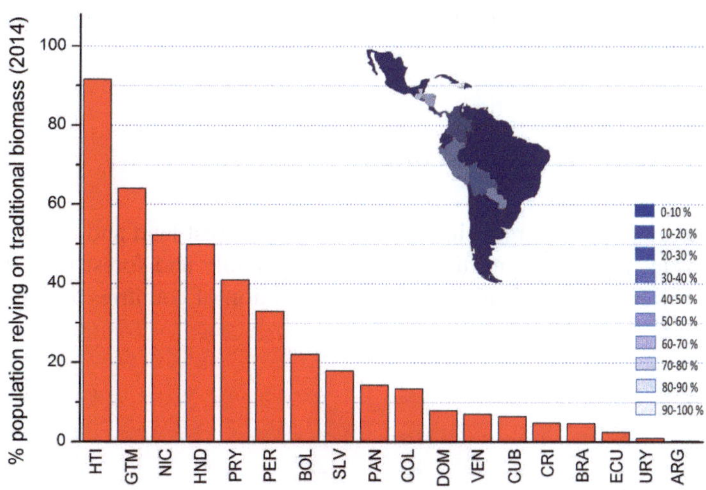

Fig. 5.11 Percentage of population relying on traditional biomass in Latin America in 2014 [1]

nutritional habits. In Africa, the variations in fuel consumption per country can be mostly attributed to the GDP per capita (Figs. 2.7 and 2.8), but also to the dominance of rural areas, the availability of fossil fuels (Figs. 3.5–3.7) and the ease of access to forest and agricultural waste, as it will be exposed below.

Case Example 5.6

Rwanda's policies for reducing the impact of using biomass for cooking:

The Rwandan government devised a strategy in 2010 to increase the efficiency and reduce the environmental impact of using biomass for cooking. Key components were: building capacity among equipment manufacturers and importers in order to make available modern appliances for the use of biomass; developing a quality label; promoting the use of these modern appliances; and launching a long-term publicity and awareness campaign to encourage households, institutions and businesses to adopt the new equipment [31].

In Latin America the percentage of population relying on traditional biomass is much lower than in Africa mostly because the GDP per capita is higher in most countries. In this case, the evolution of forest and arable land shows opposite tendencies in some countries (e.g., Haiti), as they rely most on traditional biomass but shows the lowest levels of forest and arable land. It can be attributed to the fact that in densely populated but poor countries there exists a substantial tendency to deforestation.

5.3.2 *Forest and Arable Land*

Thus, some countries are reducing their arable and forest land and, on the contrary, other countries are increasing the percentage of arable and forest land (Figs. 5.12 and 5.13) [6]. In Africa it is remarkable the evolution of countries as Egypt, a quite

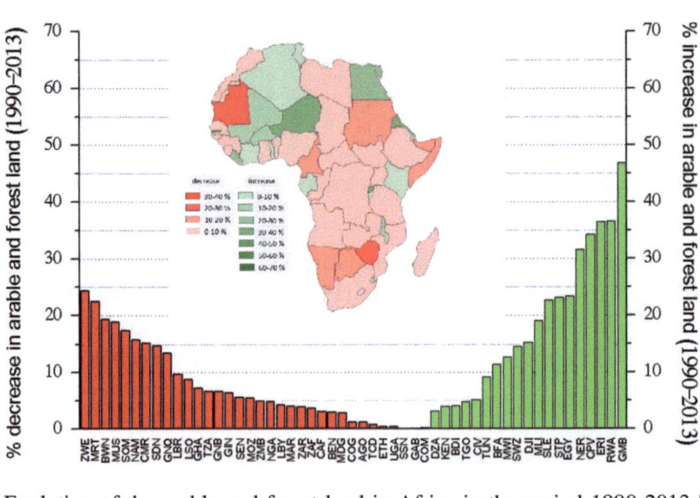

Fig. 5.12 Evolution of the arable and forest land in Africa in the period 1990-2013 [6]

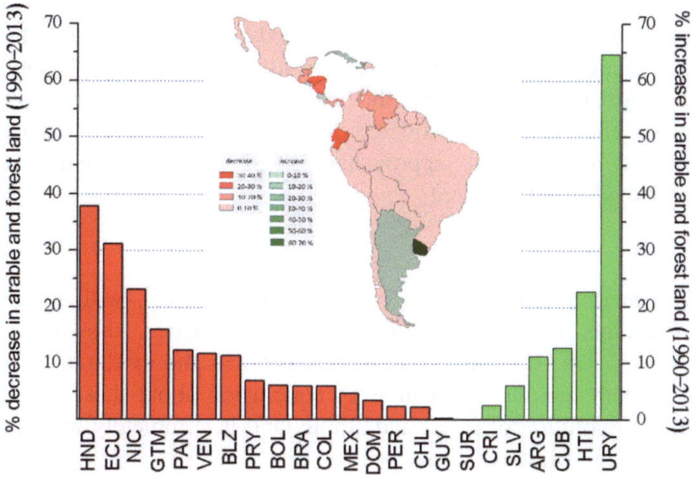

Fig. 5.13 Evolution of the arable and forest land in Latin America in the period 1990–2013 [6]

deserted country, and how increases the percentage of arable and forest land. It is also remarkable countries as Gabon, where the arable and forest land per capita is very high and also shows an increase in percentage of arable and forest land from 1990 to 2013. On the contrary, countries as Zimbabwe shows a negative evolution.

Case Example 5.7

Positive impacts of fuelwood sourcing in Maun, Botswana:

Fuelwood consumption is frequently considered as a cause of deforestation and, consequently, environmental degradation, and leading to energy insecurity among low-income households, especially where the resource is commercialized. Such views have influence among policymakers, often resulted on repressive forestry regulations and, consequently, generated a great deal of friction between state and local actors. Also, governments in many developing countries are financially, bureaucratically and legislatively ill-equipped to appropriately monitor and control diverse and dynamic patterns of resource use across vast areas of countryside. In contrast, however, a number of researchers have demonstrated that wood energy dependence is often not a significant cause of deforestation and can provide important livelihood opportunities. Deforestation should be more associated to other phenomena as agricultural growth, peri-urban commercial and residential development, fire and climate change. Also, many urban fuelwood supplies originate more from farm clearance and maintenance than cutting of forest areas. A case study in Maun, Botswana [32], under unregulated harvesting shows that impacts are significantly mitigated by collector's strong preferences for source sites with abundant dead wood, low competition, relationship to other activities, and recognized access rights. Also, fuelwood vending is found to provide critical support to rural incomes and a key source of low-cost urban energy. Finally, the benefits of fuelwood use and avoidance of negative effects were considered to be more effectively achieved by decentralized management approaches based on local institutions and understandings of the resource.

In Latin America there are also countries, as Uruguay, with a very positive tendency and other countries, as Honduras and Ecuador, showing a worrying behaviour (Fig. 5.13). It is also easy to observe that the evolution in arable and forest land in both regions is negative. A conclusion about these data at country level is that the evolution of forest and arable land is completely influenced by the policy in those countries.

If a low carbon energy technology based on biomass is planned to be introduced in a country, it is very important to consider the forest and arable land per capita. In Africa there are some countries showing large values, also close to deserts, as the population densities are small (Fig. 5.14). Also there are some dramatic situations, where biomass technologies are very difficult to implement, as in Djibouti.

In Latin America the data show similar results, and countries with low forest and arable land (i.e., Haiti and El Salvador) (Fig. 5.15) but not as critical as in Djibouti. Consequently, in which countries biomass can substantially contribute to a low carbon energy future? In our opinion, in those countries with high forest and arable land per capita and a positive evolution of total forest and arable land surface: Argentina and Gabon are the best candidates in both regions for implementing large penetration of low carbon energy technologies based on biomass.

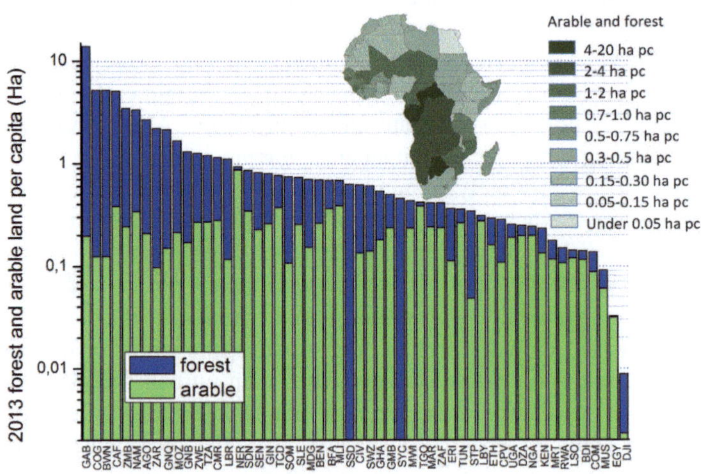

Fig. 5.14 Forest and arable land per capita in Africa in 2013 [6]

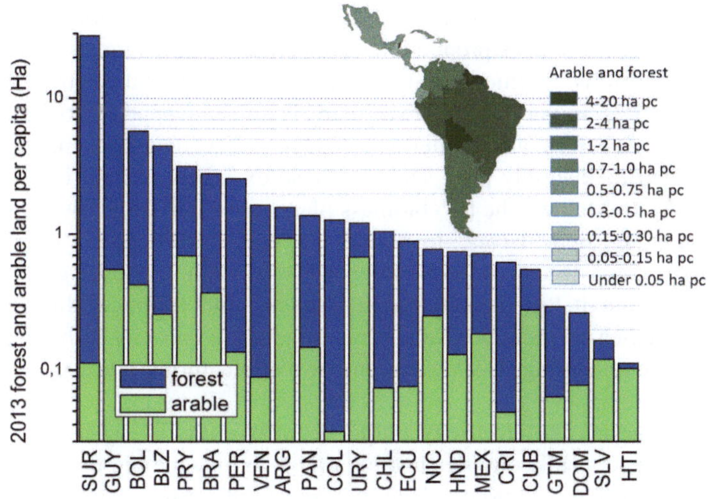

Fig. 5.15 Forest and arable land per capita in Latin America in 2013 [6]

5.3.3 *Electricity from Biomass*

To consider the penetration of low carbon energy technologies based on biomass in Africa and Latin America it is important to show the electricity output from these technologies (Fig. 5.16). For appropriate comparisons we include data from the country showing the largest absolute electricity output from biomass (USA) and the country showing the largest penetration of electricity from biomass in the power

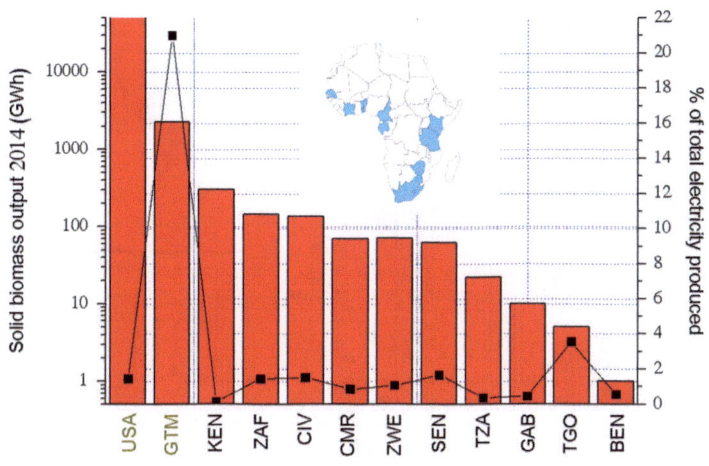

Fig. 5.16 Electricity output from biomass and share of total electricity produced in different African countries in 2014 [7]

system (Guatemala). The IEA shows electricity output from biomass only in 10 African countries, but also production of electricity from biomass and waste is detected in Mauritius, Reunion and Uganda. The share of total electricity produced in those countries is very small, except in Togo, and appears to be related to agriculture activity. Kenya also offer expertise in project development, engineering, O&M and feedstock supply [19]. There are also examples of co-firing wood-chips with coal [15]. In Malawi, the two biomass plants are captive and owned by a sugar cane producer. In Sierra Leone, in May 2014, the 32 MW Addax biomass plant, running on bagasse feedstock and adding 15 MW to the national grid, was commissioned, though it has recently faced financial and technical difficulties [19]. Also, South Africa offers expertise and manufacturing capacity in biomass project development, engineering, O&M, equipment manufacturing and feedstock supply [19].

In Latin America the electricity output from biomass is larger than in Africa, important in some small countries and significative in some big ones as Brazil and Chile (Fig. 5.17). This technology is used in most Latin American countries (except Venezuela, Paraguay and other smaller). This means that the production of electricity from biomass is capable to offer an alternative in African and Latin American countries enrolled in a future low carbon energy strategy if the incentives are appropriate. Moreover, Brazil and Mexico offer expertise and manufacturing capacity in all the biomass production chain: project development, engineering, O&M, equipment manufacturing and feedstock supply [19].

In terms of evolution of the electricity generated by biomass and waste in Africa (Fig. 5.18), only Kenya shows activity in the whole period 1990–2014 but most countries start producing electricity after 1990. The production profile for many of these countries is very flat, showing no significant upward evolution. These profiles

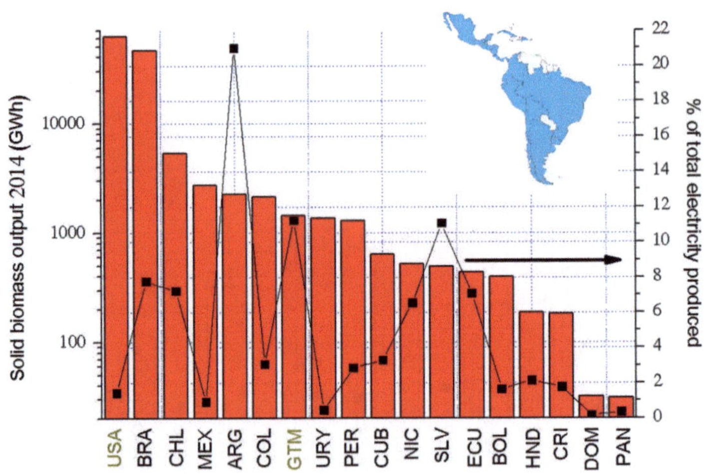

Fig. 5.17 Electricity output from biomass and share of total electricity produced in different Latin American countries in 2014 [7]

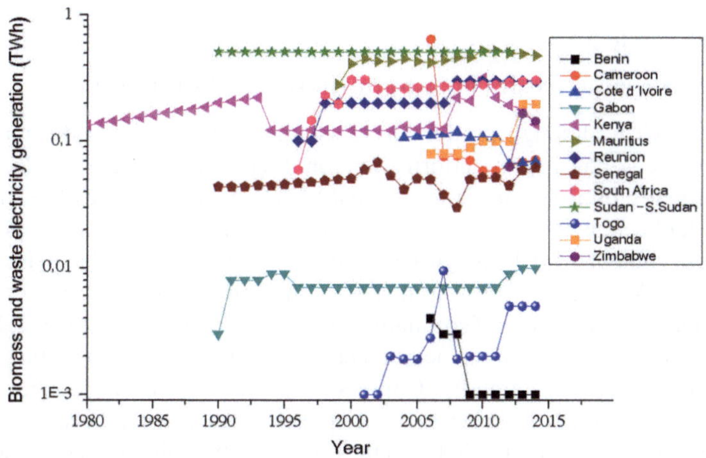

Fig. 5.18 Electricity output from biomass and waste per country in Africa in the period 1990–2014 [7, 9]

can be associated to a very incipient development and not stable in a long term in several African countries. It is also remarkable to mention the case of Mauritius, a small island, and its role as one of the largest producers in Africa.

On the other hand, in Latin America the upward tendency is common, and Brazil is leading the regional production (Fig. 5.19). Moreover, in Latin America the use of biomass for producing electricity is very common in many countries and can be attributed to huge water resources and a the development of large farms.

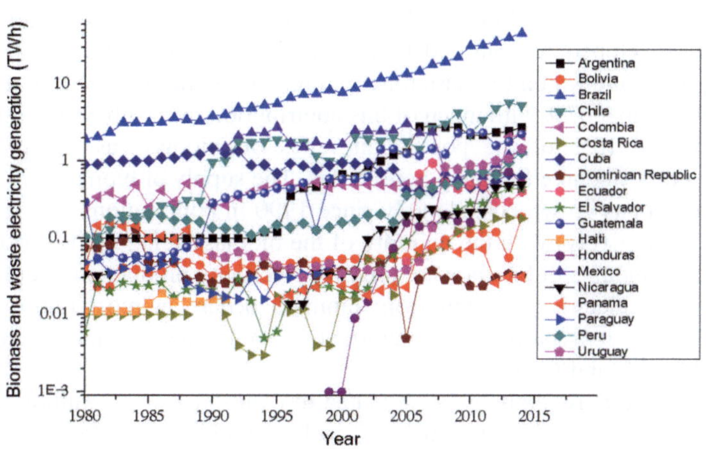

Fig. 5.19 Electricity output from biomass and waste per country in Latin America in the period 1990–2014 [7, 9]

Case Example 5.8

Fuelwood characteristics of fast-growth species in Costa Rica:

In Costa Rica, although production of energy with wood-based lignocellulosic residues has begun, information on fuel characteristics of the most commercially used species is not available yet. A paper studies the calorific value, fuel value index, effect of the chemical properties, extractives and moisture content of 10 fast-growing species (rotation periods of less than 20 years) which are the most used species in forest plantations in Costa Rica (Terminalia amazonia, Terminalia oblonga, Vochysia guatemalensis, Bombacopsis quinata, Alnus acuminata, and Swietenia macrophylla [native species] and Tectona grandis, Cupressus lusitanica, Acacia mangium and Gmelina arborea [exotic species]). The species were selected because of their capacity to produce energy, but also the quality of their timber and capacity to adapt to abandoned areas. One meter long logs of selected trees free of damage caused by insects or fungi were cut and then painted at the end to prevent loss of moisture. No consistency was observed with regard to which type of wood (sapwood or heartwood) had the highest gross calorific value (16.3–20.6 MJ kg^{-1}). The calorific values and fuel value indexes were affected by lignin, nitrogen, carbon, ashes content and extractives in sodium hydroxide, ethanol-toluene and dichloromethane [33].

5.4 Conclusions and Future Perspectives

Biomass has been a persistent fuel used by mankind from earliest civilizations through the industrial revolution and persists today as a predominant fuel for many of the less developed African and Latin American countries. Nearly 38% of the world population relies on biomass for cooking, largely in rural areas, but in Africa this value rises to a 67% with much of this contribution from sub-Saharan Africa. In Latin America, 15% of the population relies on biomass as it represents an affordable energy source for many households. The supply of biomass in Africa and Latin America has grown considerably since 1990 in both regions (greater than 81 and 42%, respectively) with the majority of the biomass being derived from solids. Very limited amounts of biomass are used for electricity production in either region although some electricity output from biomass occurs in most Latin American countries, where it is particularly important in some small countries and can be significant Brazil and Chile.

The use of biomass globally is estimated to contribute to 4.3 million premature deaths annually and is largely attributed to household (i.e., indoor) air pollution resulting from the traditional use of solid fuels. More than 577,000 premature deaths are in Africa and 74,000 in Latin America's low- and middle-income countries (2012). Exposure is particularly high among women and young children, who spend the most time near the domestic hearth with many premature deaths attributable to ischaemic heart diseases, strokes and chronic obstructive pulmonary diseases in adults, and acute lower respiratory infections in children under 5 years.

Biomass utilization in Africa varies significantly between Maghreb countries with almost no reliance on traditional biomass for energy consumption, and Sub-Saharan countries, where most countries have a strong dependence on traditional biomass. These variations are attributed to lower GDP per capita, the dominance of rural areas, ease of access to forest and agricultural waste and the limited availability of fossil fuels in sub-Saharan countries. In Latin America where GDP per capita is higher, the reliance on traditional biomass is much lower than in Africa.

Then, modernising biomass technology is not only beneficial for the economy but will also improve human health, create social benefits and reduce environmental damage. Clean cook-stoves and other modern alternatives to traditional biomass use reduces demand for energy because their efficiency in converting biomass to heat is much higher. Many countries have well-established cookstove programmes in most of Southern, Central and East Africa as well as some in West Africa, reaching penetrations of efficient cookstoves above 36% in Kenya and 50% in Rwanda. As carbon-offset projects, 2 million efficient stoves have already been installed in Africa [34]. Briquettes from agro crop residues can be a modern renewable alternative to charcoal, providing equivalent convenience of use and income-earnings opportunities. Combined with sustainable forestry management, the use of plantations could mean an opportunity to produce wood pellets for use in domestic power production or for export. Moreover, ethanol (as it is explained in Chap. 11) is a promising renewable fuel now deployed in some countries and can be used for cooking, either in its liquid form or as a gel, also to replace paraffin in lighting.

References

1. IEA (2016) World energy outlook 2016
2. IEA (2014) Africa energy outlook
3. IEA (2016) World energy outlook. Chapter 15: energy for cooking in developing countries
4. WHO (2015) Health and environment. http://apps.who.int/gho/data/node.wrapper.ENVHEALTHHAP?lang=en&menu=hide. Accessed 30 Nov 2015
5. WHO (2015) Indoor air. http://www.who.int/indoorair/en/. Accessed 30 Nov 2015
6. World Bank data (2015) http://data.worldbank.org/indicator/AG.LND.FRST.ZS. Accessed 30 Nov 2015
7. IEA Energy Access Database (2015) http://www.worldenergyoutlook.org/resources/energydevelopment/energyaccessdatabase/. Accessed 30 Nov 2015
8. Stecher K, Brosowski A, Thran D (2013) Biomass potential in Africa. IRENA
9. RESOURCE. International Renewable Energy Agency (IRENA). http://resourceirena.irena.org/gateway/dashboard/. Accessed 30 Nov 2015
10. Larkin S, Ramaje J, Scurlock J (2004) Bioenergy, Cap. 4. In: G. Boyle (ed) Renewable energy, Oxford
11. Twidell J, Weir T (ed) (2006) Renewable energy resources. Taylor and Francis

12. IEA (2008) Energy technology perspectives
13. Blankenship RE et al (2011) Comparing photosynthetic and photovoltaic efficiencies and recognizing the potential for improvement. Science 332:805–809
14. IEA (2009) Bioenergy—A sustainable and reliable energy source
15. IRENA (2015) Off-grid renewable energy systems: status and methodological issues
16. Mapelli F, Mungwe JN (2013) Modern energies services for cooking: from improved cook-stoves to domestic and community biogas based systems. In: E. Colombo, S. Bologna, D. Diego Masera (eds) Renewable energy for unleashing sustainable development, Springer. ISBN 978-3-319-00283-5
17. OECD/IEA (2010) Energy and development methodology
18. MacCarty N, Still D, Ogle D (2015) Fuel use and emissions performance of fifty cooking stoves in the laboratory and related benchmarks of performance. Energy Sustain Dev 14:161–171
19. Climatescope (2015) The clean energy country competitiveness index. The Multilateral Investment Fund, UK Department of International Development, Power Africa and Bloomberg New Energy Finance
20. Jones D, Ryan CM, Fisher J (2016) Charcoal as a diversification strategy: The flexible role of charcoal production in the livelihoods of smallholders in central Mozambique. Energy Sustain Dev 32:14–21
21. Biolite http://www.biolitestove.com/. Accessed 27 Aug 2015
22. US EPA (2015) http://www.epa.gov/burnwise/pdfs/dirtysecretbrochure2011.pdf. Accessed 30 Nov 2015
23. Masera O, Edwards R, Armendariz C, Berrueta V, Johnson M, Bracho LR, Riojas-Rodriguez H, Smith KR (2007) Impact of Patsari improved cookstoves on indoor air quality in Michoacan, Mexico. Energy Sustain Dev 11:45–56
24. Hiemstra-van der Horst G, Hovorka AJ (2008) Reassessing the "energy ladder": household energy use in Maun, Botswana. Energy Policy 36:3333–3344
25. Guerrero-Lemus R, Martínez Duart JM.(2012) Renewable energy and CO_2: cost analysis, environmental impacts and technology trends. Lecture notes in energy, vol 3. Springer, London. ISBN 978-1-4471-4385-7
26. Global Alliance for Clean Cookstoves (2014) Cookstoves—The Issues. http://carbonfinanceforcookstoves.org/about-cookstoves/introduction/. Accessed 1 September 2015
27. Allesina G, Pedrazzi S, Sgarbi F, Pompeo E, Roberti C, Cristiano V, Tartarini P (2015) Approaching sustainable development through energy management, the case of Fongo Tongo, Cameroon. Int J Energy Environ Eng 6:121–127
28. NEEDS (2008) New Energy Externalities Developments for Sustainability
29. Bird N et al (2011) Using a life cycle assessment approach to estimate the net greenhouse gas emissions of bioenergy. IEA Bioenergy: ExCo 2011:03
30. République Gabonaise. Contribution prévue déterminée au niveau national—Conférence des Parties 21–31 Mars 2015
31. OECD/IEA (2010) Energy poverty. How to make modern energy access universal. Special early excerpt of the World Energy Outlook for the UN General Assembly on the Millennium Development Goals
32. Hiemstra-van der Horst G, Hovorka AJ (2009) Fuelwood: The "other" renewable energy source for Africa? Biomass Bioenergy 33:1605–1616
33. Moya R, Tenorio C (2013) Fuelwood characteristics and its relation with extractives and chemical properties of ten fast-growth species in Costa Rica. Biomass Bioenergy 56:14–21
34. Global Alliance for Clean Cookstoves (2014) Clean Cookstoves and fuels: a catalog of carbon offset projects and advisory service providers—2015 (2nd edition) http://cleancookstoves.org/binary-data/RESOURCE/file/000/000/381-1.pdf. Accessed 30 Nov 2015

Chapter 6
Photovoltaics

Ricardo Guerrero-Lemus and Les E. Shephard

Abstract Solar photovoltaic (PV) production in Africa and Latin America combined comprise less than 3% of the total global photovoltaic energy production in spite of the solar resources located across these regions. However, increased solar PV modularity, reduced costs of both large utility-scale and smaller residential-scale systems, limited operation and maintenance costs, and conversion efficiencies independent of power capacities is expected to contribute to future broader scale PV adoption in both Africa and Latin America. Reported prices of PV systems vary widely and depend on many factors including system size, location, customer type, connection to an electricity grid and technical specifications. In this chapter it is shown that while PV capacity in Africa and Latin America is far below many other global regions, there are some countries experiencing rapid growth, and that other nations are investing in solar supply chain capabilities. The low cost of the simplest solar-powered lamps produce savings on kerosene that makes them a smart investment. Also, solar-powered pumps are cost-effective, reliable, and can save many hours of labour daily in rural off-grid areas, where water hauling is traditionally done by hand by women and children. These pumps are durable and immune to fuel shortages needed for conventional pumping systems, and they may cost less than diesel-powered generators in many African and Latin American areas. In conclusion, PV is growing rapdily in many African and Latin American countries because of its modularity and drop in costs, offering a hopeful future not only to many countries that need affordable electricity to grow, but also for small rural communities that would like to be connected to the global community.

Keywords Energy · Africa · Latin America · Electricity · Photovoltaics · Inverters · Solar cells · Energy efficiency · Silicon · PV modules · Solar irradiance · Semiconductors · Electric grid · Batteries · Payback

R. Guerrero-Lemus and L.E. Shephard, *Low-Carbon Energy in Africa and Latin America*, Lecture Notes in Energy 38,
DOI 10.1007/978-3-319-52311-8_6

6.1 Introduction

Solar energy can be used to produce electricity by converting solar radiation directly into an electric current through the so-called solar cells or photovoltaic cells. The modularity of this technology, showing almost the same conversion efficiency independently of the power capacity, and also the gap narrowing in terms of cost between large and small PV plants, makes them very suitable for grid-connected and also off-grid rural areas in Africa and Latin America (Fig. 6.1).

Moreover, there are no moving components in the photovoltaic systems, so the operation and maintenance costs are low. Globally, the latest data available (2015) estimates a 1.2% PV energy share of the world electricity production [1]. If we consider the PV energy generation deconvoluted in regional values, non negligible values are obtained in Latin America from 1991 and in Africa from 2000, but PV generation in Africa has surpassed Latin America from 2010 to present (Fig. 6.2). Thus, in 2015 is estimated 2.1 TWh PV generation in Africa, 2.5 TWh PV in Latin America and 178.3 TWh worldwide.

We can expect the current trend of Africa surpassing Latin America in terms of PV electricity generated as the global horizontal radiation in the former is substantially larger than in the later (Fig. 6.3). Also the solar irradiation in Africa is more homogeneously distributed between countries than in Latin America.

6.2 Technology State of the Art

92% of PV technologies (2014) are based in silicon wafers and 8% on thin films [4], and this dominance is expected to continue in the medium term. A silicon solar cell consists in a semiconductor pn junction which is illuminated by light from the Sun. Figure 6.4a shows an scheme of a solar cell in which the emitter (n type in this case), directly faces the Sun. The emitter has to be quite thin, since the incident light has to reach the interface between the p-type and the n-type material, and it is known that silicon is a strong optical absorber. On the top of the emitter an antireflection coating has to be placed to minimize losses of light by direct reflection on silicon.

As it is well known from solid state electronics, an internal electric field arises at the area surrounding the pn interface or spatial charge region. Therefore, if the incident photons from sunlight, with energies higher than the semiconductor bandgap, reach the pn interface they will create electron-hole pairs (see Fig. 6.4a). The electrons and holes thus created will move in opposite directions by the action of the field, resulting in a current which will be the sum of the currents due to both types of carriers (recall that the holes are charged positively).

From the above concepts, it is convenient to emphasize that a solar cell under illumination behaves like an electric generator, or fem, and therefore, contrary to the case of an ohmic resistance, the output current and voltage should have different

Fig. 6.1 Shop selling PV modules in Tanzania. *Source* Simone Tuchella and Engineers without borders

signs. The solid curved line in Fig. 6.4b represents the I–V characteristic of a typical solar cell. If the cell is in open circuit (I = 0), the corresponding voltage is maximum and it is called the open circuit voltage V_{oc}. On the contrary, if the terminals of the cell are shorted (V = 0), then the current is called the short circuit current I_{sc}, and its value is maximum.

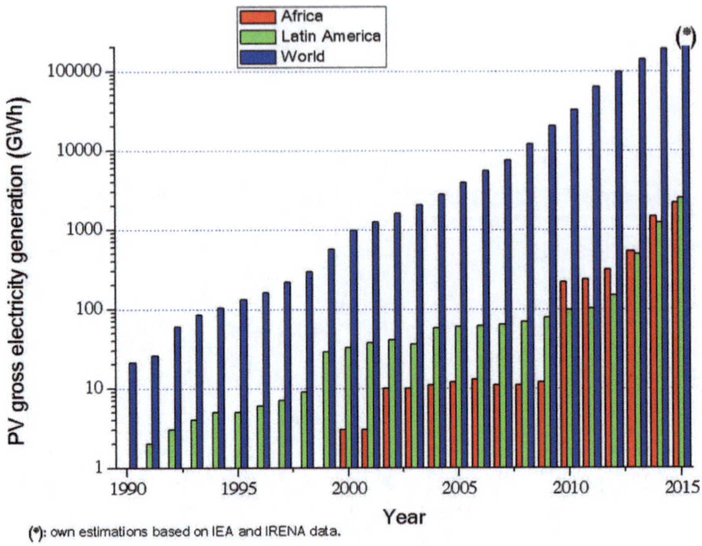

Fig. 6.2 Evolution of the annual production of electricity from PV systems in the period 1990–2015 [2, 3]

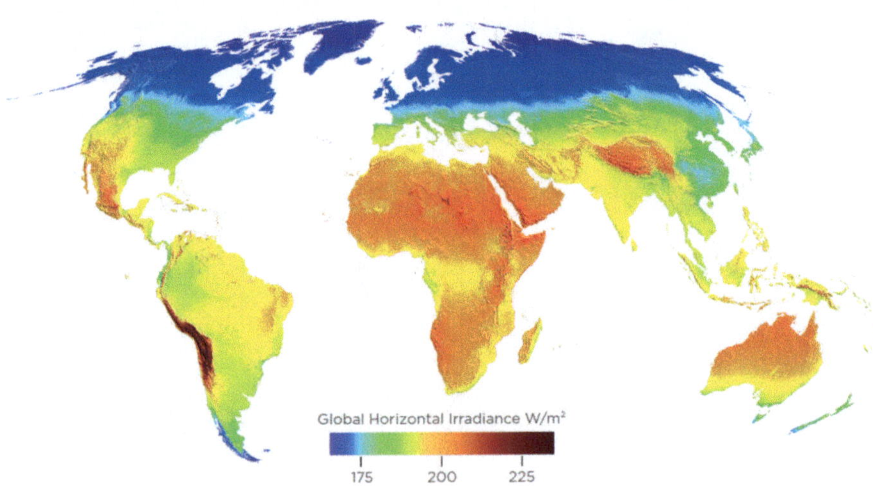

Fig. 6.3 Global horizontal irradiance. *Source* Vaisala Inc. 2015 ©

Evidently, it is usually desired that the output power is maximum (P_{mp}), as it is the case in power generation. Therefore, the product $I \cdot V$ should be maximum, which in terms of the diagram of Fig. 6.4b means that the operating point Q should be chosen so that the area of the rectangle subtended by Q should be maximum, i.e., $P_{mp} = I_{mp} \cdot V_{mp}$. The proper location of the operating point is obtained by choosing

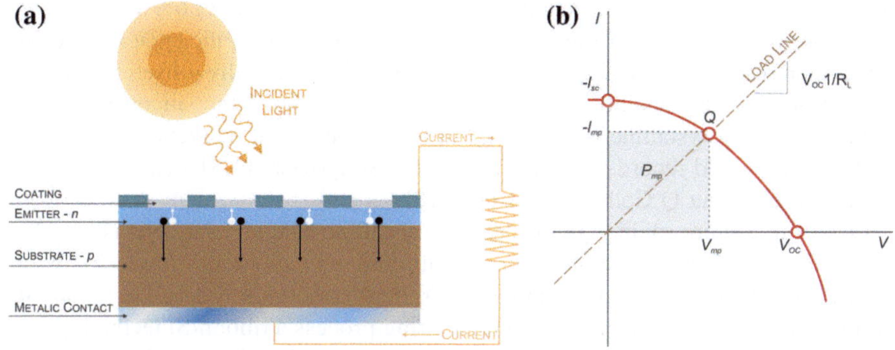

Fig. 6.4 **a** Scheme of a silicon solar cell; and **b** I–V characteristic of a typical solar cell

the load external resistance R in such a way that the load straight line intersects the solar cell I–V characteristic at the point Q at which $P = P_{mp}$.

Each silicon-based solar cell has a current density (~30 mA/cm^2) and voltage (~0.5 V) output very low, so it must be connected in series and in parallel to produce the necessary electrical power for the different applications. According to the Shockley-Queisser (S-Q) detailed-balance model, the limiting photovoltaic energy conversion efficiency for a single-junction solar cell is 33.7%, for an optimum semiconductor band gap of 1.34 eV. Modules are assembled joining several cells in series and in parallel. To produce these modules the solar cells are encapsulated under stringent conditions to have a durability of 20–25 years or more under atmospheric conditions.

On the other hand, there are three different generations of photovoltaic cells at different stages of technological development: (i) first generation, it is a mature technology based on crystalline silicon, (ii) second generation, it can be considered a quasi-mature technology and it is based on semiconductor thin films, and (iii) third generation, based on many physical phenomena (mainly quantum effects), usually sun-radiated by concentrator systems (making it necessary to use mechanical structures to track the sun path), currently under a intense research and development activity.

6.2.1 *Wafer-Based Solar Technology*

They are based on crystalline silicon wafers 180–200 μm thick either single crystal (c-Si: 25% of the market in 2015) or multicrystalline (mc-Si: 68% of the market in 2015) [5] with large grain sizes (about 1 cm side). Material usage for silicon cells has been reduced significantly during the last 10 years from around 16 g/W_p to less than 6 g/W_p due to increased efficiencies and thinner wafers [5]. The pn (or pin) junctions are formed on the wafer and covered by antireflection coatings and electric contacts. These cells dominate the PV market and the maximum efficiency of these cells converting sun radiation to electricity reached a 25.6% world record, very close to its theoretical limit. The mc-Si solar cells have efficiencies somewhat lower than for c-Si, reaching a 20.8% world record, usually offset by cheaper production costs but mainly depending on the evolution of the silicon monocrystalline and multicrystalline price per kg in the international markets. For large PV modules, the world records are 24.1% for c-Si (Sunpower, 11,310 cm^2) and 19.5% for mc-Si (Hanwha Q Cells, 14,661 cm^2) [6].

One of the future savings in the cost of crystalline silicon solar cells is based on the thinning of silicon substrates, but also introducing new processing technologies (e.g., PERT, HIT and bifacial technologies). Furthermore, in order to eliminate the cutting and polishing steps in the manufacturing process a ribbon-Si technology has been successfully developed. The most advanced ribbon-Si technique is called "Edge-defined film-fed growth" where molten silicon is solidified through a carbon die forming a tape ("ribbon") and then introduced in a manufacturing process to produce solar cells.

6.2.2 Thin Film Solar Technology

This technology reached a 8% of the market in 2015 [5], and it implies substantial savings of semiconductor materials, as the pn junction some micrometers thick is deposited on a substrate of under a superstrate mainly based on glass, stainless steel or polymeric material (rigid or flexible). Also thin transparent conducting oxides (e.g., ITO or ZnO) are used as top electric contacts. Although their efficiency is lower compared to Si-based solar cells, as discussed below, and their production costs are lower too, the market share is lower because of the need of a larger surface to obtain the same power, and also durability concerns. There are different thin film solar cells:

(i) Amorphous silicon solar cells (a-Si:H): the semiconductor is deposited after the decomposition of silane gas (SiH_4) in a plasma-enhanced chemical reactor. Commercial a-Si:H modules appeared in the 1980s but their current market penetration is still small because of the low efficiency of solar cells (world record: 13.6%) and the initial decrease of the electrical output when they are exposed to UV radiation (about a 20%). For large a-Si:H modules (14,322 cm^2) the world record efficiency is 12.3% [6].

(ii) CdTe solar cells: these cells are considered the less expensive of the market in terms of cost per kWh and are mainly produced by chemical deposition in an aqueous solution and/or evaporation of CdTe on a substrate. A major drawback of these cells is due to strict environmental controls for toxic elements like cadmium. Shortages in raw materials also limit production of tellurium. A 21.0% maximum efficiency has been achieved for CdTe solar cells and 18.6% for CdTe modules (7,038.8 cm^2) [6].

(iii) CIGS solar cells: the first CIGS cells were produced in the early 1980s by evaporation of the constituent elements Cu(In,Ga)(S,Se). However, there are other deposition techniques being used or tested as chemical bath deposition, sputtering and/or electrodeposition. Although a 21.0% CIGS cells efficiency is the highest reached in thin film technology, the CIGS module (9703 cm^2) record efficiency reaches a 15.7%. Problems arise due to the scarcity of In and the toxicity of Cd and, consequently, Cd-free CIGS modules have been fabricated obtaining good results (17.5% eff.) but in a small area (808 cm^2) [6].

(iv) Dye-sensitized solar cells: these cells are based on a transparent conducting oxide (ITO), a photoelectrode (dye and TiO_2), a redox electrolyte (tri-iodide) and a counter-electrode (Pt). The TiO_2 is prepared on a colloidal solution and, subsequently sintered and submerged in the dye solution and, subsequently pasted to a superstrate with TCO. Durability is a main concern with these devices, mainly due to photo-corrosion. The best efficiency has reached a 11.9% [6].

(v) Perovskite-based solar cells: are usually an extension of the dye-sensitized solar cells where the active layer is methylammonium lead trihalide ($CH_3NH_3PbX_3$, where X is an halogen ion such as I^-, Br^-, Cl^-), with an optical bandgap between 2.3 and 1.6 eV depending on the halide content. Also perovskite active layers have been considered in tandem structures for CIGS and c-Si solar cells. The most important feature is the low cost and the evolution of the efficiency from a mere 3.8% in 2009 to a 19.7% in 2015, making this the fastest-growing solar conversion efficiency to date. However, concerns about the use of lead and the long term stability of this active layer makes it under strong research activity currently.

(vi) CZTSSe: these cells are based on $Cu_2ZnSn(S,Se)_4$ semiconductors, showing similar optoelectronic properties than CIGS and CdTe but also avoiding the use of toxic (Cd) and scarce (In) materials. First produced in the 1980s, they avoid vacuum processes and the main challenge is an adequate crystallization of the semiconductor. Best cell efficiencies are low, reaching a 9.8% [6].

(vii) Organic solar cells: these cells are based on organic molecules that can show n-type and p-type semiconductor properties, and also extremely large absorption coefficients for being built as thin film solar cells. The development of organic solar cells began more intensively in 2001, reaching maximum values of efficiency still low (11.2%) [6], also low (8.7%) for organic PV modules (802 cm^2) [6]. The main attraction of these cells is their low production costs, but raises important durability problems and the need to increase their efficiency to be commercially competitive.

6.2.3 Third Generation Solar Cells

Such cells are based on different and, sometimes, novel optoelectronic concepts that allow or are expected to increase the efficiency of the cells in the future. Notable among these are the multijunction monolithic structures applied in the world's efficiency record solar cells, primarily with two p–n junctions configurations (mid 80s), and recently reaching the current world absolute record of 46.0% (2015) based on GaInP/GaAs, GaInAsP/GaIn junctions configurations, 5.2 mm^2 area and 508 (suns) concentration intensity. The world record efficiency reaches a 38.8% efficiency for a 5-junctions solar cell (1.021 cm^2) measured under standard test conditions (non-concentration).

Other strategies to increase the efficiency of solar cells and considered under the third generation concept are based on (i) intermediate gaps; (ii) photon up-conversion, down-conversion and down-shifting; (iii) silicon nanostructures and other quantum wells; (iv) hot carriers; (v) thermo-photovoltaics and thermo-photonics; (vi) surface plasmons; and (vii) impact ionization.

6.2.4 Efficiencies and Required Areas

The evolution of the solar cells efficiency produced in laboratories from 1976 to the present [7] shows a dramatic rise in efficiency in recent decades, especially in multijunction solar cells, but also it is being observed in organic and perovskite-based cells. In the case of monocrystalline silicon solar cells the efficiency is close to its theoretical limit. The energy yield for most PV technologies increase with solar concentration ratios where areas with high direct normal irradiation exists, but excessive heating can be detrimental to the efficiency and lifetime of the solar cells. Organic and amorphous silicon cells are generally heat-sensitive to be used in concentrators, but conventional monocrystalline silicon cells can operate efficiently at lower concentrations (1–100 suns) without needing active cooling mechanisms. In fact, the Si-based solar cells efficiency record (27.6%) was reached at 92 suns, and for thin films-based solar cells (CIGS) the efficiency record (23.3%) was reached at 15 suns. On the other hand, GaAs and multijuntion solar cells are better used in medium-high concentration systems (100–300 suns and above). Thus, the GaAs-based solar cells efficiency record (29.1%) was reached at 117 suns and the multijunction solar cells efficiency record (46.0%) was reached at 508 suns [6]. Mainstream concentration technologies utilized are parabolic dish collectors and Fresnel lenses, and high concentration systems (300–1000 suns) clearly dominate the market.

However, commercial solar cells and modules usually show quite lower efficiencies compared to world records, and average commercial values (Table 6.1) are more useful for estimations about area per kilowatt-peak (kW_p) of capacity

Table 6.1 Approximate commercial values of efficiencies and area required per kW_p for solar cells and solar modules (based on ISE-FhG data) [5]

	Thin film	mc-Si	Cz-Si	HJT/IBC
Efficiency (%)	13.8–15.7	15.8–16.8	15.8–17.7	18.5–21.3
m^2/kW_p	6.8	6.1	6.0	5.0

HJT Heterojunction technology; *IBC* Interdigitated back contact installed. In the last 10 years, the efficiency of average commercial wafer-based silicon modules has risen from about 12–16% and, at the same time, CdTe module efficiency has risen from 9 to 13% [5]. It can be appreciated the advantage that provides the crystalline silicon technology in terms of area required per kW_p compared with thin film technology. With concentrator technology, module efficiencies of up to 38.9% have been reached in the lab and 27–33% efficiencies are commercially available [5].

Case Example 6.1

Protection of PV systems against theft in South Africa and Zimbabwe

In South Africa, communities susceptible to solar panel theft resort to mounting their panels flat on the ground in front of their houses so they can be looked after during the day and taken indoors at night for safe keeping, or use security lights to illuminate the perimeter of the PV plant to provide security. In Zimbabwe, some households have resorted to shielding their panels with steel bars, which causes shading of the panels [8]. Consequently, the power losses due to the usage pattern of these systems are higher than the values specified for the system. A case study is performed in Thlatlaganya village within Limpopo province in South Africa. Economic analysis indicates that the battery life is also negatively affected. The authors consider that the measures adopted by locals to protect the PV modules are a reflection of lack of user education and the high cost of alarm systems and radio frequency identifiers, and propose an alternative bench-rack mobile system for mounting solar panels. Data on the PV usage pattern was obtained through semi-structured interviews with 88 PV users [9].

On the other hand, PV system performance has strongly improved. Before 2000 the typical performance ratio was about 70%, while in 2015 it is in the range from 80 to 90% [5]. The energy payback time of PV systems is dependent on the geographical location: mc-Si PV systems in Northern Europe need around 2.5 years to balance the input energy, but less than a year in the north of Africa [5]. Also other innovations can reduce the share of BoS costs for PV systems, as click-and-plug panels on residential rooftops, and new robots to keep systems clean and detect any faults [10].

Case Example 6.2

Ghana's PV development:

Until late 2013, when Ghana's first utility-scale PV plant became connected, renewable energy production became solely from distributed PV generation, which currently amounts to about 5 MW across thousands of small installations. This distributed capacity was built through both on- and off-grid development aided by Ghana's Energy Development Access Project (GEDAP) and has contributed to the country's high electrification growth rate, which stood at 72% in 2012, up from just over 60% in 2009. A 10-year fixed tariff for wind, solar, hydro, biomass and biogas projects took effect in 2013. New levies on petroleum products to fund the scheme and other energy sector investments were introduced in 2015. Interest in Ghana's FiT sharpened after the rates were raised in September 2014. Included in the new regulations was a cap put on the size of solar PV projects to maintain grid stability. However, developers have raised concerns over the duration of the power purchase agreement (PPA) as they only guarantee rates for 10 years, and the credit worthiness of the Electricity Company of Ghana, which buys 72% of all power [11].

6.2.5 Inverters

Inverters are DC/AC converters and are required when the PV power is supplied to AC appliances. To effectively perform pulse width modulation (PWM) control for the inverter, insulated gate bipolar transistor (IGBT) and metal oxide semiconductor field effect transistor (MOSFET) are mainly used for switching. The efficiency range for all inverters varies normally from 94.5 to 98.5%. There are different configurations that should be described:

(i) Single stage/central inverter: a single power processing stage take care of all tasks of Maximum Power Point Tracking (MPPT), voltage amplification and grid side current control. The global efficiency of the generating system is reduced due to the fixing of a common operating point for all PV modules.
(ii) Multi-stage inverter: each string of modules is connected to a double- or a single-stage inverter and a DC/DC boost converter is used for the MPPT and the DC/AC inverter controls the grid current.
(iii) Multi-string multi-stage inverters with high frequency transformer: lower power DC/DC converters are connected to individual strings and to the central inverter. Each PV string has its own MPPT.

The installation conditions (indoor or outdoor), ambient temperature, requirements of water, dust-proofing, noise level, protective functions (DC over/underpower, DC over/undervoltage, DC overcurrent, AC over/undervoltage, frequency increase and drop, and detection of DC or AC grounding faults) and electromagnetic compatibility (EMC) should be strongly considered. Also standby power consumption of the inverter is a very important parameter, particularly for off-grid PV applications. The use of micro-inverters to ensure that one faulty panel does not affect the performance and output of the rest, and new inverters to limit the amount of power exported to the grid from PV panels and to smooth power ramps derived from short-term oscillations of the sun irradiation due to cloud shading are being considered.

Also utilities want export-limiting equipment to respond to demand changes on the grid in less than one second. The system could also switch equipment on or off, channel any surplus renewable power into batteries and operate "private lines", which are independent distribution networks which might set up by a group of neighbouring consumers and prosumers to share locally produced solar power. It is estimated that 1 GW_p of DC/DC converters, also called "power optimizers", have been installed in 2014 [5]. New features for grid stabilization and optimization of self-consumption; storage units included in the inverter; and the utilization of innovative semiconductors (SiC or GaN) which allow very high efficiencies are the new trends in inverters technology [5].

6.2.6 Costs

The sharp decline in price that PV experiences from recent years opens important possibilities to develop PV systems in African and Latin American locations with limited or no financial incentives. Also, as electricity from the grid is very expensive compared to PV in many African and Latin American countries, such sharp declines can produce grid defection if batteries are available and cost effective, as it will be explained below.

Reported prices of PV systems vary widely and depend on many factors including system size, location, customer type, connection to an electricity grid and technical specifications. System prices for off-grid applications are significantly higher than for grid-connected applications because of the additional requirement of storage batteries and/or other associated equipment. More expensive grid-connected system prices are usually associated to building integrated PV because of smaller size and more costly BOS. Therefore, system prices below 1 USD/W_p for large-scale PV systems seems to be now common in very competitive tenders [12] and the range of prices tend to converge (Fig. 6.5), with the lowest prices decreasing at a reduced rate while the highest prices are reducing faster. Thus, average price for modules in 2015 ranged from a low USD 0.52/W in India to USD 0.72 in Japan [13].

Therefore, considering highly irradiated areas in Africa and Latin America (2000–2400 kWh/year), current PV systems price range (1.0–2.5 USD/W_p) and using Table 6.2 [14], the price of electricity could potentially range between 0.04 and 0.09 USD/kWh. However, economies of scale, lack of expertise, trading barriers and other considerations could increase the final cost of electricity in these regions.

However, in terms of levelized cost of electricity (LCOE), the International Renewable Energy Agency (IRENA) establishes a range of USD 0.131–0.264/kWh in Africa with an average value of 0.190/kWh, and USD 0.084–0.216/kWh in South America with an average value of USD 0.110/kWh (Fig. 6.6) [2]. Thus, in

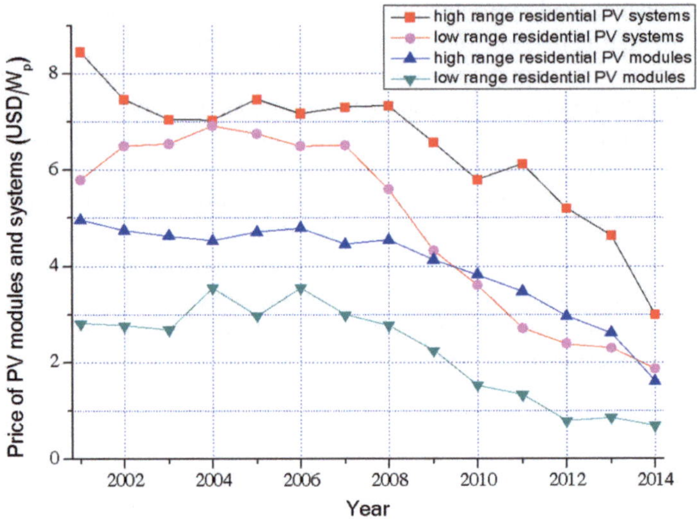

Fig. 6.5 Evolution of PV modules and small-scale system prices in selected reporting countries from 2001 to 2014 [12]

Table 6.2 Relation between PV investment costs (EUR/kWp), solar irradiation (kWh/year) and estimated cost of electricity [14]

\kWh EUR/kWp\	2400	2200	2000	1800	1600	1400	1200	1000	800
5000	0.1788	0.1951	0.2146	0.2384	0.2682	0.3065	0.3576	0.4292	0.5365
4600	0.1645	0.1795	0.1974	0.2194	0.2468	0.2820	0.3290	0.3948	0.4935
4200	0.1502	0.1639	0.1803	0.2003	0.2253	0.2575	0.3004	0.3605	0.4506
3800	0.1359	0.1483	0.1632	0.1812	0.2039	0.2330	0.2718	0.3262	0.4077
3400	0.1216	0.1327	0.1461	0.1621	0.1824	0.2085	0.2432	0.2918	0.3648
3000	0.1073	0.1170	0.1288	0.1431	0.1609	0.1839	0.2146	0.2575	0.3219
2600	0.0930	0.1014	0.1116	0.1240	0.1395	0.1594	0.1860	0.2232	0.2790
2200	0.0787	0.0858	0.0944	0.1049	0.1180	0.1349	0.1574	0.1889	0.2360
1800	0.0644	0.0702	0.0773	0.0858	0.0966	0.1104	0.1288	0.1546	0.1931
1400	0.0501	0.0546	0.0602	0.0668	0.0751	0.0858	0.1001	0.1203	0.1502
1000	0.0358	0.0390	0.0431	0.0477	0.0536	0.0613	0.0715	0.0860	0.1073
600	0.0215	0.0234	0.0260	0.0286	0.0322	0.0368	0.0429	0.0515	0.0644
400	0.0143	0.1560	0.0172	0.0191	0.0215	0.0245	0.0286	0.0343	0.0429

average and compared to other regions, production of electricity from PV is favourable in South America but not yet in Africa. Thus, in 2015 record low PPA prices were set in Peru (USD 0.048/kWh) and Mexico (USD 0.040/kWh) [15].

Case Study 6.3

The lowest bid for PV worldwide in 2016: Chile

The solar PV cost reduction spiral reached its lowest point before sending this book to press at an auction in Chile in August 19, 2016. A Spanish developer Solarpack Tecnologica won contracts to sell power from a solar plant high in

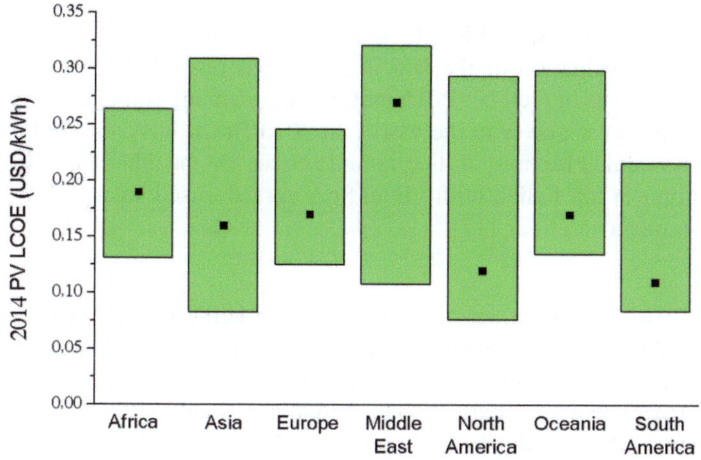

Fig. 6.6 Levelized cost of electricity from PV in the different global regions (2014) [2]

the Andes for USD 29.10/MWh. The 120 MW plant to be built in Chile's sunny Atacama desert surpassed the previous record-low for solar PV brokered in Dubai in May, and was priced at almost half the cost of coal power in the same event. Overall, Chile's auction contracted 12.4 TWh of new generation with 52% awarded to renewables at an average price of USD 45.1/MWh, according to BNEF estimates. The 20-year contracts paid in Chilean pesos were awarded to 11 companies, including Mainstream Renewable Power and Endesa Chile—a subsidiary of Enel. Yet some power-market players doubt the projects can deliver, given the auction's record-low power purchase agreements, because project owners in Chile are exposed "substantially to fluctuating power prices", according to BNEF. Low power demand in remote regions and discrepancies in selling power at different points in the distribution network can mean that even projects with contracts secured in the regulated market are exposed to spot price fluctuations [16].

Case Study 6.4

Brazil auctions and currency depreciation, and Mexico:

Given the drought, and the necessity to add new capacity to the country's mix, in 2014 the government contracted a total of 4 GW of biomass, solar, small hydro and wind projects through auctions. This was particularly important for the photovoltaic sector, which was included for the first time in the country's federal-level tenders, taking 1 GW of contracts in 2014, at average prices of USD 87/MWh, at the time of the auction. However, the subsequent devaluation of the Brazilian real has pushed down this price to USD 56/MWh as of October 2015, putting these among some of the cheapest contracts for solar power in the world. In 2016 Brazil has cancelled the only auction scheduled because of its economic crisis. On the other hand, In March 2016, a unit of top Italian utility Enel SpA agreed to sell solar in Mexico for $35.50 a megawatt-hour [17].

In June 2016 in Zambia, an African record low USD 0.0602/kWh was reached, and in May 2016, an auction of 800 MW of solar PV in Dubai attracted a bid of USD 0.0299/kWh. Other prices reached for utility-scale solar PV in other African and Latin American markets are USD 0.075/kWh in South Africa, USD 0.065/kWh in Chile and USD 0.080/kWh in Brazil [13]. Notwithstanding, this evolution in prices decline can be slowed in these and other countries because of high financing costs and delays in expanding high-voltage transmission lines to connect the solar plants to power demanding areas.

Case Example 6.5

The lowest bid for PV in Africa:

In June 2016, Zambia managed to bag the lowest bid for solar in Africa, at USD 0.0602 per kWh, through a World Bank organised auction programme. A 45 MW plant will be built by First Solar, the biggest US panel producer, and France's Neoen. The auction in Zambia was conducted through the "Scaling Solar" programme, developed to help introducing solar in a country using competitive auctions and the endorsement of the World Bank. That support helps overcome international banks' concerns about political risk and makes these emerging markets more appealing to developers. It also includes standardised contracts, eliminating the often lengthy process of negotiating power-purchase deals one at a time. Senegal and Madagascar are also participating in the programme, which may eventually spur development of 850 MW of solar capacity in the three power-deprived countries. That would be roughly USD 1 billion in investments.

Finally, in the past there has been some controversies about the energy payback time of PV cells and systems. As the efficiencies of the systems involved in manufacturing PV have substantially increased and also the efficiencies of the PV systems, this issue is not a concern currently. Thus, recent studies conclude that the energy payback time of assembled modules shipped to the customer are 1.09 and 0.93 years for monocrystalline and multicrystalline silicon, respectively, both under southern European conditions. CO_2 emission rates strongly depend on the scenario considered, ranging from 18 g/kWh in southern Europe to 60 g/kWh in northern

Germany under worst-case conditions for useful lifetime and degradation [18]. The energy payback time for a large PV system located in a highly irradiated area in Africa (Ouarzazate, Morocco) can be below 1 year. Also the CO_2 emissions associated to the PV activity (from construction to disposal) are under 30 g-CO_2-eq/kWh [19].

6.2.7 *Pico-Solar Products*

The 1.2 billion people living without access to the power grid spend about USD 27 billion annually on lighting and mobile-phone charging with kerosene, candles, battery torches or other fossil fuels [20]. Solar-powered portable lights and home kits can offer an improved service at a lower cost, and many companies are now actively focusing in this area. Pico-solar products can be defined as those that demand less than 10 W power, and more than 44 million pico-solar products had been sold worldwide as of the second half of 2015 [20].

Case Example 6.6

An off-grid power kiosk in rural Zambia:

The off-grid kiosk is expected to be placed in Filibaba, Zambia in September 2015 for recharging mobile phones (35/day) and other electronic devices for a fee. Also customers can rent portable battery kits (7–17 Ah, 12 V_{DC}) with external LED bulbs to provide light to their homes. Electricity will be available to the school/community center for 6 h in the evening after the energy kiosk closes. The kiosk is powered by a 1.8 kW PV system and four 12 V, 220 Ah lead-acid batteries (10.56 kWh total energy storage capability). Filibaba is a farming settlement of over 2000 inhabitants without access to the power grid and located about 23 km of the Chigola main town center. A face-to-face survey showed an average family of seven members living in a three bedroom home and spending USD 4.65/month to light their homes via candles, kerosene lanterns, or other non-electric means. In addition, they paid USD 2.66/month to charge mobile phones, traveling an average distance of 9.9 km. 79% of the population was willing to pay more than USD 8/month for access to electricity (over twice their current expenditures). As the kiosk is not able to provide services to all the Filibaba's population, authors consider the best strategy a transparent and equitable system (e.g., lottery) for determining who is eligible. The load profile was based on data from another energy kiosk in Zambia (5.9 kWh/day). The project has been motivated by a previous work in the village of Filibaba by one of the authors devoted to study the potential use of biogas for cooking, and represents a collaboration between NGOs from USA, Zambia and IEEE Smart Village [21].

Running a simple wick kerosene lantern for a year can cost up to USD 54 in Kenya, compared to a simple portable solar light conservatively priced at USD 13 per unit, offering a payback period of about 3 months [20]. Solar lanterns account for 97% of reported sales. Modern LEDs have already replaced incandescent light bulbs in all new off-grid solar products, saving on the battery size and allowing manufacturers to scale the PV cell down from 2.5 W to less than 1 W for the same performance. 98% of the batteries used are lithium based, and only a 2% are run on either lead-acid or nickel-metal hydride. The functionality of the solar lanterns has increased to include phone charging as a standard feature. Low-cost generics comprise at least half the wider pico-solar market, showing that mass manufacturers are entering what was once a niche market.

Case Example 6.7

Pay-as-you-go approaches for solar home systems in Kenya:

Pay-as-you-go (PAYG) finance approaches are emerging in areas where finance for sustainable energy technologies like solar home systems (SHSs) previously consisted mostly of hire-purchase and microfinance models. Reports in 2015 suggest that PAYG solar companies are now active in at least 20 Sub Saharan countries. In Kenya these finance approaches have seen the most significant emergence largely due to the success of mobile payments and the sharp decline in PV costs. In Kenya there are around six companies offering payment plans for SHSs based on real-time monitoring via machine-to-machine technologies and PAYG payment plans. The analysis is focused on the largest three providers at the time of the study (M-KOPA, Mobisol and Azuri), accounting for about 250,000 customers in the region in mid 2015. All companies offer a one-stop-shop model where they integrate provision of the SHS technology (usually 8–10 W PV systems), finance and after-sales service. Moreover, the three companies developed their business models based on close attention to people's existing payments for energy, particularly buying kerosene and paying for mobile charging. All customers interviewees emphasised provision of energy services to the rural poor (e.g., light and mobile charging) as their business core motivation. Some of these companies experimented with different deposit sizes to secure a sense of ownership, and offer opportunities to upgrade their systems once they pay off their initial system. For after-sales services, some companies rely on locally available technicians or remote assessment before sending technicians to villages. Financing costs is a critical concern, and also political uncertainty (e.g., reintroduction of import duty on solar products) [22].

Solar home systems are a little more complex than a portable lantern, as most products on the market offer at least some degree of customisation and a variety of appliances (e.g., 4 h of light, 4 h of TV, 6 h of radio and 1 mobile-phone charging per day). Although batteries and PV panels costs are expected to decrease, the major reduction is expected to come from the switch to DC appliances, enabling the use of smaller PV panels and batteries. In 2014, standard appliances required 121 W of PV and 125 Ah of battery capacity, compared to just 27 W of PV and 28 Ah of battery capacity to get the same level of service from super-efficient appliances [23]. Also the DC appliances are expected to become cheaper as the production volume increases [24]. Indeed, DC radios are popular and widely available, generally priced around USD 5–10, and consuming around 1 W of power. 19″ DC TV offers the best trade-off between price, power consumption (10 W), shareability and a retail price of about USD 104 in 2015 [24]. Fans are the second-highest priority in appliances after TVs, and 12 W 12″ rotor DC fans can be obtained in the market for about USD 18. Prices for a 50 litre DC compressor fridge using 45 W in 2015 are around USD 300 each. DC refrigerators and fans are about 50% more expensive than AC ones, but the former are more efficient. Other appliances such as irons, water heaters or washing machines for laundry are unlike to become commercially relevant in the short term [24].

Case Example 6.8

Appliances adapted for African rural areas:

There are some remarkable examples of real pico-solar products currently offered in the African market. An Ugandan company called SolarNow offers a USD 200 low voltage television set that runs on the direct current (DC) used by solar systems. A British designed fridge called Sure Chill needs only a few hours of power a day to maintain a constant 4 °C. Finally, a company in South Africa has just launched solar powered ATMs for rural areas with intermittent mains power [25].

6.3 Regional and National Perspectives on Technology

6.3.1 Evolution on Electricity Produced from PV

PV capacity in Africa and Latin America is far below the values shown in other regions but there are some countries there rapid growth is currently detected. In Africa the leading country is South Africa but only from 2014 as Egypt is almost stagnant from 2010 (Fig. 6.7). This is attributed to the recent blackouts and lack of supply in South Africa and instabilities in Egypt. Many other countries in Africa are growing in PV capacity because of the decrease in costs and the huge solar resources in this region, but the role of PV is yet very small. Then, South Africa became the first African PV producer in 2015 with 1122 MW installed, mostly ground mounted. Egypt is in second position, announcing a FiT program targeting

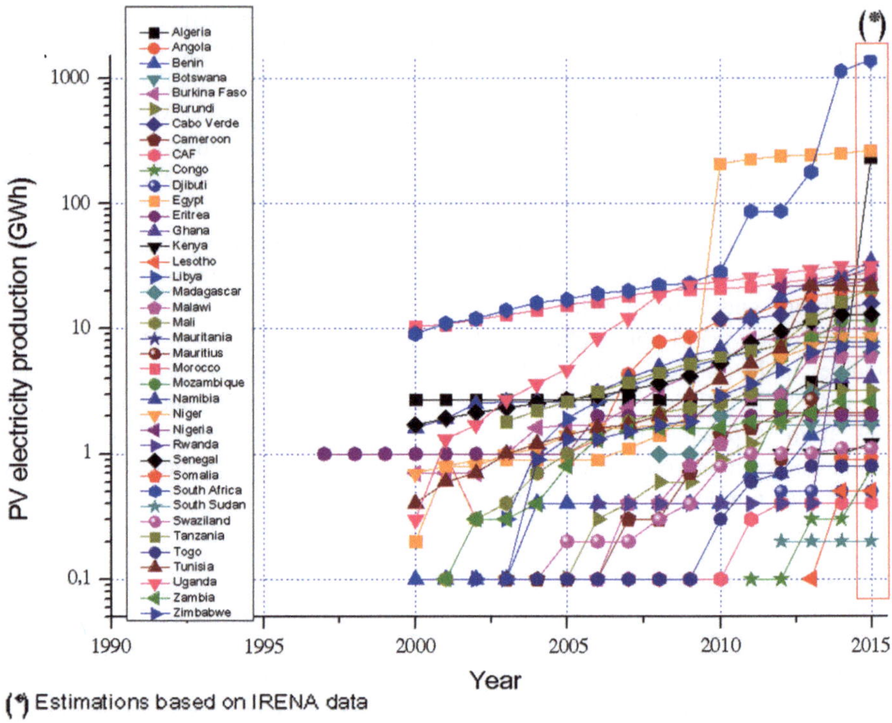

Fig. 6.7 PV gross electricity production evolution in the African countries. *Source* IEA and IRENA

2.3 GW of installations between 50 kW and 50 MW, and 300 MW for installations below 50 kW. In Rwanda, a 8.5 MW plant has been inaugurated at the beginning of 2015. In Algeria, a new FiT scheme has been set up in 2014 for ground-mounted systems above 1 and 400 MW have been planned. Also, in 2014 Nigeria signed a memorandum to install at least 1.2 GW of PV in the coming years. In other several African countries (Ghana, Mali, Ivory Coast, Burkina Faso, Cameroon, Gambia, Mauritania, Benin, Sierra Leone, etc.) plants from 11.5 to 155 MW have been announced to replace or complement existing diesel generation [4]. Moreover, Kenya shows expertise in project development, engineering, O&M and inverters [11]. Mozambique is one of the few countries in Sub-Saharan Africa to have a solar supply-chain facility (the module assembly plant near Maputo aims to produce components for the national market [11]). South Africa, Brazil and Mexico offer expertise and manufacturing capacity in project development, engineering, O&M, modules, inverters and balance of plant [11]. In 2014, Rwanda commissioned the largest PV project in sub-Saharan Africa outside South Africa (USD 23.7 million 8.5 MW PV plant developed by GigaWatt Global, constructed by Scatec Solar, which is also an equity investor joining Norfund, FMO, OPIC and the Emerging

Africa Infrastructure Fund [11]), and Tanzania should soon have its first utility-scale solar plant, a 5 MW project by developer NextGen [11].

Case Example 6.9

Medium size PV-diesel-battery system for an isolated power system in Namibia

The settlement of Tsumkwe is situated in the north of Namibia and has about 250 permanent households. It is not connected to the national grid, and it is not expected to be connected in the next 20 years. This settlement was originally a South African military post with a micro-grid (11 kV) with two diesel generators provided electricity. At Namibia's independence in 1990, the government established a school, a clinic and a police station in the village. The electricity system has been poorly maintained and severe pollution of diesel has contaminated the soil around the generators and the fuel tank. Originally electricity was provided 24 h per day, however, increasing cost of diesel limited the production to 10–14 h per day (5 am to 1 pm and 5–10 pm), producing a negative impact in the community. The cost of electricity reached USD 0.85/kWh in 2010, and the regional government therefore subsidized the tariff, charging USD 0.14/kWh to households and USD 0.27/kWh to institutions and businesses, resulting in an annual deficit of USD 170,000. The solar-diesel hybrid system consisted in a 202 kWp PV array (50% of daily electricity demand estimated), 766 kWh battery array, 3 diesel generators (2 × 150 kWA and 1 × 350 kVA) showing a peak output of 150 kW, and the microgrid was renovated. An energy efficiency campaign was launched replacing electric stoves and hot plates with LPG, introducing CFL light bulbs and solar water systems. The total cost of the project was USD 1.4 million. All domestic and business customers connected to the hybrid system are equipped with prepayment methods. Community members were informed about how to efficiently use the electrical appliances. However, at the end of the project: (i) the integration of the PV and diesel was not working, which resulted in manual operation; (ii) expertise for troubleshooting and repairing the system was not sufficient; (iii) there is not long term strategy for increased demand due to increased population; (iv) the cost efficient tariff is USD 0.60/kWh; and (v) no revenues are devoted to operation and maintenance. Successful implementation of the system needs to solve these problems for a long-term supply of electricity in the settlement [26].

In Latin America a similar trend is based on the huge increase in PV capacity in Chile acting as the main player from 2014 (Fig. 6.8) and almost completely driven by utility-scale projects developed. In contrast to Africa, in Latin America the main PV player has always been Mexico, but internal instabilities make investment a risky activity in many areas compared to the safety rules of Chile. Also it is

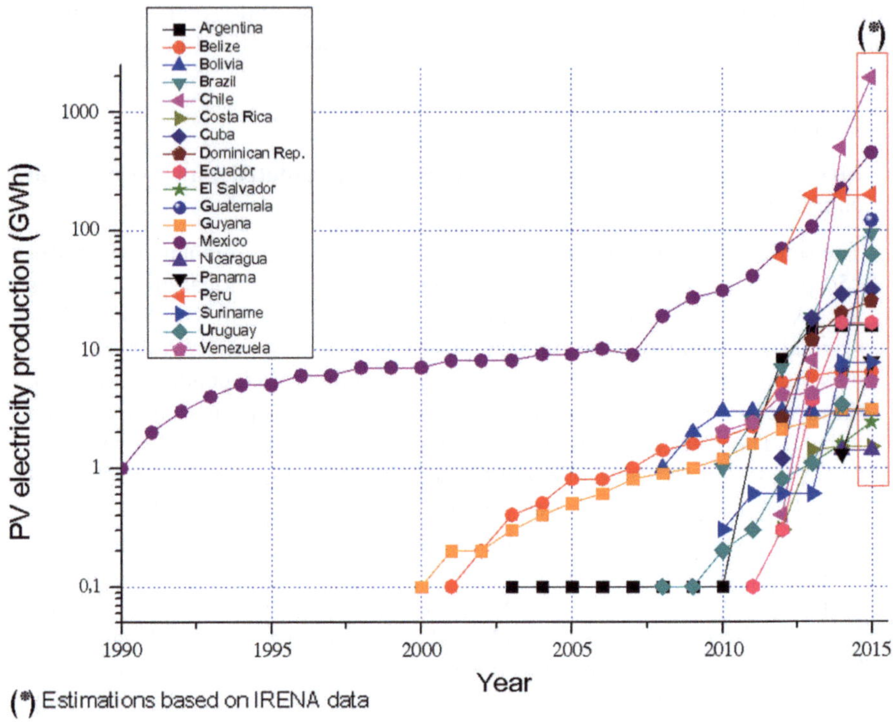

Fig. 6.8 PV gross electricity production evolution in the Latin American countries. *Source* IEA and IRENA

important to mention the case of Peru, where a recent surge in PV capacity makes the trend remarkable. Many other Latin American countries shows some PV capacities but very small. Then, Chile is leading the PV market in Latin America, reaching 848 MW in 2015 because of high electricity prices and high solar irradiation in some areas. 103 MW of PV systems were installed in Mexico in 2015, increasing the total capacity in the country to 203 MW and, from the beginning of 2015, more than 200 projects have been approved with a total capacity of 5 GW, as the Mexican government has announced a target of 600 MW of PV systems in 2018 and 6 GW of self-consumption by 2024 [22]. Mexico has accelerated depreciation policies for PV systems, net metering below 500 kW also for condominium and a Solar Villages programme for rural electrification [4]. Brazil has started to include PV in auctions for new power plants and more of 1048 MW were granted in 2014 (~98 USD/MWh), has now in place a net metering system but with limited results, and the government has set up a 3.5 GW target for PV in 2013. Other Latin American countries (Argentina, Perú, Honduras, Ecuador and Uruguay) are significantly promoting PV electricity generation but with more limited results [4]. Indeed, Honduras has launched a feed-in tariff scheme offering contracts of USD 180/MWh for the first 300 MW of PV commissioned before July 2015 and USD 150/MWh after. Then, on the second half of 2015, a total of 387 MW of PV capacity had been commissioned in Honduras, making it the second largest solar market in Latin America.

Case Example 6.10

A small PV powered reverse osmosis system for water purification in a remote Mexican community

A small photovoltaic reverse osmosis system (PVRO) for cleaning water (1 m^3/day) in a remote village in the Yucatan Peninsula of Mexico (village of La Mancalona, about 450 people) was designed, fabricated and deployed. The community residents were indigenous subsistence farmers and bee-keepers. Many times they were forced to drink brackish groundwater, rain-water from cisterns, or water found in open ponds, streams or rivers, as they cannot afford commercially purified water. All usual sources of water can contain unsafe amounts of contaminants. The purified water from the closest town costs about USD 1.50 for a 20-liter bottle. The MIT Field and Space Laboratory developed the technology. The system is based on a PV system (400 W nominal power), batteries, UV pretreatment, pre-filter, high pressure pump and reverse osmosis. A plan was developed in cooperation with the community aimed at making the system self-sustaining in the long-term. Methods and materials were developed to permit the community members to operate and maintain the system themselves (classroom and hand-on training in Spanish and using Spanish manuals). The economic results show that the capital costs (components, shipping and installation) are about USD 10,000, but commercial prices would be substantially lower. The production cost of water was USD 0.17 per 20-liter bottle with a production of 1000 litre/day. Some challenges were detected during the study about the assembly of the system without power for tools, and the complete understanding of the water sources [27]. A similar case study has been detected in a rural community in northern Tanzania, but devoted to avoid high fluoride concentrations in drinking water, which produce severe dental and skeletal fluorosis [28].

6.3.2 *Electricity Share from PV*

In terms of share of electricity produced by PV, the last available data are from 2014, and show values very small compared to the country (Germany) where the share of PV is the highest worldwide (Fig. 6.9). These results also reflect the

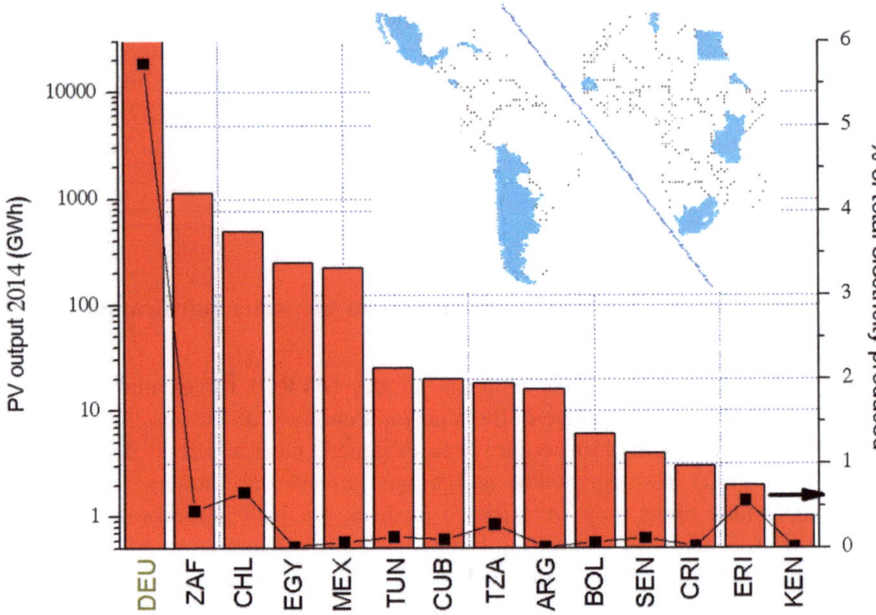

Fig. 6.9 Electricity output from PV and share of total electricity produced in different African and Latin American countries in 2014 [1]

current small role of PV in the African and Latin American countries. South Africa is leading the PV output in 2014, but it is expected that Chile will lead it in 2015, also in terms of share of total electricity produced. However, Chile is a very complex country as it has 4 power grids and some of the PV systems are directly supplying electricity to mines and related activities in the north of the country, isolated from the main power grids.

Case Example 6.11

Chile's PV plants selling electricity in the spot market:

Chile is currently is the leader between African and Latin American countries in solar installations. But more important is that most of this development has been propelled by merchant projects and the PV plants have been developed to sell electricity in the spot market. Also, it is interesting to mention that in order to improve the competitiveness of renewable projects in the auctions, generators bid to supply electricity during a certain time blocks (11 pm to 8 am; 8 am to 6 pm; 6–11 pm), a characteristic very favourable to solar energy systems. However, a recent fall of prices in the wholesale market has made developers shift to new alternatives. But for households and small businesses, in 2014, a net metering regulation was published. Under the regulation, retail electricity consumers who have renewable energy or cogeneration installations smaller than 100 kW will be able to connect to the national grid, deliver surplus generation and obtain credit for electricity delivered [11].

6.4 Conclusions and Future Perspectives

Solar photovoltaic production in Africa and Latin America combined comprise less than 3% of the total global photovoltaic energy production in spite of the homogenous distribution of solar resources across the regions. However, increased solar PV modularity, reduced costs of both large utility-scale and smaller residential-scale systems, limited operation and maintenance costs, and conversion efficiencies independent of power capacities will likely contribute to future broader scale PV adoption in both Africa and Latin America. Reported prices of PV systems vary widely and depend on many factors including system size, location, customer type, connection to an electricity grid and technical specifications. System prices for off-grid applications are significantly higher than for grid-connected applications because of the additional requirement of storage batteries and/or other associated equipment. In highly irradiated areas in Africa and Latin America (2000–2400 kWh/year), current PV system prices range from 1.0 to 2.5 USD/W_p with a commensurate cost of electricity ranging between 0.04 and 0.09 USD/kWh. However, economies of scale, lack of expertise, trading barriers and other considerations increase the final cost of electricity in these regions.

While PV capacity in Africa and Latin America is far below many other global regions, there are some countries experiencing rapid growth. In Africa, South Africa has invested significantly in solar PV because of recent blackouts and lack of supply. Other nations are investing in solar supply chain capabilities. In Latin America new solar PV capacity is being driven largely by utility-scale solar deployments, and Chile is now considered a leader among all Latin American nations.

The simplest solar-powered lamps cost around USD 8 [25]. That can be a high investment for poor people, but the saving on kerosene makes it a good investment. In 2013 Africa had around eight million solar systems installed to power public lighting [30]. These solar-powered systems can also be used to charge a mobile phone (charging in kiosks in Papua Guinea can reach USD 200/kWh).

Solar-powered pumps can save many hours of labour daily in rural off-grid areas, where water hauling is traditionally done by hand by women and children. In Egypt, 9% of PV applications are for water pumping [29]. These pumps are durable and immune to fuel shortages needed for conventional pumping systems.

Two phases of the Regional Solar Programme (RSP), implemented by the Permanent Interstate Committee for Drought Control in the Sahel, have resulted in the deployment of solar pumping stations systems in nine countries, providing improved access to water and electricity for 2 million people [29]. Also PV costs less than diesel-powered generators in many African and Latin American areas. However, limitations may arise when pumping head and/or water volume requirements are large or highly seasonal.

In conclusion, PV is growing rapidly in many African and Latin American countries because of its modularity and drop in costs, offering a hopeful future not only to many countries that need affordable electricity to grow, but also for small rural communities that would like to become connected to the global community.

References

1. Ren 21 (2015) Renewables 2016. Global status report. Key findings
2. International Energy Agency database (2015) http://data.iea.org/. Accessed 30 Nov 2015
3. IRENA (2015) Resource. http://resourceirena.irena.org/gateway/dashboard/. Accessed 30 Nov 2015
4. IEA (2015) Report IEA-PVPS T1-27:2015. Trends 2015 in photovoltaic applications
5. ISE-FhG (2015) Photovoltaics report. Fraunhofer Institute für Solare Energiesysteme 17 November 2015. https://www.ise.fraunhofer.de/de/downloads/pdf-files/aktuelles/photovoltaics-report-in-englischer-sprache.pdf. Accessed 30 Nov 2015
6. Green MA, Emery K, Hishikawa Y, Warta W, Dunlop ED (2016) Solar cell efficiency tables (version 48). Prog Photovolt: Res Appl 24:905–913
7. NREL Best Research-Cell Efficiencies Chart. http://www.nrel.gov/ncpv/images/efficiency_chart.jpg. Accessed 30 Nov 2015
8. Mulugetta Y, Nhete T, Jackson T (2000) Photovoltaics in Zimbabwe: lessons from GEF solar project. Energy Policy 28:1069–1080
9. Azimoh CL, Wallin F, Klintenberg P, Karlsson B (2014) An assessment of unforeseen losses resulting from inappropriate use of solar home systems in South Africa. Appl Energy 136:336–346
10. NREL (2015) SAPC best practices in PV operations and maintenance. www.nrel.gov/docs/fy15osti/63235.pdf. Accessed 30 Nov 2015
11. Climatescope (2015) The clean energy country competitiveness index. The Multilateral Investment Fund, UK Department of International Development, Power Africa and Bloomberg New Energy Finance
12. IEA (2015) Trends 2015 in photovoltaic applications. Survey report of selected IEA countries between 1992 and 2014. Photovoltaics Power Systems Programme. Report IEA-PVPS T1-27:2015
13. IRENA (2016) Letting the light. How solar photovoltaics will revolutionise the electricity system
14. Schreieck R (2015) PV production forum. Market session. In: 25th EU PVSEC—Valencia 7 Sept 2010
15. The Economist (2015) Follow the sun. Solar energy. Available 16 Apr 2016
16. Bloomberg New Energy Finance (2016) Week-in-review. Available 23 Aug 2016
17. Bloomberg New Energy Finance (2016) Week-in-review. Available 21 June 2016
18. Wetzel T, Borchers S (2015) Update of energy payback time and greenhouse gas emission data for crystalline silicon photovoltaic modules. Prog Photovolt: Res Appl 23:1429–1435

19. Ito M, Lespinats S, Merten J, Malbranche P, Kurokawa K (2016) Life cycle assessment and cost analysis of very largescale PV systems and suitable locations in the world. Prog Photovolt: Res Appl 24:159–174
20. Bloomberg New Energy Finance and Lighting Global (2016) Off-grid solar market trends report 2016
21. Louie H, Shields M, Szablya SJ, Makai L, Shields K (2015) Design of an off-grid kiosk in rural Zambia. In: Global humanitarian technology conference (GHTC), 2015 IEEE
22. Rolffs P, Ockweel D, Byrne R (2015) Beyond technology and finance: pay-as-you-go sustainable energy access and theories of social change. Environ Plan 47:2609–2627
23. Phadke A, Jacobson A, Park WY, Lee GR, Alstone P, Khare A (2015) Powering a home with just 25 Watts of solar PV: super-efficient appliances can enable expanded off-grid energy services using small solar power systems. Lawrence Berkeley National Laboratory. LBNL-175726
24. Global LEAP, Clean Energy Ministerial and Dalberg (2016) Global LEAP off-grid appliance market research: final report
25. The Economist (2015) Africa, a brightening continent. Special report: energy and technology. Available 17 Jan 2015
26. Klintenberg P, Wallin F, Azimoh LC (2014) Successful technology transfer: what does it take? Appl Energy 130:807–813
27. Elasaad H, Bilton A, Kelley L, Duayhe O, Dubowsky S (2015) Field evaluation of a community scale solar powered water purification technology: a case study of a remote Mexican community application. Desalination 375:71–80
28. Shen J, Mkongo G, Abbt-Braun G, Ceppi SL, Richards BS, Schäfer AI (2015) Renewable energy powered membrane technology: fluoride removal in a rural community in northern Tanzania. Sep Purif Technol 149:349–361
29. IRENA (2015) Africa 2030: roadmap for a renewable energy future
30. IRENA (2015) Off-grid renewable energy systems: status and methodological issues

Chapter 7
Solar Thermal Energy for Heating, Cooling and Power

Ricardo Guerrero-Lemus and Les E. Shephard

Abstract This chapter shows that solar thermal energy is growing moderately around the world in part because solar photovoltaics is acting as a very competitive alternative in terms of costs and O&M. There is a significant penetration of solar thermal systems for heating water in some African and Latin American countries but challenges remain to produce the technologies involved locally with local qualified manpower, particularly in the vacuum technology and hot water plumbing sectors. No solar cooling systems for direct building air conditioning have been reported in Africa and limited opportunities are reported in Latin America with no detailed information on capacity and/or location. Some solar sorption refrigeration have been reported in Africa and Latin America since the late 1970s, but mostly devoted to research and development activities and not to wide-scale application. Finally, solar cookers hold significant promise in both regions but to date have limited wide scale adoption in either region because of dependence on weather conditions, cultural habits and maximum temperature limitiations by with many of these devices, which are not compatible with basic food recipes in many areas. Concentrating solar power (CSP) has tremendous promise globally but continues to grow at a measured pace, as Africa and Latin America have very limited CSP, with 2 GWh and 8.6 GWh, respectively. CSP plants require water for cooling and for the condensation processes which can be problematic in many desert areas of Africa and Latin America where the sun direct radiation is high enough to produce high power outputs.

Keywords Energy · Africa · Latin America · Electricity · Solar thermal · CSP · SHC · Solar cookers · Thermal insulation · Thermal storage · Solar collectors · Parabolic troughs · Linear Fresnel systems

7.1 Introduction

Solar energy is not only used to produce electricity by converting solar radiation directly into an electric current but also to produce heat. This heat can be directly used for different purposes (e.g., heating water and food, low-temperature heating

R. Guerrero-Lemus and L.E. Shephard, *Low-Carbon Energy in Africa and Latin America*, Lecture Notes in Energy 38,
DOI 10.1007/978-3-319-52311-8_7

processes in industry, or cooling spaces) in solar heating and cooling (SHC) systems or converted into electricity, mainly through concentrated solar power (CSP) technology. The direct use of heat also offers modularity for applications at different levels (household, buildings, industry, etc.), but CSP is mostly considered for large power units.

The ability of SHC technologies to capture solar energy in the Earth's surface continues to grow worldwide, reaching by the end of 2015 an installed capacity of 435 GW_{th}, corresponding to a total of 622 million m^2 of collector area [1]. The vast majority of the total capacity in operation in 2014 is in China (289.5 GW_{th}) and only small amounts are place in Latin America (10.0 GW_{th}) and Africa (2.1 GW_{th}). The nominal solar thermal capacity of solar collectors surface is conventionally transformed into energy applying the conversion factor 0.7 kW_{th}/m^2 [1]. Compared to previous years, the global market growth rate 2013/2014 of a 1.8% tended to flatten out as the market growth in the period 2013/2014 was 1.9% and 6.7% in the period 2012/2013 (Fig. 7.1) [1]. Thus, the market for new collector installations is for the first time shrinking, a trend that seems to continue in 2015. In Latin America the growth between 2012 and 2013 has been 1.0%, and in Africa 5.0% [1]. The amount of heat from the sun that can be captured by solar collectors range from 300 to 800 $kWh/m^2/year$ depending on the design and location [2] and the world technical potential for solar water heating on the basis of available roof area and the irradiation per region was estimated to amount to 123 EJ/year in the long term [3].

The solar thermal collectors can be divided into water flat-plate collectors (FPC), evacuated tube collectors (ETC), unglazed water collectors, and glazed and

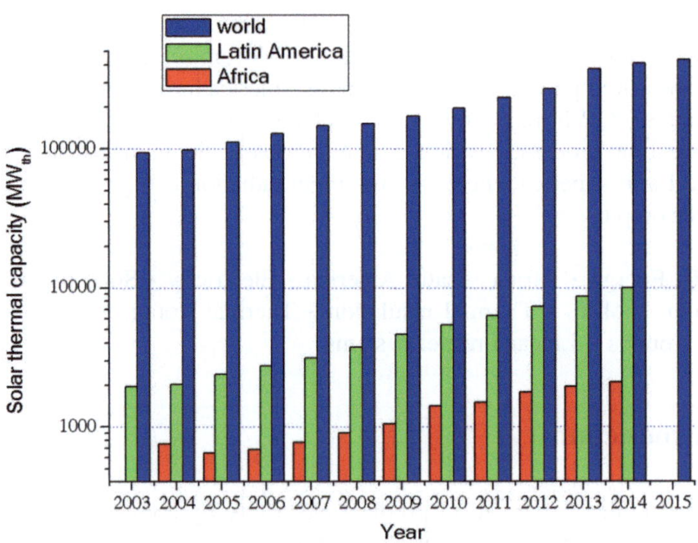

Fig. 7.1 Evolution of the annual solar thermal capacity in the period 2003–2015 (*source* SHC-IEA)

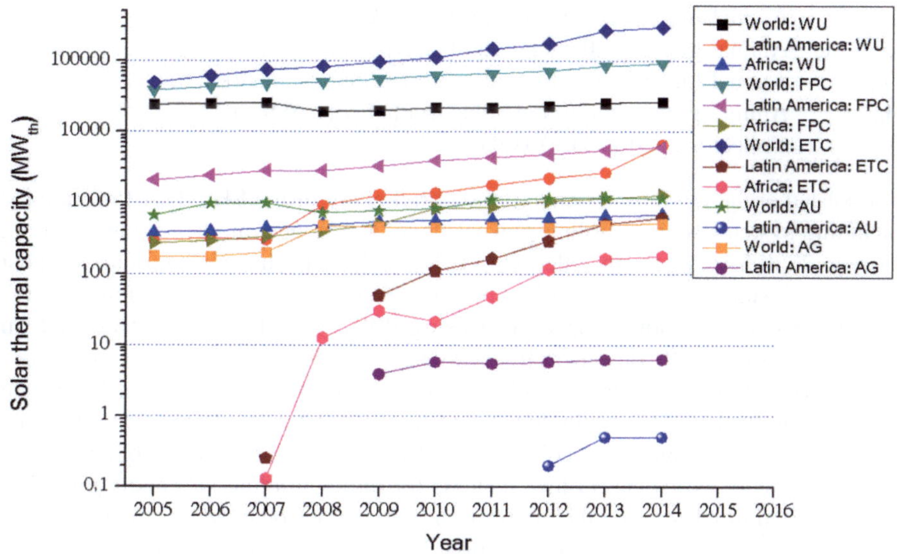

Fig. 7.2 Evolution of the global solar thermal capacity from 2005 to 2014 for different collector technologies. *WU* Unglazed water collector; *FPC* Flat plane collector; *ETC* Evacuated tube collector; *AG* Air glazed collector; and *AU* Air unglazed collector (*source* SHC-IEA)

unglazed air collectors, being ETC followed by FPC the main contributors to solar thermal energy production (Fig. 7.2) [1]. However, in Africa and Latin America the main contributor is FPC followed by the water unglazed collector, and this difference can be attributed to the higher difficulties for installing and maintaining compared to ETC.

The SHC Program of the IEA and major solar thermal industry associations have agreed on a common and simple calculation method to estimate the annual solar collector energy output in kWh [4]. By calculating the amount of energy produced by solar thermal systems, worldwide solar thermal energy can be easily compared with other (renewable) energy sources:

– Unglazed collectors for pool heating:

$$Q_{uc} = 0.29 \cdot H_0 \cdot A_a \tag{7.1}$$

– Glazed collector in DHW (domestic hot water systems):

$$Q_{gc,DHW} = 0.44 \cdot H_0 \cdot A_a \tag{7.2}$$

– Glazed collector in combi-system:

$$Q_{gc,c} = 0.33 \cdot H_0 \cdot A_a \tag{7.3}$$

Q_{uc}	Annual unglazed collector output (kWh)
$Q_{gc,DHW}$	Annual glazed collector output (kWh)
$Q_{gc,c}$	Annual glazed collector output in combi systems (kWh)
H_0	Annual global horizontal solar irradiation (kWh m^{-2})
A_a	Collector aperture area (m^2)

For solar thermal cooling systems, only a brief study of the SHC-IEA Program [5] including systems with a cooling power above 20 kW and another report counting close to 2200 SAC plants installed worldwide in 2014 [6] have been found. The list counts 81 installed large scale solar cooling systems (eventually including systems which are currently not in operation) using different technologies: 56 based on absorption chillers; 10 on adsorption chillers; and 17 on DEC (desiccant evaporative cooling) systems. The overall cooling capacity of the solar thermally driven chillers was 9 MW. The technology used in 53% of the total area was flat plate collectors, 37% vacuum tube collectors, 7.3% compound parabolic collectors and 2.6% air heaters. No solar cooling systems for direct building air conditioning has been reported in Africa, and the applications are mainly focussed on small-size cold boxes for food or vaccines preservation [7]. In Latin America there are some companies offering solar thermal air conditioning but no statistics about capacity and/or location has been detected.

Special mention should be made to solar cookers as a potential solution to introduce cleaner ways of cooking in Africa and Latin America and to avoid carbon emissions, deforestation and time to collect biomass. For immediate food consumption temperatures above 70 °C are needed. These temperatures can be achieved by using solar concentration combined with a radiation absorber where the food is placed and a greenhouse enclosure to retain the heat (Fig. 7.3) [8]. A high concentration requires continuous orientation of the concentrator, but low concentrations, as in solar ovens, requires orientation every 30’.

Passive solar systems (thermal mass, Trombe wall, solar chimney, etc.) for utilization in bioclimatic architecture are not addressed in this book.

As discussed above, CSP technology converts solar heating to electricity. CSP uses mirrors as optical elements to concentrate solar energy, converting it into thermal energy at medium (300–600 °C) and high ($\geq$600 °C) temperatures, usually carried by steam or hot air to move turbines according to a specific thermodynamic cycle (Rankine, Brayton, Stirling, etc.) and, consequently, to produce electricity. Therefore, unlike PV solar cells which “directly” produce electricity from the sun, for CSP the sun energy is transformed into thermal energy in an intermediate step before producing electricity. When the temperatures reached are high enough ($\geq$600 °C) the CSP systems could be considered for solar fuel production, mainly hydrogen. The different CSP technologies are classified in relation to the characteristics of the mirrors system: (i) parabolic troughs; (ii) parabolic dishes; (iii) flat mirrors, called heliostats, focused on a receptor located in a tower; and (iv) linear Fresnel reflectors, where no solar tracking is needed (Fig. 7.4). To adjust the power demand from the grid to the power output of the CSP plant and to levelized falls in electricity production due to momentary cloudy weather,

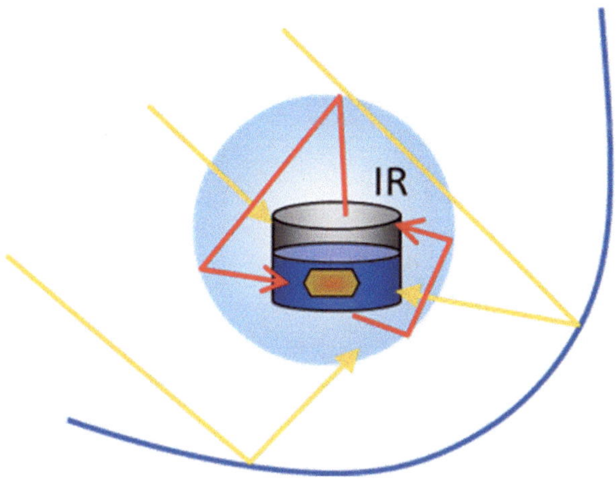

Fig. 7.3 Standard solar cooker scheme, composed by a solar concentrator, radiation absorber where the food is placed and a greenhouse enclosure to retain the heat

increasing number of CPS plants are associated to thermal storage systems, which are much more efficient than electricity storage systems.

Figure 7.5 shows the evolution of the world electricity output from solar thermal energy from 1990 to 2014 [9].

On a global scale, electricity from CSP has grown substantially since 2010, mostly in Spain, but also more recently Africa shows electricity production from CSP, as the solar irradiation conditions are favourable in this region. Thus, CSP is operating in some African countries (Algeria, Egypt and Morocco) since 2011 and new plants in South Africa and Chile are expected to add electricity output from 2015 [10].

The CSP capacity installed or under construction worldwide and in operation in 2015 consists mainly of parabolic troughs (92 projects), and to a lesser extent on tower systems (22 projects), linear Fresnel reflectors (12 projects) and parabolic dishes (2 projects) [11].

The solar energy that CSP plants exploit is the direct component normal irradiance (DNI), which is the energy received on a surface tracked perpendicular to the sun's rays. This is because the diffuse component cannot be optically focused. CSP developers typically define a bottom threshold for DNI of 1900 kWh/m^2/year–2100 kWh/m^2/year [12]. DNI measurements provide only a first approximation for a CSP plant's electrical output potential. In practice, variations in direct sunlight over the course of a day is more important as below a certain threshold CSP plants have no net production due to constant heat losses in the solar field. Optimal DNI is usually found in arid and semi-arid areas with reliably clear skies, which typically lay at latitudes from 15°–40° North or South (Fig. 7.6). Closer to the equator the atmosphere is usually too cloudy and wet in summer, and at higher latitudes than about 40°, values of the DNI are small, and the skies are usually too cloudy. DNI is

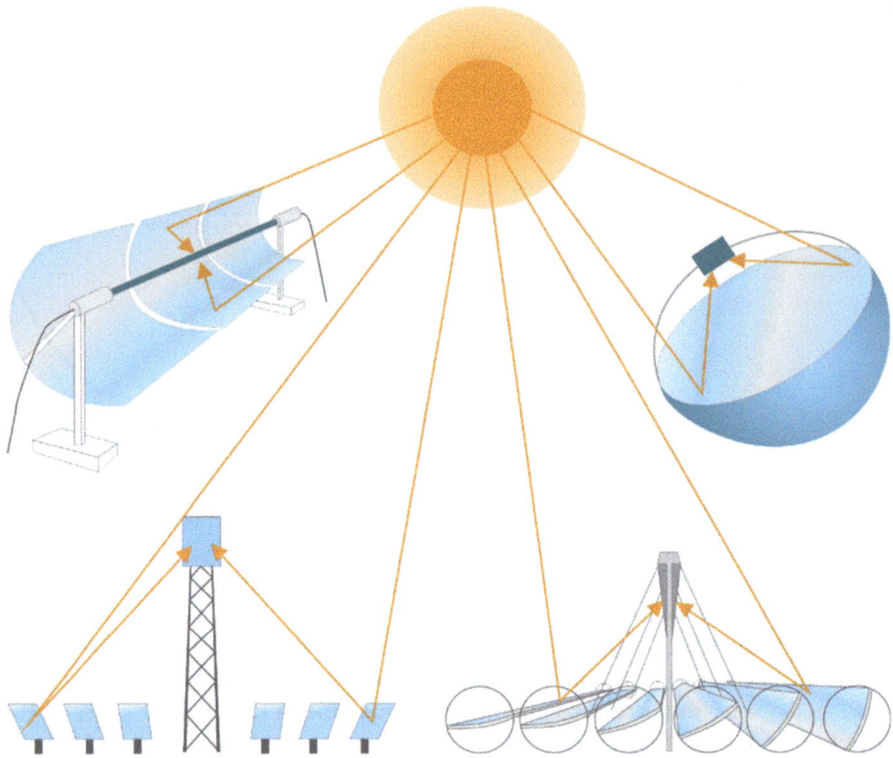

Fig. 7.4 Different CSP technologies: (i) parabolic troughs; (ii) parabolic dishes; (iii) tower; and (iv) linear Fresnel reflectors

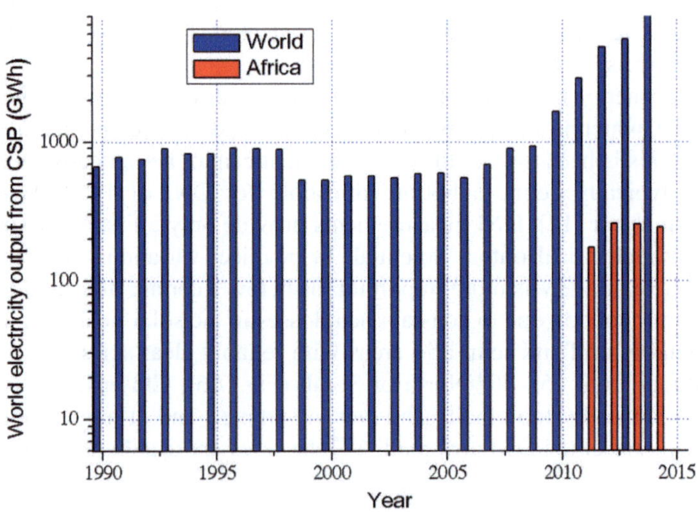

Fig. 7.5 Evolution of the world's, African and Latin American electricity output from solar thermal energy in the period 1990–2014 [9]

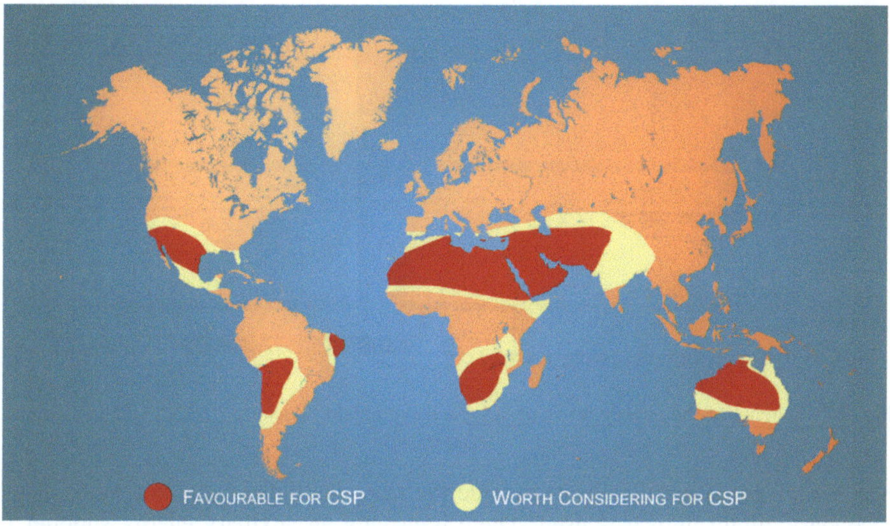

Fig. 7.6 Optimal geographical areas for installing CSP plants [11]

also significantly larger at higher altitudes, where absorption and scattering of sunlight are much lower.

7.2 Technology State of the Art

7.2.1 Solar Thermal Fundamentals

The system most widely used to collect thermal energy from the sun is the flat-plate collector, whose operating principle is based on the greenhouse effect (Fig. 7.7). The main effort is focused on achieving maximum radiation absorbed in the collector and, simultaneously, minimizing the thermal energy losses. These thermal energy losses are produced:

- From convection: reduced by a glass on the absorber, where an air-vacuum sheet is present in the space between the glass and the absorber.
- From conduction: it is necessary to use thermal insulation between the absorber and collector housing and to cover the fluid transport tubes.

Moreover, to get a maximum absorption energy in the absorber it is necessary to use anti-reflecting coatings.

Thus, the heat absorbed by the solar collectors is transferred to a liquid carrier flowing in the primary circuit. The thermal energy is transferred to sanitary water, which circulates in a secondary circuit, through a heat exchanger (Fig. 7.8). The

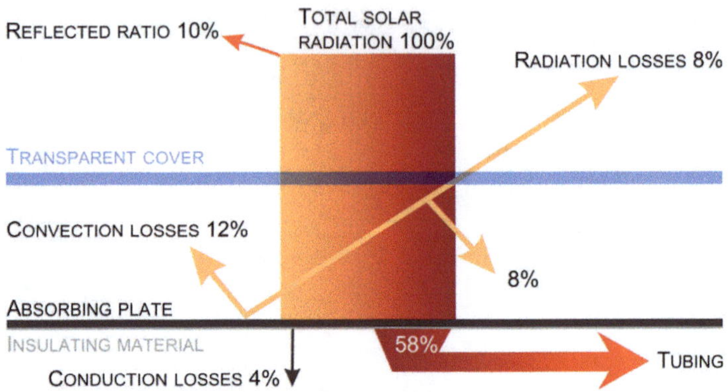

Fig. 7.7 Scheme of the energy flows on a typical flat solar collector

heat exchanger is usually located inside the tank that collects the hot water of the secondary circuit until it is used. There is usually an auxiliary energy supplier to increase the water temperature of the secondary circuit when it is not high enough to be consumed. This auxiliary energy supplier is usually an electrical resistance.

Although the solar thermal technology for hot water production can be considered mature, it is estimated that the materials traditionally used in this technology (copper and aluminium) can be replaced by new ones (polymers) that offer improved performance, durability and cost reduction.

Other research activities aim to improve the thermal insulation and self-cleaning properties of the windows situated at the top of the collectors, raising the systems efficiency. Moreover, it is detected a progressive substitution of traditional materials by others which offer higher resistance against corrosion processes, overheating and

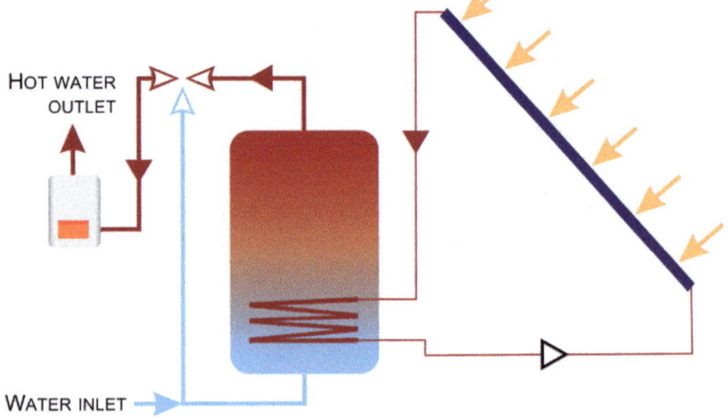

Fig. 7.8 Scheme of a basic solar thermal system

can offer more versatility to manufacturing non-rectangular collectors, which can improve the integration of solar collectors in some outer areas of the buildings receiving sun light.

7.2.2 Cooling and Air Conditioning

Although various technologies exist to convert solar heat into cooling there are four dominant technologies being used today:

- Closed chillers, which produce chilled water using liquid sorption materials, are the most common. The dominating material pair uses water as a refrigerant and a LiBr aqueous solution as sorbent. Another pairing being used is ammonia as refrigerant and water as sorbent. LiBr-water systems also are available as single-effect machines, operating at temperatures starting at 75 °C, and as double-effect machines, with driving temperatures of 130 °C and above and, consequently, using concentrating solar collectors.
- Closed chillers using solid sorption materials and water as a refrigerant. In particular, machines in the small capacity range have been developed in recent years and are entering the market in increasing numbers. Silica gel is used as sorbent in most cases, but also zeolite coatings applied on heat exchanger metal surfaces, which show a promising potential. A completely different category of systems uses salt as sorbent material, which is completely dried during regeneration thereby providing a kind of thermo-chemical storage. Thus a mismatch between availability of driving heat (e.g., coming from solar collectors) and the operation of the cooling system (e.g., during night) can be overcome.
- Open cooling cycles using solid sorption materials and rotating wheels with a matrix impregnated with the desiccant material. Various combinations of the desiccant rotor with sensible heat exchangers and evaporative coolers (humidifiers) exist, which are adapted to different climatic conditions.
- Open cooling cycles using liquids as sorption materials, having achieved a status of early market deployment and a number of installations have been realized. Open cooling systems using a liquid desiccant have two main advantages: (i) cooling of the sorption process can be realized easier in comparison to systems with solid sorption materials; and (ii) the concentrated solution can be stored. This provides an attractive method of storage that allows overcoming mismatches between the availability of driving heat and the need of conditioning the ventilation air.

Until recently the smallest thermally driven chiller available on the market had a capacity of 35 kW and only one manufacturer existed. Currently, there are small capacity systems for residential or small commercial applications starting at a rated cooling capacity of about 8 kW [13]. However, the market penetration for solar thermal cooling systems remains still small after many years. The main shortcoming from a technical perspective is related to shortcomings in the proper design and energy management of systems that result in a high overall electricity consumption of auxiliary components. A particular area where faults are made is related to the heat rejection subsystem, which often has not received sufficient attention in the past. Another error often made is that many systems were far too complex and as a result created non-optimal control and large maintenance efforts. Today, systems are available on the market that require a relatively high electricity demand and, thus, do not save much energy. Even worse, these systems lead to a negative energy balance, that is, they consume more (primary) energy compared to a conventional system. The overall energy balance for solar thermal systems will improve as other needs such as heat for space heating or sanitary hot water are included. An overview of market availability for thermally driven cooling systems based on sorption technology is provided elsewhere [14].

In Africa and Latin America it is important to use materials that can be obtained locally. Thus, low cost refrigerators have been tested in operation with activated carbon/ethanol to store vaccines in a temperature range of 2–8 °C. Activated carbon can be obtained by using coconuts shells and ethanol can be found in alcoholic beverages available locally. When solar energy is not available, the refrigerator could operate with biofuels or wood fire [15].

7.2.3 Solar Thermal Collectors

Collectors are usually divided between the so-called flat-plate (glazed or unglazed) and vacuum tubes (with direct or indirect flow) (Fig. 7.9). The fluids most commonly used as heat carriers in the primary circuit are water, mineral and synthetic oils, water-glycol mixtures and water-glycerine mixtures. The vacuum tube systems work better in cloudy and cold environments since convection losses are reduced up to 35%. However, they are less robust than the flat-plate collectors, especially as vacuum is lost from the tubes with ageing.

The standard configuration of the collectors is rectangular in geometry, but it is increasing the design of flat-plate collectors with different geometries to improve integration in buildings as well as the development of hybrid thermal solar—PV in order to provide both heat and electricity to buildings.

Another technology which is increasing in interest is based on high-temperature solar thermal collectors. These are considered when high temperature working-fluids (above 130 °C) are required, and use mirrors or lenses to concentrate sunlight on the absorber. Concentrating solar technologies require clear skies and sufficient

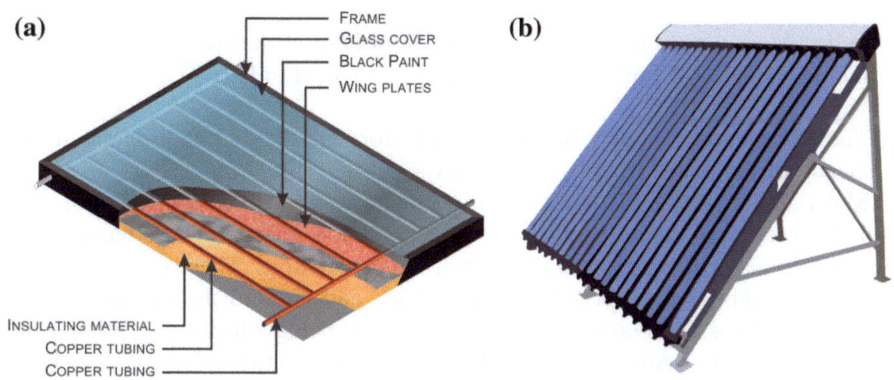

Fig. 7.9 **a** Scheme of flat-plate collector and **b** image of the vacuum tube collector

DNI to reach high levels of performance. The applications requiring such high temperatures are, mainly, the solar cooling and air conditioning systems, but also the combi-systems.

Case Example 7.1

Real solar adsorption refrigeration system working in Bou-Ismail, Algeria:

The aim of this work has been to characterize a low-maintenance device with a new tubular absorbent configuration since it is the most important element to study for enhancing the efficiency of a solar absorption refrigeration system. The device has been built with an adsorbent-adsorbate silica-gel-water pair and tested in the weather conditions of Bou-Ismail, Algeria. The system consists of three major components: (i) a solar collector (12 matt black painted Cu tubular adsorbent 90 × 70 cm bed, reflecting surface, single glazed cover and 4.5 kg of 2–5 mm micro-pore silica-gel material); (ii) an air-cooled condenser (9/7 mm diameter copper tube); and (iii) an evaporator (0.8 l volume 40/38 mm copper tube inside a 15 l cold chamber insulated with 150 mm thick polyurethane). During the day the system follows a process of isosteric heating of the adsorbent and a subsequent isobaric desorption of the adsorbate steam, condensed and stored in the evaporator. During the night the system follows a process of isosteric cooling, where the adsorbent regained the ability of adsorbing the adsorbate. Experimental maximum temperature at 33 °C average ambient temperature ranges between 95 and 117 °C, condensing temperature between 45 and 53 °C, and +5 °C for minimum evaporator temperature. The solar COP ranges from 0.083 to 0.09 (receiving around 19 MJ/m^2). The device is expected to be applied for medical storage. These results show the capability of African research centres to develop quite complex solar thermal technology. A detailed description of the device and the building process can be obtained in [16].

The key parameters for solar energy collectors are the output temperature and efficiency of the collected solar energy, exposed in Table 7.1.

Many applications require hot water available anytime at the consumption place (tap, shower, etc.), which involves using recirculation systems. These recirculation systems are also being applied in solar thermal installations. These systems usually demand an additional supply of power, so they add costs to the operation of the installation. Thus, programmable thermostats are being offered to optimize the cost-demand ratio. Another area where advanced control systems are being demanded is in hybrid systems, i.e., combining solar heat and biofuels (or fossil fuels) to guarantee hot water supply.

7.2.4 CSP Technology

The main differentiating technology in CSP is the solar collector, as the heated fluid is always used for producing steam which moves a steam turbine for producing electricity. There are four types of solar collectors:

- Parabolic troughs: represent the most mature technology. The mirrors have typical concentration ratios between 50 and 200 suns and, consequently, the heating carrier temperature is not high enough to consider this technology as suitable for thermo-chemical solar fuel production. The mirror arrays can be more than 100 m long with a curved surface 5–6 m across. The structure can rotate on an axis parallel to the north-south direction. The movement is controlled by a computer algorithm that contains the position of the sun throughout the day. The thermal fluid (mainly, a special oil or deionized water) circulates through a tube along the cylinder axis and it is usually made of a metal, e.g., stainless steel covered by a highly absorbing optical film and insulated in an evacuated glass envelope. This tube is surrounded by a highly transparent Pyrex glass or quartz. The conversion efficiency to thermal energy can reach 60%, while the electricity conversion efficiency can range between 8 and 35%

Table 7.1 Output temperature and efficiency of the collected solar energy

	Output temperature (°C)	Efficiency (%)
Unglazed flat-plate collector	20–40	40–50
Glazed flat-plate collector	60–110	70–75
Evacuated tube collector	90–110	75–80
High-temperature collector	>150	75–80

Table 7.2 Approximate values of current efficiencies and areas required for CSP technologies

	Tower	Trough	Dish	Fresnel	Storage
Peak efficiency (%)	20–35 (concepts)	15	25–30	8–10	98
Area needed (m^2/kW)	47.5	23	4	15	18 (7 h)
Solar fuels	Yes	No	Yes	No	

(Table 7.2) [14]. Troughs technology has been successfully tested in several CSP plants for many years, using a variety of transfer fluids: molten salts, mineral oils, water-vapor, etc.

- Tower systems: also known as central receiver systems (CRS), the flat mirrors or heliostats focus the solar energy onto a receiver located on top of a tower. Each heliostat has a typical surface around 40–120 m^2 and a two-axis tracking system to focus all the mirrors on the receiver. The typical solar concentration in the receiver is about a thousand suns. Then, the heated fluid temperature at the receiver can be high enough to consider this technology as suitable for solar fuel production. The heated fluid can be used directly to move turbines, or/and be directed to a thermal exchanger for storing its thermal energy.
- Parabolic dishes: these collectors focus solar radiation onto a small punctual area located on the parabolic focus. The concentration ratio ranges between 3000 and 10,000 suns, and can heat the carrier fluid to temperatures above 600 °C. Then, the heating carrier temperature can be high enough to consider this technology as suitable for solar fuel production. With these high temperatures the highest electricity conversion efficiencies of all CSP technologies can be obtained (Table 7.4). Due to its geometry, two-axis orientation tracking systems are needed to obtain maximum solar radiation parallel to the axis of the dish mirror. The dish technology is not widespread and is usually associated with relatively small plants, between 10 and 20 kW, and with Stirling heat engines (or micro-turbines) placed at the focal point. This design eliminates the need for a heat transfer fluid and for cooling water.
- Linear Fresnel: in this case, no tracking system is needed to concentrate the sun radiation because the Fresnel mirrors direct the sunlight to the heat exchanger ideally at any incidence angle. However, the electricity conversion efficiency is the lowest of all CSP systems (Table 7.4). Then, the heating carrier temperature is not high enough to consider this technology as suitable for solar fuel production and also introduces many difficulties for efficient thermal storage. This low conversion efficiency is expected to be compensated by a significantly lower production cost than any other CSP technology and less area per kW needed than parabolic troughs and tower systems.

CSP plants require water for cooling and condensing processes, and this resource is difficult to obtain in desert areas where the sun direct radiation is high enough to produce high power outputs. CSP water requirements are about 3000 l/MWh for parabolic trough and linear Fresnel plants, compared to about 2000 l/MWh for a conventional coal plant and only 800 l/MWh for combined-cycle natural gas plants. Tower CSP plants water needs are less than for trough plants, but depend on the efficiency of the technology. Dishes are cooled by air [12]. Dry cooling (with air) is an alternative for all CSP technologies, but it is more costly and reduces efficiencies. Thus, dry cooling installed on trough plants in hot desert areas reduces annual electricity production by 7% and increases the cost of the produced electricity by about 10%. These values are somewhat lower for solar towers than for parabolic troughs [12].

None of the CSP technologies can be considered in a competition stage, since in almost all countries incentives or specific bids for PPAs are required for CSP plants. However, all technologies are in a market phase, considering that all of them have plants over 1 MW currently in operation. Then considering the power capacity currently in operation worldwide (exposed above), parabolic troughs are ahead in terms of development, followed by the tower systems. Linear Fresnel and dish technologies are also entering in the market stage.

7.2.5 Thermal Storage

Thermal energy storage (TES) systems can be charged from heating or cooling environments and hold this energy over time. The usual device with zero energy consumption for hot water storage is a thermosyphon. Located above the collectors, the thermosyphon uses convection currents to storage the heated water (Fig. 7.10). These systems are cheap and can store heat for days or even one week or two at acceptable cost. But they are bulky and therefore are not an ideal solution for long-term storage.

Thermal energy storage can be categorised based on the underlying physical principles of the storage technique (Table 7.3): (i) sensible heat, making use of the

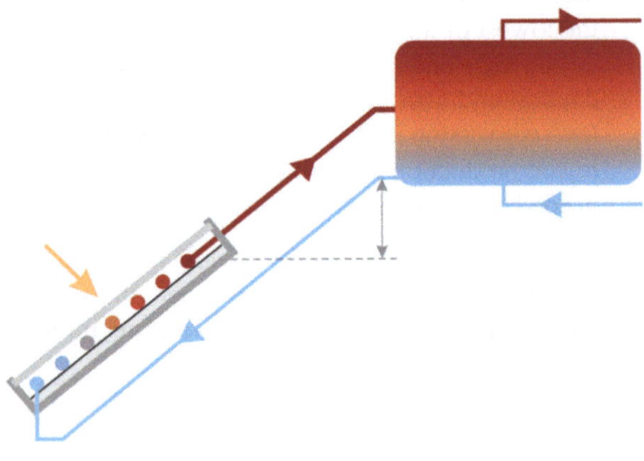

Fig. 7.10 Diagram of a solar collector with thermosyphon

Table 7.3 Low-medium temperature thermal storage categories [17]

Sensible (water, ground, concrete, etc.), e.g., hot/chilled water tanks, underground thermal energy storage	**Latent** (inorganic salts, paraffins, etc.), e.g., ice storage, PCM panel sand modules	**Thermochemical** (chemical reactions), e.g., silica gel + water, zeolite + water	**Absorption/adsorption** (Open/closed systems) inorganic salts + water, ammonium + water

Table 7.4 Energy capacities, power, efficiency and storage time of thermal energy storage technologies [17]

TES technology	Capacity (kWh/t)	Power (kW)	Efficiency (%)	Storage time
Hot water tank	20–80	1–10,000	50–90	Day-year
Chilled water tank	10–20	1–2000	70–90	Hour-week
ATES low temp.	5–10	500–10,000	50–90	Day-year
BTES low temp.	5–30	100–5000	50–90	Day-year
PCM-general	50–150	1–1000	75–90	Hour-week
Ice storage tank	100	100–1000	80–90	Hour-week

ATES Aquifer thermal energy storage; *BTES* Borehole thermal energy storage

heat capacity of a material (vast majority of systems); (ii) latent heat, utilising the phase-change properties, either melting or evaporation; (iii) thermochemical heat, storing energy by way of chemical reaction; and sorption heat, using water vapour uptake by a sorption material.

The key parameters of thermal energy storage systems are their capacity, power rating (ability to discharge), efficiency (losses over time and with charge/discharge) and cost. The current values for these parameters are exposed in Table 7.4, (except for costs, described in the next section).

Sensible heat storage systems (e.g., hot and chilled water) and some latent heat stores (e.g., ice storage) are mature technologies. However, developments in advanced phase change materials (PCM) and chemical reactions are creating new application possibilities described below.

When thermal storage is included in the CSP plant design, the size of the solar field must be increased in order to capture the extra energy required to be stored when sunlight is available. A large variety of materials for thermal storage are used in different CSP plants, mainly thermophysical storage materials: (i) molten salts, which consist of a eutectic mixture of salts (typically a 60% $NaNO_3$/40% KNO_3, with ranging work temperatures of 260–565 °C); (ii) mineral oils, which should be non-toxic or flammable and are therefore costly; (iii) water/steam for direct transfer of steam to the turbine (the storage time is less than for molten salt for the same storage volume); and (iv) air, as it has very low cost and easy handling. The thermophysical approaches rely on changes in a system's physical state and use sensible heat (by increasing the temperature of the storage material), latent heat (absorbed at a constant temperature, as in a phase change), or both, and thermal insulation is needed to minimize heat losses to the environment.

Thermochemical storage materials are the alternative option, where chemical reactions reversibly store energy without a need for insulation. Despite the advantages of enabling high energy density and insulation-free long-term storage, thermochemical technologies have yet to be widely used. On the other hand, thermophysical storage relies primarily on liquids and solids and offers high volumetric energy density but low gravimetric energy density. Thermochemical systems generally contain at least one gas-phase component, affording light weight but requiring large volume in the absence of mechanical compression.

Installation of thermal buffer storage (usually filled with water) is also common in solar cooling systems. Sizes range from small buffers to overcome short term fluctuations, up to large buffer to save solar power for hours, and applied on the hot and/or cold side. In a few applications, ice storage has been applied to increase the storage density [6].

In this section it is important to mention the *solar multiple* concept, as the ratio of the actual size of a CSP plant's solar field compared to the field size needed to feed the turbine at design capacity when solar irradiance is at its maximum (about 1 kW/m^2). Plants without storage have an optimal solar multiple around 1.1–1.5 (up to 2.0 for linear Fresnel reflectors), depending primarily on the amount of sunlight the plant receives and its variation throughout the day. Plants with large thermal storage capacities may have solar multiples of up to 3–5 [12].

A last aspect to be considered is the attractiveness to ensure continued electricity production from CSP plants, which can also require biomass and/or fossil fuel input (e.g., natural gas) to provide energy for the turbines when sun radiation and storage is not available.

7.2.6 Solar Cookers

As it has been exposed in Chap. 5, cooking with biomass by many Africans and Latin Americans is an activity associated to serious health concerns due to inhaling the smoke and deforestation. Also, hours spent gathering wood or biomass fuel can be more useful devoted to education and work. Books completely devoted to solar cookers and multitude of aspects many of them difficult to find elsewhere (e.g., heat transfer theory, cooking recipes, cooking tools and heat-storage kitchens) can be found elsewhere [18]. Some technical progress has been made to reduce such concerns but the health concerns are associated to any biomass combustion without adequate ventilation. An alternative is to introduce cleaner ways of cooking, such as with solar ovens.

The three most common types of solar cookers are (Fig. 7.11) [19]: (i) box cookers, cooking at moderate to high temperature typically between one to three hours, high durability, high thermal isolation, and they are the most widespread; (ii) parabolic cookers, using a bowl shaped reflector to focus the light more directly

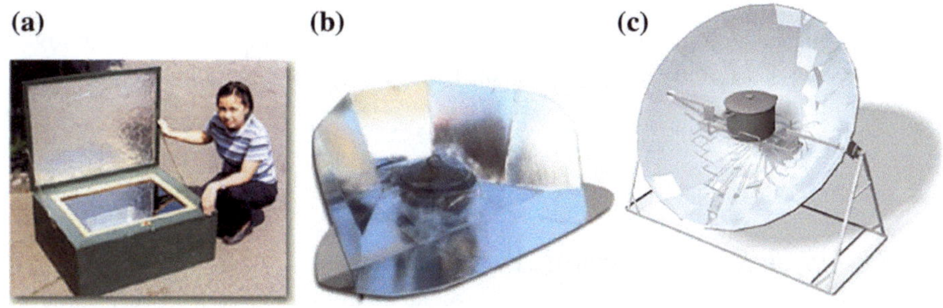

Fig. 7.11 The three most common types of solar cookers: **a** box cookers; **b** panel cookers; and **c** parabolic cookers (*source* Wikipedia Commons)

onto the cook pot, sensible to wind turbulences, and typically do not require a greenhouse enclosure because of the high cooking temperatures obtained; and (iii) panel cookers, incorporating elements of box and parabolic concentrator cookers, and offering low durability.

Solar-cooking systems face challenges because the most durable stoves are expensive and difficult to manoeuvre or repair, and the cheap ones are insubstantial and easily stolen. Also, there are concerns about the use of solar cookers in sunny regions but with large rainy seasons, and for basic foods (e.g., tortilla in Latin America) that require higher temperatures than those obtained with these devices. Concentrated solar cookers with parabolic mirrors are vulnerable to wind damage and require constant repositioning to catch sunlight. Also there are difficulties to develop financially sustainable markets for solar ovens [20]. Solar cookers can also be useful to other purposes, as the sterilization of instruments in hospitals.

Case Example 7.2

Solar cooking experiences in Central America:

The Central American Solar Energy Project (CASEP) was launched in 1991 putting women from the local communities at the heart of the programme to facilitate the integration of the technology. Women constructed the stoves from plywood, glass, aluminium and metal, and use hammers, hand saws, shears and sealant guns to build them. The ovens were built collectively to increase the spirit of solidarity. Ongoing support was provided for at least two years by a NGO (CASEP) staff in each country. Finding and retaining skilled female leaders is the biggest challenge. The difficulty to adapt lifestyles to cook with sunlight and longer cooking times depending on weather conditions is also a challenge [20]. Thus, in a solar-oven project funded by CASEP in Nicaragua surveys indicated that users continued to cook on biomass-burning stoves and savings were not significant, after reading data from thermometers inside the solar ovens. In fact, the training did not result in self-sustaining community action after the project ended [21]. In parts of

Latin America preparing tortillas accounts for more than 50% of cooking fuel usage [22], but solar cooker designs are not hot enough for tortilla preparation.

7.2.7 Other Solar Thermal Applications

The most common applications of solar thermal technologies are those related to water heating and air conditioning, but there are other useful applications that can be valuable for Africa and Latin America. Indeed, a number of industrial sectors use a significant proportion of their process heat at temperatures <120 °C and, therefore, can consider solar thermal energy for their needs. Examples of these sectors are transport equipment, machinery, mining and quarrying, food and tobacco, and textiles and leather. Characteristic processes are washing, leaching, cooking, drying, preheating of boiler feed water and space heating in industrial buildings [23].

On the other hand, water treatment and seawater desalination can come from distillation or reverse osmosis technology, with the first preferred for more saline water. Distillation heat requirements can be totally or partially provided from solar thermal technologies [23].

Case Example 7.3

Sun dryers in Togo:

Many food products are dried for food preservation to avoid product losses and promote trading. Dryers can be classified in terms of type (industrial and small scale), energy, solar energy collection (direct, indirect or mixed), physical state of products (solid, paste, liquid), operation mode (batch or continuous), drying time (ultra-short < 1 min, 1 min < short < 60 min, long > 60 min) product movement, number of stages, drying temperature, operating pressure, drying medium, heat input type, and ventilation. A thermo-economic analysis was carried out to determine which dryers are most successfully adopted. The survey was mainly carried out for agricultural products dryers in Togo (37), but extended to Benin (6) and Burkina Faso (6). Although there are many types of dryer technologies in the scientific literature, the variety of technologies really used in the field was limited and based on local construction materials. Sun drying was the most used, representing 70% of the total dryers surveyed, using surfaces as asphalt roads, cemented areas or tables, sometimes covered to reduce insect and bird contamination. The gas dryers found (gas is subsidized to fight against desertification) are the Atesta model built with wood on a cement base, offering a resulting quality of dried fruit suitable to export (solar energy variability and irregularity do not enable the standard quality required). Traditional sun dryers are widely used from 1 to 4000 kg per cycle, costing EUR 0–7.63 per m^2 drying area and operating cost under EUR 1.53/year; low power solar dryers from 1 to 1000 kg per cycle, costing EUR 9.11–238.55 per m^2 of tray and EUR 0.76–15.3/year were used in the Sahel and Sudan-Sahel tropical climate; and gas dryers are used in all climatic zones at around 100 kg per cycle, with a high drying power, costing EUR 107.52–181.75 per m^2 of tray and EUR 486–1867 per year. The analysis showed that for the most widely used dryers, the drying cost was EUR 0.01/kg for direct exposure (sun drying) and EUR 0.07/kg for the Atesta gas dryer, amounting to 7–8% of the achieved profit, respectively. Considerations about the product to be dried, product added value, local energy availability and cost, labour skills and availability, local material availability, and user needs and investment capacity are also needed to choose a dryer properly [24].

7.2.8 Thermal Insulation

An increase in energy demand for cooling and heating in buildings is being witnessed also in many African and Latin American countries. In many of these countries it is very unusual to be under comfort conditions (22–26 °C, 40–70% relative humidity). Solutions can be proposed ex-ante by an adequate architecture and orientation of the building, where the rule is to avoid direct incidence of the sun in hot seasons and promote it in cold seasons. It can be partially obtained by site location and orientation, and the simple orientation of the building to the south if it is placed in the northern hemisphere and to the north in the southern hemisphere is the general recommendation. Additional avoidance of the direct sun incidente can be obtained by fixed or mobile sunblinds, natural ventilation and proper window-to-wall ratios.

On the other hand, the thermal insulation of the building envelope is the most significative solution for buildings just in place. Architectural and physical properties of buildings, such as thermal mass, structural materials, and shape, are the most important parameters that define the thermal load. An optimum thickness and properties of insulation offers minimum total cost [25]. Moreover, some studies show that modern habitats in many African and Latin American countries are uncomfortable compared to traditional buildings [26]. The location of the insulator in the envelope of the building is also important to avoid other materials to store undesirable heat due to external weather conditions.

The insulation properties of materials are defined mostly by their thermal conductivity, which is dependent on the density, porosity, and the moisture content. Also, spectral reflectance and surface emissivity should be considered. Thermal conductivity varies with operating temperature. The highest thermal insulation can be obtained with vacuum, but it is difficult to manage in buildings. Silica aerogel, polyurethane and polystyrene (in different configurations) are very popular and cost-effective materials for building thermal insulation. Also double glass windows with air between the two glasses are very useful for diminishing thermal conductivity.

Another option is what constitutes a cool roof: a surface that reflects sunlight and emits heat more efficiently than other roofs. Research studies shows that the roof surface temperature can exceed the temperature of other surrounding surfaces covered with vegetation by 20 °C [27]. Different alternatives are: (i) evaporative cooling through roof ponds (e.g., roof with moist soil shaded by 10 cm of pebbles); (ii) high albedo roof surfaces (light and bright colours) preferred to vegetal roofs; and (iii) rounded or curved roofs with the optimal orientation of the vault facing north-south [28].

7.2.9 Costs

Solar thermal costs vary quite markedly depending on weather conditions, more or less complex installations, labour costs, capacity and how mature is the market, varying between under USD 200/m^2 and above USD 1000/m^2 [29]. The collector has an investment weight of 30–50% of the total cost, and the accumulation system between 20 and 25%. The cost of the regulation system is usually between 4 and 7% of total investment. O&M costs are estimated in USD 28–50/kW year [14]. Also the cost is sensitive to variation in prices of some raw materials (mainly copper, aluminum and the steel) showing substantial volatility in the last decade. For example, considering the average solar irradiance of Spain, the stand-alone solar thermal heating systems cost usually is more expensive than heat produced from natural gas, but competitive with retail electricity prices [30]. Consequently, this technology is very competitive in Africa and Latin America against electricity, and also in many areas against natural gas. For district heating the cost also depends on the system capacity and collector technology, but is above of stand-alone systems because additional investment in the distribution network. Costs are expected to decrease substantially in future and also experience-curves for solar thermal heating systems indicate that costs are reduced by 20% when the cumulative capacity doubles in a country [14].

Case Example 7.4

PROSOL program to promote solar water heating in Tunisia:

In 1994, Tunisia recorded, for the first time, a deficit in its energy balance, partially solved doubling the Algeria-Italian gas pipeline in 1995. But from 1999, the energy balance became again in deficit and rising due to the steady rise in energy demand coupled with a stable indigenous supply. Tunisia has very advanced regulatory instruments to promote renewable energies, starting in the early 1980s. One of such instruments is the program PROSOL, established in 2005 to promote solar water heating (SWH) systems. Previous attempts failed because the mandate to buy the SWHs to a unique local public manufacturer called SERPET (and 200 installers), supplying equipment with many technical problems that produced a rapid decline in the market activity; and the exhaustion of funds designed to finance the subsidies. The main goal of the PROSOL program was the installation of 222,000 m^2 of solar collectors in the period 2005–2011 and a more stable market. The main features of this program are: (i) the SWH must be acquired from an eligible supplier; (ii) the supplier submits a loan application to a participating Tunisian bank; (iii) once the loan is approved, the installation takes place; and (iii) the customer obtain a subsidy of 20% of the cost (ceiling of 200 dinars/m^2) and repays the loan over a 5-year term, through the electricity bill. The subsidy has a participation from the Italian Government (80 dinars of a 300 l SWH cost). Due to PROSOL, the installed surface of SWHs jumped from 7000 m^2 in 2004 to 80,000 m^2 in 2010, and the local industry grew to offer 45 eligible suppliers containing 8 local manufacturers, 1100 authorized installers and 206 SWH system models, and exports (mainly to North African countries) increased from 25,334 dinars in 2006 to 495,917 dinars in 2010. The authors also consider that the program can be improved if the government designs sub-strategies for each Tunisian region for considering some specificities, the promotion of technicians capable to maintain the systems, and the lack of clear procedures for suspension of subsidies in case of non-compliance from suppliers [31].

The generation of solar cooling requires a higher investment, being able to reduce costs if the facility is designed to provide cooling in summer and heating in winter, in a way that improves the balance between supply and demand. The front costs for solar air conditioning (SAC) are between 2 and 5 times higher than a conventional state-of-the-art system depending on local conditions, building requirements, system size and selected technical solution. In recent studies, investment costs of cooling systems range from USD 2000 to 5000 per kW_{cold} and even higher in some particular cases [6]. This large range is due to different sizes of systems, different technologies, application to different sectors, and other boundary conditions. Collectors represent 20–30% of the total investment, and the cooling system a 20–25% [17]. A recent trend is to develop of solar cooling kits mainly for small capacities (<35 kW), and prices excluding installation dropped below USD 5000/kW in 2013 [6]. Under promising conditions payback times of ten years or less can be obtained. Of course, these costs cannot be extrapolated to rural areas in Africa and Latin America because of the lack of maintenance and the ambient conditions expected. Tentative costs for small-size solar refrigeration technologies adapted to Africa have been exposed elsewhere.

Case Example 7.5

Solar thermal refrigeration in Kenya:

The ISAAC solar power icemaker is a successful example of technology proven in rural Africa [32]. It produces up to 50 kg of ice per sunny day that is capable of chilling up to 100 l of milk. Three solar icemakers were installed (Oct 2008) in each of two separate rural communities in the coast of Kenya and two profitable dairy co-operatives were established [33]. The ISAAC used in this project have a 2.4 × 4.8 solar collector. Milk is chilled by making an ice bath and immersing milk in milk cans into the ice bath. The icemakers were shipped to Kenya and installed by a US company with the assistance of the technicians from the co-operatives and other men from the villages. Installation is a very straightforward process of preparing the foundation, putting components in position, piping components together, installing the collector, and charging with refrigerant. Installation is an important part of training the operators. After the equipment was installed, the operators were trained about how to operate the icemakers. After a few days, the assessment of the operators was that the equipment is "easy to operate". Rainy and overcast days were a problem for the dairy because the refrigeration depends upon solar energy. The co-operatives started by restricting milk collection when there is no ice, which is the day after a rainy day. Another strategy was to store extra ice from sunny periods in an insulated box for rainy days. It was also possible to buy ice from where it was available. Another option was to have a back-up ice making system. The co-operatives also needed to develop their marketing plan. It induced businesses for milk production, milk collection, packaging and sale for cooperatives, and the production of yogurt and mala that are sold at a higher price, generating additional profits. It is estimated that, when produced in-country where wages are low and transportation costs can be minimized, the 11 square meter ISAAC can be produced for less than USD 7000 [34]. Co-op net monthly income varied between USD 3290 and 1263 compared to USD 180 without ice production. The project generated rural income, rural jobs, alleviates poverty, and contributes to food and energy security. The project has enormous potential for replication.

Systems using electrically driven vapor compression cycles combined with photovoltaic systems may be an alternative to use solar energy for air-conditioning or refrigeration. Photovoltaic modules have experienced a continuous cost reduction due to a huge growing market during the last two decades. In particular, as the PV cost has dramatically decreased in recent years, most probably the economic outcome will be in favor of a photovoltaic-based cooling system.

On the other hand, investment costs for CPS plants have remained high, from USD 4000 to 9000/kW, depending on the solar resource and the capacity factor, which also depends on the size of the storage system and the size of the solar field, as reflected by the solar multiple. However, costs are expected to decrease as CPS deployment progresses, following a learning rate of 10% [35]. The cost of electricity produced by CSP plants depends on the direct component of solar irradiation on the geographical location where it is built and the technology selected, and on many other factors that can be included in a deeper CSP cost analysis [36]. However, recents PPA (Noor 1 CPS plant at Ouarzazate in Morocco and another one in USA) secured at USD 190/MWh [35]. Moreover, LCOE of CSP plants can be further reduced as in some cases the technical lifetime of a CSP plant can be significantly greater than 30 years. For example the owner of the oldest SEGS plants in California, which are about 30 years old, is considering significant refurbishment, including adding thermal storage, to expand their lifetime by 20 years and negotiate a new PPA with the company that buy the electricity (Southern California Edison).

In regional terms, the levelized cost of electricity (LCOE) has been recently established by the International Renewable Energy Agency (IRENA) in a range of USD 0.170–0.310 kWh in Africa and South America (Fig. 7.12) [10]. Thus, in average and compared to other regions, production of electricity from CSP is quite favourable in South America and Africa.

There is also adequate to consider the investment costs of CSP technologies. Thus, the investment costs for large parabolic trough plants typically range from USD 4.3 to 8.5/W, depending on different parameters exposed in Table 7.5 [12]. Plants without storage systems and located in highly sun irradiated areas are on the lower range of investment costs. Investment costs for parabolic trough are expected to decrease by 20% when CSP plants are scaled up to 200 MW.

Tower investment costs are more difficult to estimate, but are generally higher than for trough systems. However, increasing efficiency from 15 to 25%, mass production of flat mirrors and technology maturity could bring the investment costs down by 40–75% [37]. Dish investment costs are about twice those of parabolic troughs [37].

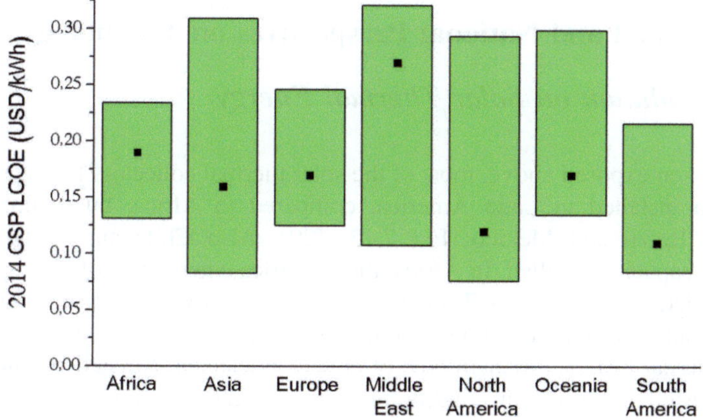

Fig. 7.12 Levelized cost of electricity from CSP in the different global regions (2014) [10]

Table 7.5 Investment costs of a 50 MW trough plant associated to a 7h thermal storage system

Concept	(%)	Concept	(%)
Solar field	30	Project finance	6
Allowances	14	Heat transfer fluid	5
Storage	9	Power block	5
Project management	8	Project development	3
Balance of plant	8	Grid access	3
Civil works	7	Miscellaneous	2

Table 7.6 Solar thermal storage costs for different technologies [37]

TES technology	USD/kWh$_{th}$
Hot water tank	0.10–0.13
Chilled water tank	0.10–0.13
PCM-general	13–65
Ice storage tank	6–20
Thermal-chemical	10–52

Finally, the storage cost for the solar energy storage systems exposed in Table 7.5, are shown in Table 7.6, considering a cost reduction of 50–75% for PCM centralized storage systems [17].

For thermal storage costs, some studies [38] indicate that they introduce an extra cost for electricity production ranging at 2.6–7.5%, depending on the technology used. Other studies [39] indicate that the technology using PCM is 2.5% cheaper compared to the technology based in two tanks storage, while the technology using storage in concrete passive systems is 11.9% more expensive. These cost values vary not only in terms of construction costs but also on how the plant is operated. Then, operational procedures as defocusing of the mirrors, adjusting the storage procedure based on the weather conditions at the beginning of the day, or the optimal time to discharge the storage systems are important issues to be considered to reduce storage costs.

7.3 Regional and National Perspectives on Technology

7.3.1 Evolution on Solar Thermal Energy

As it has been exposed above, most of the solar thermal collectors for heating water have been installed in Latin America, compared to Africa, but mainly in two countries: Brazil and Mexico (Fig. 7.13) [29]. Also Chile and Uruguay show additional capacity installed. In Africa, the countries where most of the capacity has been installed is South Africa, Tunisia and Morocco, but also Mauritius, Zimbabwe, Namibia and Mozambique shows additional capacity installed. In terms of technology, the flat plate is the main type of collector selected. No statistics about solar cooling systems or solar air conditioning disaggregated in terms of countries has been found in the literature.

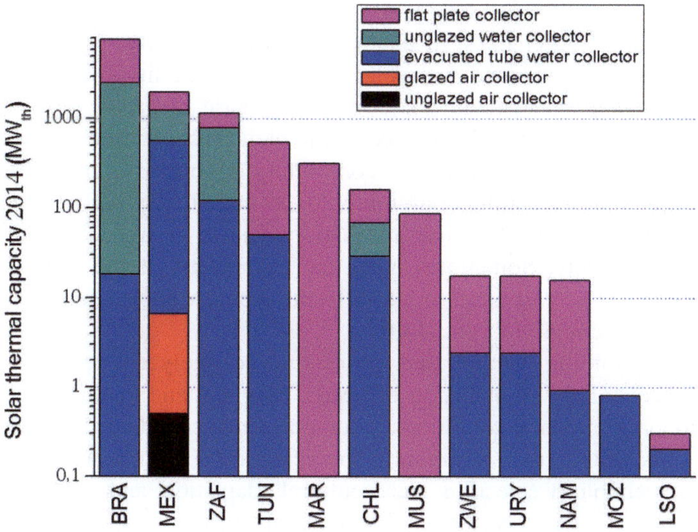

Fig. 7.13 Solar thermal capacity in different African and Latin American countries in 2014 [29]

Case Example 7.6

Solar water heater technology transferred to rural communities in Argentina:

Many times renewable devices are abandoned by rural communities because of a non-adequate process of technology transfer between suppliers and these local communities. An analysis of case studies in the Andean communities of Cabreira, San Juan and Las Capillas (Province of Salta, Argentina) where solar water heaters were transferred, identified main factors to be considered. The demand for water heaters was a request originated from one community after obtaining funding for the building and improvement of bathrooms. Another community demanded heating water also for a communal laundry. The solar water heaters emerge as a viable option because the availability of solar resources and the lack of other energy sources (natural gas, electricity and firewood shortages). In order to get the communities to learn and get involved with the technology, the following activities were conducted: audio-visual workshops, and direct observation and testing of three types of solar water heaters, for some families. For the selection of technology the residents valued, after the test, the water storage capacity, light equipment, materials resistant to wind and solar radiation, durability over time, price, mobility, maintenance within the community, ease of installation, speed of heating the water, affordable and accessible parts, easiness of being placed on the roof, and aesthetics. Also, the community appointed a group of young people with technical skills in plumbing (who had already participated in the construction of toilets) for their installations and to be trained in the use of solar water heaters. Sometimes, after the installation of the water heaters, technicians had no resources to return to the area and most equipments were broken after a couple of years [40].

The determining factors that condition the analyzed processes were: (i) solar technology (satisfaction of needs, easy use and maintenance, adaptation to environmental conditions, good performance, adequate information, and replicability); (ii) community participation (identification of actual demand, technology selection, socio-cultural adaptation, and training of local technicians); (iii) involvement of technicians (accompanying the process, fulfill the commitment, opening-up, and ability to communicate); (iv) inter-institutional coordination (socio-technical association, defined roles, good communication, previous contact, leadership and creative responsiveness to unforeseen events); (v) logistics and monitoring of the project (flexible planning, respect for the dynamics of social processes, continuous technical advising, post-installation monitoring, evaluation of the process and results, and local technical autonomy); and (vi) structure of the community (good organization, positive experience in previous projects, local empowerment, and community values).The analysis of interactions showed a strong interdependence among factors and particularities depending on the case analysed, so it requires the adaptation of technologies and methodological approaches from the point of view of researchers and technicians. Among the most influential factors, the participation of the community represents the largest contributor to others. At the same time this factor has a high dependence on other factors such as inter-institutional coordination, logistics and monitoring of the project [40].

Solar sorption refrigeration have been reported in Africa (Burkina Faso, Egypt, Kenya, Morocco, Nigeria and South Africa) and Latin America (Argentina, Brazil, Cuba and Mexico) since the late 1970s, but mostly devoted to R&D activities [15, 41]. They are mainly devoted to small-size cold boxes for food and vaccines preservation, and no direct building air conditioning based on absorption or adsorption has been detected.

7.3.2 Evolution on CSP

For CSP technology the capacity evolution in Africa is slow and concentrated in four countries: Algeria, Egypt, Morocco and South Africa (Table 7.7) [10]. Most of the projects are in the pipeline but not in operation. The technology in operation is parabolic trough, but there are some projects considering linear Fresnel and tower technologies. Also, an increase in power capacity per plant is expected to rise substantially with the new projects and most of them plan to use storage systems (mainly molten salts) and dry cooling.

Table 7.7 CSP plants expected and in operation in Africa [10]

Country	Name	Tech	Year[a]	Cost	Capacity (gross)	Transfer/storage (hours)	Cooling
DZA	Hassi R'mel	P	2011	315 M€	25MW$_e$	None	Dry
EGY	Kuraymat	P	2013	–	20MW$_e$	None	Wet
MAR	Ait Baha	P	2014	–	3MW$_e$	Air (1 atm)/stones (5)	–
MAR	eCare	L	2014E	–	1MW$_e$	Water/steam	Dry
MAR	IRESEN	L	2017E				
MAR	Ain Beni Mathar	P	2010	–	20MW$_e$	None	Wet
MAR	Noor I	P	2015	1042 M€	160MW$_e$	Dowtherm A/molten salt (3)	Dry
MAR	Noor II	P	2017E		200MW$_e$	molten salt (7)	–
MAR	Noor III	T	2017E		150MW$_e$	molten salt (8)	
ZAF	Bokpoort	P	2016	565 M$	54.5MW$_e$	Dowtherm A/molten salt (9)	Dry
ZAF	Ilanga I	P	2017E		100MW$_e$	Thermal oil/molten salt (4.5)	
ZAF	Kaxu Solar Park	P	2018E		100MW$_e$	Thermal oil/ molten salt (4.5)	
ZAF	Kaxu Solar one	P	2015	–	100MW$_e$	Thermal oil/molten salt (3)	Dry
ZAF	Khi Solar One	T	2016	–	50MW$_e$	Steam/steam (2)	Dry
ZAF	Redstone	T	2018E		100MW$_e$	Molten salt/ molten salt (12)	Dry
ZAF	Xina Solar One	P	2016E	–	100MW$_e$	Oil/molten salt (5)	–

E Expected; *L* Linear Fresnel; *P* Parabolic trough; *T* Tower

[a]Year starting (or expected) into operation. As some expected starting dates are passed, it can be derived that some delays are being produced

Case Example 7.7

CSP in Tunisia and interconnection with Europe:

Tunisia is separated from Europe by a distance of 140 km at the Sicily Channel, offering an ideal position to transfer renewable power directly to the European markets. The majority of the Tunisian regions is connected to the grid, and the transport network covers the majority of Tunisian territory. The country has also a huge potential constituted by fossil aquifers to the south, but also dry cooling technologies can be used. A 50 MW CSP plant based on parabolic trough in the Tataouine region at the south of Tunisia is simulated and compared with a reference CSP plant Andasol in Spain. About 40% of the country territory (at the center and south) are desert, and with slope less than 2%. The energy production simulated in Tunisia is quite higher than the reference in Spain. However, the plant becomes economically competitive in Tunisia when the majority of the plant components (e.g., collectors structure, mirrors and storage system) can be manufactured locally. Also, plans for a 200-km long 1000-MW HVDC submarine cable linking El Hawaria in Tunisia and Partanna in Sicily added to the existing Transmed gas pipeline linking Algeria, Tunisia and Italy should be also considered [42].

In Latin America no CSP plant is in operation, and the projects are only being executed in two countries: Chile and Mexico (Table 7.8) [10]. Moreover, most of the CPS capacity is expected only in Chile and based on the two main technologies: parabolic trough and tower systems.

Table 7.8 CSP plants expected and in operation in Latin America (December 2016) [10]

Country	Name	Tech	Year[a]	Cost	Capacity (gross)	Transfer/storage (hours)	Cooling
CHL	Pedro de Valdivia	P	2015E	2610 M$	360 MW_e	Thermal oil/molten salt (10.5)	Dry
CHL	Acatama I	T	2018E	–	110 MW_e	Molten salt/molten salt (17.5)	–
MEX	Agua Prieta II	P	2014E	–	14 MW_e	None	–

E Expected; *L* Linear Fresnel; *P* Parabolic trough; *T* Tower
[a]Year starting (or expected) into operation. As some expected starting dates are passed, it can be derived that some delays are being produced

Case Example 7.8

Integration of CSP in the Brazilian electric power system:

The study evaluates four types of CSP plants with parabolic troughs (simple, hybridized and with storage 6 h-indirect and 12 h-direct) at two sites: Bom Jesus de Lapa and Campo Grande, and expanded to other Brazilian sites. The CSP simulations are made based in the System Advisor Model (SAM) and MESSAGE. Findings indicate that the least-cost expansion of the Brazilian electricity system should be based on hydroelectric and thermoelectric plants fueled by natural gas and sugarcane bagasse, and CSP would not be chosen. However, if CSP is forced to into the model because of an hypothetical mandatory use in the Brazilian grid, solar energy would replace more hydro resources than fossil resources. Also, the CSP plants would lead to energy independence for the Northeastern region of Brazil, which may even become an exporter of electricity to the Southern and Southeastern regions. This exportation means additional revenues for poor semi-arid regions where Brazil's best DNI is found [43].

Figure 7.14 shows the evolution of the electricity output from CSP in the African countries where these plants are in operation. Currently the evolution is almost flat but it is expected that South Africa and Chile could be added to this graph very soon.

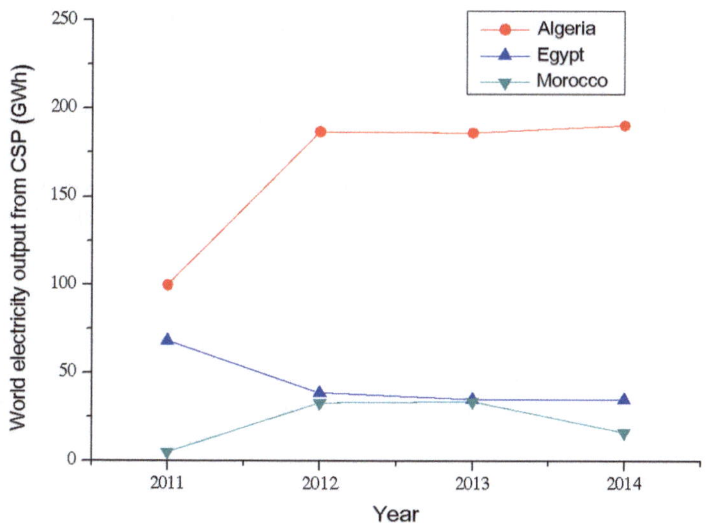

Fig. 7.14 Evolution of the installed CSP power capacity in the different African and Latin American countries

Also, it is important to mention the use of concentrated solar thermal systems integrated into coal-fired power plants, as in Egypt and Algeria. The solar thermal systems inject the heat into the conventional power-generation process to reduce the amount of fossil fuel needed to be burned in power plants up to a 50%. Authors show that this strategy is economically preferable to installing carbon-capture and sequestration equipment for mitigating an equivalent amount of GHG emissions [44].

7.4 Conclusions and Future Perspectives

Solar thermal energy is growing moderately around the world because photovoltaics is acting as a very competitive alternative in terms of costs and O&M. There is a significant penetration of solar thermal systems for heating water in Brazil and Mexico with limited capacity in Chile and Uruguay. In Africa, South Africa, Tunisia and Morocco have developed solar hot water capacity with minor development in Mauritius, Zimbabwe, Namibia and Mozambique. The challenge is to be able to produce the technologies locally with qualified manpower, particularly in vacuum technology and hot water plumbing sectors. No solar cooling systems for direct building air conditioning has been reported in Africa and limited opportunities are reported in Latin America with no detailed information on capacity and/or location. Solar sorption refrigeration has been reported in Africa (Burkina Faso, Egypt, Kenya, Morocco, Nigeria and South Africa) and Latin America (Argentina, Brazil, Cuba and Mexico) since the late 1970s, but mostly devoted to research and development activities and not wide-scale application. Solar cookers hold significant promise in both regions but to date have limited wide scale adoption in either region because of the dependence on weather conditions, cultural habits and limitations on the highest temperatures reached by many of these devices, which are not compatible with basic food recipes in many areas.

Concentrating solar power (CSP) has tremendous promise globally but continues to grow at a measured pace on a global scale with total installed capacity exceeding 375 GW. Africa and Latin America have very limited CSP, with 2 and 8.6 GWh, respectively. Current global capacity emphasizes parabolic troughs with lower capacity associated with tower systems, linear Fresnel reflectors, and parabolic dishes. The main differentiating technology in CSP is the solar collector that heats a fluid used to produce steam that drives a turbine for producing electricity. CSP plants require water for cooling and for condensation processes which can be problematic in desert areas where the sun direct radiation is high enough to produce high power outputs.

References

1. Mauthner F, Weiss W, Spörk-Dür M (2014) Solar heat worldwide. Markets and contribution to the energy supply 2014. Solar heating & cooling programme. International Energy Agency. 2016 Edition
2. IEA (2008) Energy technology perspectives
3. Hoogwijk M, Graus W (2008) Global potential of renewable energy sources: a literature assessment. Background report. Ecofys. http://www.ecofys.com/files/files/report_global_potential_of_renewable_energy_sources_a_literature_assessment.pdf. Accessed 13 Feb 2016
4. SCS-IEA (2011) New method to calculate solar thermal output. Solar Update 54
5. Sparber W, Napolitano A, Melograno P (2007) Overview on worldwide installed solar cooling systems. 2nd international conference solar air conditioning, Tarragona—Spain, October 2007
6. IEA (2015) Cooling position paper. Task 48 quality assurance and support measures for solar cooling. Solar heating and cooling programme. http://www.iea-shc.org/data/sites/1/publications/IEA-SHC_Solar-Cooling-Position-Paper_Sep2015.pdf. Accessed 12 Feb 2016
7. N'Tsoukpoe KE, Yamegueu D, Bassole J (2014) Solar sorption refrigeration in Africa. Renew Sustain Energy Rev 35:318–335
8. Lecuona A (2014) Cocinas solares, cómo reducir la pobreza y el consumo energético. Universidad de La Laguna
9. International Energy Agency database (2015) http://data.iea.org/. Accessed 30 Nov 2015
10. IRENA (2015) Resource. http://resourceirena.irena.org/gateway/dashboard/. Accessed 30 Nov 2015
11. NREL data (2015) http://www.nrel.gov/csp/solarpaces/by_technology.cfm. Accessed 30 Nov 2015
12. OECD/IEA (2010) Technology roadmap. Concentrating solar power
13. Henning H-M (2011) Solar air conditioning and refrigeration. 2010 Annual Report. SHC-IEA 2011
14. Guerrero-Lemus R, Martínez-Duart JM (2013) Renewable energies and CO_2. Cost analysis, environmental impacts and technological trends. Lecture Notes in Energy 3. Springer. ISBN 978-1-4471-4385-7
15. N'Tsoukpoe KE, Yamegueu D, Bassole J (2014) Solar sorption refrigeration in Africa. Renew Sustain Energy Rev 35:318–335
16. Bouzeffour F, Khelidj B, Tahar abbes M (2016) Experimental investigation of a solar adsorption refrigeration system working with silicagel/water pair: a case study for Bou-Ismail solar data. Sol Energy 131:165–175
17. IEA (2012) Energy-efficient buildings: heating and cooling equipment. Technology roadmap
18. Lecuona-Newmann A (2016) Cocinas Solares. Fundamentos y aplicaciones. Herramienta de lucha contra la pobreza energética. Editorial Marcombo, Barcelona. ISBN 978-84-267-2372-7
19. Introduction to solar cooking. Solar cookers international network. http://solarcooking.wikia.com/wiki/Introduction_to_solar_cooking. Accessed 30 Nov 2015
20. Brown LS, Lankford WF (2015) Clean cooking empowers women. Nature 521:284–285
21. Bauer G, Vukelich S (2016) Solar ovens beaten by rain and tortillas. Nature 524:161
22. Masera O, Edwards R, Arnez CA, Berrueta V, Johnson M, Bracho LR, Riojas-Rodríguez H, Smith KR (2007) Impact of Patsari improved cookstoves on indoor air quality in Michoacán, Mexico. Energy Sustain Dev 11:45–56
23. IEA (2012) Technology roadmap. Solar heating and cooling
24. Boroze T, Desmorieux H, Méot J-M, Marouzé C, Azouma Y, Napo K (2014) Inventory and comparative characteristics of dryers used in the sub-Saharan zone: criteria influencing dryer choice. Renew Sustain Energy Rev 40:1240–1259
25. Naouel (2011) A study on optimum insulation thickness in walls and energy savings in Tunisian buildings based on analytical calculation of cooling and heating transmission loads. Appl Energy 88:156–164

26. Nematchoua MK, Tchinda R, José AO (2014) Thermal comfort and energy consumption in modern versus traditional buildings in Cameroon: a questionnaire-based statistical study. Appl Energy 114:687–699
27. Uemoto KL, Sato MN, John V (2010) Estimating thermal performance of cool colored paints. Energy Build 42:7–22
28. Dabaieh M, Wanas O, Hegazy MA, Johansson E (2015) Reducing cooling demands in a hot dry climate: a simulation study for non-insulated passive cool roof thermal performance in residential buildings. Energy Build 89:142–152
29. IEA (2015) 2014 Highlights. SHC task 52. Solar heat and energy economy in urban environments. Solar heating and cooling programme. http://task52.iea-shc.org/data/sites/1/publications/IEA_SHC-Task52-Highlights-2014.pdf. Accessed 12 Feb 2016
30. BCG (2011) Evolución Tecnológica y Prospectiva de Costes de las Energías Renovables. Estudio Técnico PER 2011–2020. Boston Consulting Group
31. Omri E, Chtourou N, Bazin D (2015) Solar thermal energy for sustainable development in Tunisia: the case of the PROSOL project. Renew Sustain Energy Rev 41:1312–1323
32. IRENA (2010) Africa 2030: Roadmap for a renewable energy future
33. World Bank data (2015) http://siteresources.worldbank.org/DEVMARKETPLACE/Resources/205097-1128108124421/Rural_Milk_Preservation_Paper.pdf. Accessed 30 Nov 2015
34. Energy Concepts LLC (2015) http://www.energy-concepts.com/_pages/app_isaac_solar_ice_maker.htm. Accessed 30 Nov 2015
35. OECD/IEA (2014) Technology roadmap. Concentrating solar power
36. Hernández-Moro J, Martínez-Duart JM (2012) CSP electricity cost evolution and grid parities based on the IEA roadmaps. Energy Policy 41:184–192
37. Kearney AT (2010) Solar Thermal electricity 2025. STELA
38. NEF (2009) New energy finance. Guardian cleantech summit. 23 Nov 2009
39. Muren RB, Chapman DJ, Arias DA (2010) Yearly performance modeling of concrete and phase change material thermal energy storage systems for concentrated solar power plants. Proceedings of ES2010 ASME Energy Sustainability 2010 May 17–22, Phoenix, Arizona, USA
40. Escalante KN, Belmonte S, Gea MD (2013) Determining factors in process of socio-technical adequacy of renewable energy in Andean Communities of Salta, Argentina. Renew Sustain Energy Rev 22:275–288
41. Best R, Pilatowsky I (1998) Solar assisted cooling with sorption systems: status of the research in Mexico and Latin America. Int J Refrig 21:100–115
42. Balghouthi M, Trabelsi SE, Amara MB, Ali ABH (2016) Potential of concentrating solar power (CSP) technology in Tunisia and the possibility of interconnection with Europe. Renew Sustain Energy Rev 56:1227–1248
43. Malagueta D, Szklo A, Soria R, Dutra R, Schaeffer R, Borba BSMC (2014) Potential and impacts of Concentrated Solar Power (CSP) integration in the Brazilian electric power system. Renewable Energy 68:223–235
44. Dalvi VH, Panse SV, Joshi JB (2015) Solar thermal technologies as a bridge from fossil fuels to renewables. Nat Clim Change 5:1007–1013

Chapter 8
Hydropower and Marine Energy

Ricardo Guerrero-Lemus and Les E. Shephard

Abstract In this chapter it is shown how hydropower plays an extraordinarily important role in the energy future of both Africa and Latin America. Thus, more than 20% of total global hydropower generation occurs in Africa and Latin America combined, and more than 90% of existing renewables in both regions is produced with hydropower. Also, the hydropower capacity in Africa and Latin America is expected to increase over time as several nations are expected to pursue new hydropower development or expansion of existing capacity. Moreover, it is estimated that 92% of the existing total hydropower capacity in Africa is untapped. Conversely, hydropower production in some countries exceeds 85%. Consequently, in many countries energy-intensive industry users can be located as they can consume large quantities of low-cost power produced by hydropower. On the other hand, small hydropower capabilities are likely to become more pervasive over time as environmental concerns, drought and climate change may limit efficiency of larger facilities. In particular, mini-hydro could provide many of the rural electricity. Moreover, climate change may drastically impact the hydropower and consequently, the system's capability to supply load. Indeed, two lessons can be extracted from recent Brazilian and Venezuelan hydro crisis: the need for diversification from hydropower and the importance of distributed generation. Finally, marine energy plays a very marginal role in renewable electricity production as it has virtually no role in Africa or Latin America. As technology evolves and costs decrease marine energy may play a role in select future locations however it is largely believed that more traditional renewable resources will be adopted much earlier in the energy development cycle.

Keywords Energy · Africa · Latin America · Electricity · Hydropower · Wave power · Currents · Tidal range · Salinity gradient · Thermal gradient · Micro-hydropower · Hydrokinetics · Biodiversity · Reservoir

R. Guerrero-Lemus and L.E. Shephard, *Low-Carbon Energy in Africa and Latin America*, Lecture Notes in Energy 38,
DOI 10.1007/978-3-319-52311-8_8

8.1 Introduction

Hydropower plays a central role in renewable energy production since 74.1% of world electricity generation from renewable resources is produced in hydropower plants (2014, last data available) [1]. In Africa this percentage reaches a 90.8% and a 87.8% in Latin America. This proportion is slightly diminishing over the years as other renewable energy sources such as wind, solar, etc., are gaining ground in the electricity mix. If we consider the hydropower generation deconvoluted into regional values, this technology has been steadily growing in Africa and Latin America, with Latin America having much larger capacity (Fig. 8.1). In 2015 there was an estimated 130 TWh hydropower generation in Africa, 766 TWh in Latin America and 4225 TWh worldwide. The estimated global technical potential for hydropower is about 463 EJ/year. Most of this potential is located in Africa, Asia and Latin America, while in developed countries almost all of the economic potential has been exploited. Thus, the most untapped global region is Africa, where 92% of the total potential has not yet been developed [2]. There are many developing countries where the electricity produced by this technology can exceed 85% of the country's demand as illustrated below. Hydro generation decreases drastically during the dry season and during dry years in many African and some Latin American countries.

Hydropower is derived from the available potential energy of water falling from a certain height, causing a rotational movement of the turbines or power generators

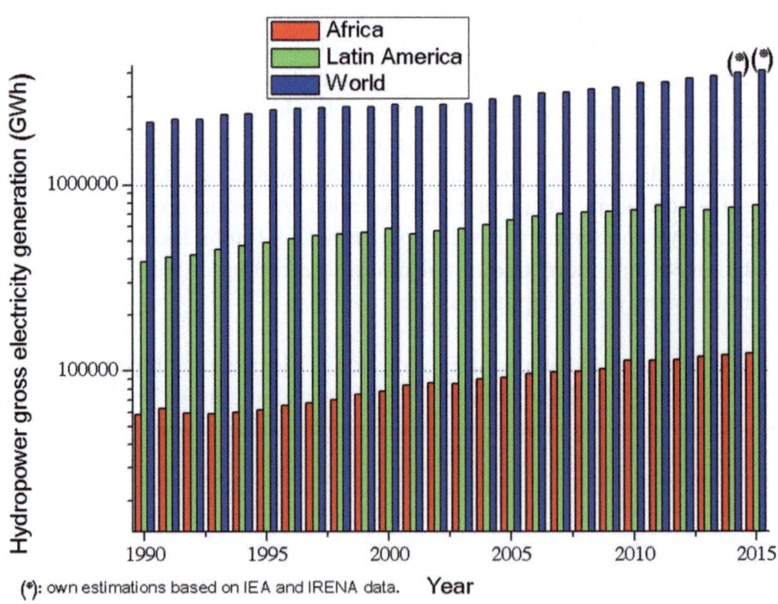

Fig. 8.1 Evolution of the percentage of hydropower generation (gross electricity) worldwide in Africa and Latin America in the period 1990–2015 [1]

that is transformed into electricity. This available power can be estimated by the product of the height between the inlet and outlet reservoirs, the efficiency of the turbine, the flow volume per second, the density of the fluid and the gravity constant.

Hydropower has a number of very significant advantages relative to other energy sources:

1. It is a well established technology today, and it has been producing electricity at competitive prices for more than a century.
2. Hydropower produces electricity with a very flexible and quick response time (in the order of seconds), so it can cover peak-power demands and production gaps and is especially useful when unmanageable energy sources (mainly wind and solar) have a large capacity share in the grid.
3. Water storage in reservoirs by pumping permits an adequate management of renewable electricity excesses when they are produced from sources. This water stored can be reconverted to electricity when it is needed.
4. The GHG emissions per kWh and energy payback show excellent performance.

Hydropower also has a number of drawbacks, mainly from the environmental point of view. These include flooding of high biodiversity areas, obstacles to fish passage in both directions of the rivers, decreasing river flow to unsustainable levels, and often, displacement of human populations. Also, large hydropower plants need long lead times for their development and construction (reservoir, tunneling and canal, powerhouse, site access and grid connections). Obviously, these drawbacks are largely diminished in the case of mini- and micro-hydropower plants.

Hydropower can be classified according to various criteria. As a first approach, the factor usually considered relates to the power of the plants (Table 8.1) [3]: (i) large hydropower systems, $P > 10$ MW; (ii) small hydropower systems, also called mini-, $1 \text{ MW} < P < 10$ MW; and (iii) micro-hydro systems, $P < 1$ MW.

A second criterion usually refers to the height, h of the head: (i) high head hydropower systems, $h > 100$ m; (ii) intermediate head hydropower systems, $10 \text{ m} < h < 100$ m; and (iii) low head hydropower systems, $h < 10$ m. Finally, hydropower systems are also classified between conventional (referring to hydropower dam and reservoir) and run-of-river systems (without reservoirs or with very small reservoirs). Most of the run-of-river systems are also mini-and micro-hydropower systems.

Water storage in reservoirs by pumping is increasingly being used to match offer and demand in a power grid, storing energy when offer is in excess and powering

Table 8.1 Hydropower capacity worldwide, in Africa and in Latin America classified in terms of the power range [3]

2015	P > 10 MW	10 MW > P > 1 MW	P < 1 MW
World	910,242	114,395	30,785
Africa	26,748	432	43
Latin America	164,115	6325	546

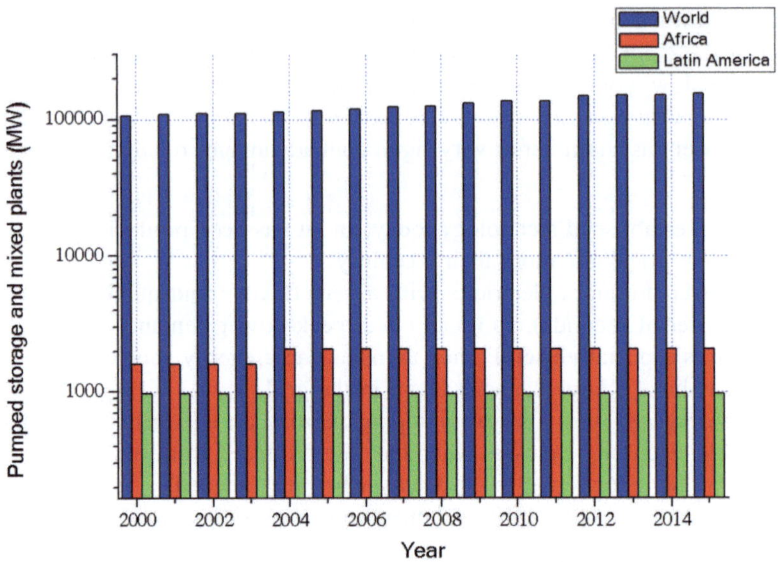

Fig. 8.2 Evolution of pump storage capacity between 2000 and 2015 worldwide, in Africa and Latin America [3]

the grid when demand is in excess. These operations are very attractive when bulk power is not easily adjustable (nuclear power plants) or the share of non-dispatchable energy resources (mainly solar and wind) in the grid is too high. World statistics show an increasing capacity, reaching 154 GW (Fig. 8.2) and 12.7% of the total hydropower capacity in 2015 [3]. However, there are only two countries in Africa showing hydro-storage capacity (Morocco and South Africa) and one country in Latin America (Argentina), and the capacity is flat from 2004. The energy conversion efficiency of recently pumped hydropower is over 80% [2]. The key performance metric for pumping storage hydropower systems is the availability factor, closely related to the rains and dry seasons in the region.

On the other hand, marine energy plays a very marginal role in renewable electricity production worldwide. Only 999 GWh were produced in 2014 (last data available) [1], mostly tidal range, that represents 0.02% of the total renewable electricity produced. No ocean electricity production is detected in Africa and Latin America but 0.1 MW of marine capacity is detected in Brazil from 2012.

Marine energy can be considered an incipient renewable energy because its worldwide capacity is at present the lowest of all renewable technologies (see Table 8.2) [4]. Also, there is not yet a consensus about the more efficient technologies to be applied to extract energy from oceanic waves and currents. Thus, in the past, large amounts of public and private funds have been dedicated without a substantial success in the short term. The energy contained in the oceans is available primarily through the exploitation of the flow of waves, ocean currents and tides, but there are also activities to extract energy from ocean thermal and salinity

Table 8.2 Estimated resources for the production of electricity from the ocean energy worldwide [4]

EJ/year	Estimated global resources
Waves	288
Tidal currents	2.88
Tidal range	1.08
Salinity gradient	7.2
Thermal gradient	36

Table 8.3 Ocean Power installed and being installed and under installation worldwide (2015) [5, 6]

	kw Installed (2015)	Presented projects
Wave power	3031	80,504
Tidal and ocean currents	7078	146,158
Salinity gradient	50	100,000
Tidal range	518,105	255,550
OTEC	220	220
Total	528,484	482,532

gradients. In addition, some studies consider the exploitation of marine biomass as a resource to be considered within the general category of ocean energy, although in our case it is considered in Chaps. 5 and 11, devoted to biomass and biofuels, respectively.

Some of these ocean energy technologies such as wave power and tidal currents are beginning from very low power capacities, although with an intense activity. On the contrary, tidal range is a well established technology but with short growth potential mainly due to technical and environmental conditions (Table 8.3).

8.2 Technology State of the Art

8.2.1 *Turbines*

There are many different types of turbines that reach an efficiency up to 92% [2] (Fig. 8.3) :

(a) Reaction (or pressure) turbines, driven by the hydrostatic pressure of water. The most commonly used turbines are the Francis, fixed pitch propeller and Kaplan turbines.
(b) Impulse (or action) turbines, driven by the kinetic energy of water. The most commonly used are the Turgo, crossflow and Pelton turbines.

The Francis turbine is by far the most used reaction turbine for heads of 10–250 m. It is also called a radial flow turbine as the water flows from the outer circumference to blades curved toward the center. This turbine is completely

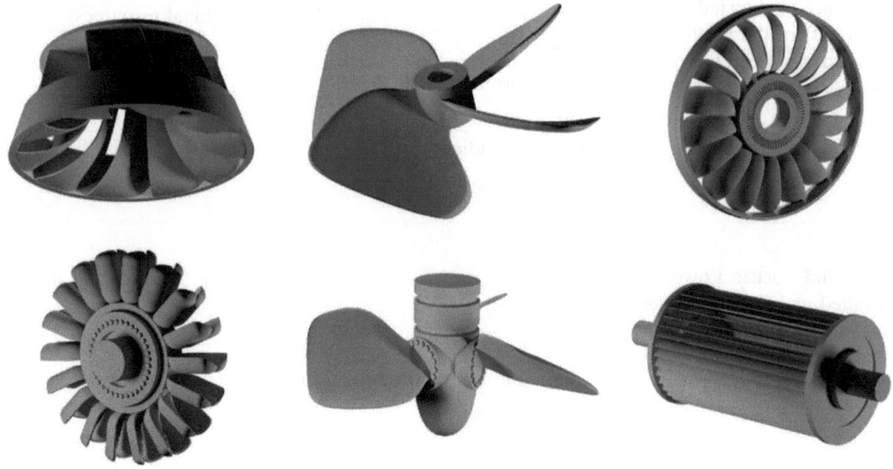

Fig. 8.3 Different types of turbines used to produce hydropower power: *top-left* Francis; *top-center* Kaplan; *top-right* Turgo; *bottom-left* Pelton; *bottom-center* propeller; and *bottom-right* crossflow [7]

submerged and its axis can be positioned both vertically and horizontally. In these turbines, the water exerts a pressure on the rotor blades and the rotational energy gained is primarily due to the pressure on the turbine.

Propeller turbines operate essentially like the propeller of a boat, but in reverse. They are normally used with low and rushing heads. These are called Kaplan turbines when the angle of incidence of the blades is automatically adjusted for flow conditions and to set the conversion efficiency to changes in electricity demand.

Among the impulse turbines, the most used is the Pelton turbine for high heads (up to 300 m). It is essentially a bladed wheel with a double ladle mounted on its edge. The high pressure water impinges on the edge that separates the two ladles. Split into two jets, the water is ejected laterally after having transferred its momentum to the wheel. Thus, the wheel can reach speeds up to 1000 rpm. An essential difference between the Pelton impulse turbine and the reaction turbines described above that operate entirely submerged and are affected by the pressure difference is that the impulse turbines operate in air at atmospheric pressure. Figure 8.4 [7] shows a practical scheme for the correlation among different turbines for several head, flow and power ranges.

8.2.2 Large Hydropower Systems

Currently, there are only three hydropower power plants that exceed 10 MW. The Three Gorges Dam in China (22.5 MW); the Itaipu in Brazil (14GW) and Guri in Venezuela (10.4 GW). These large systems are not entirely dedicated to supply

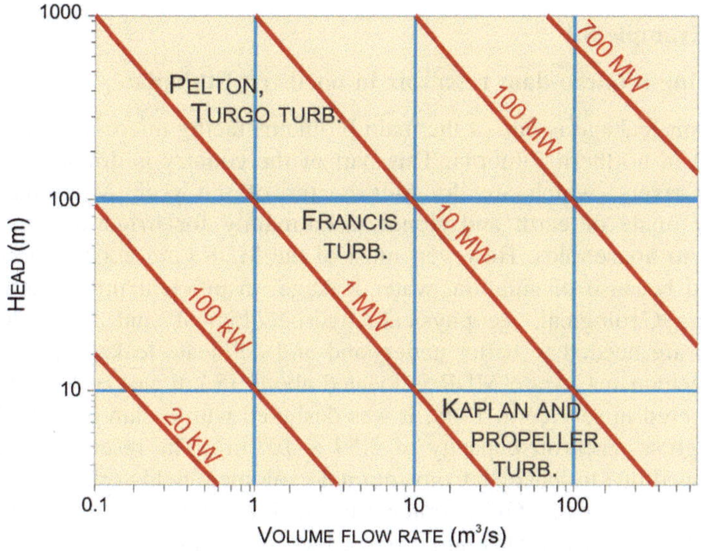

Fig. 8.4 Correlation between height, flow and power for different turbines

power to the grid, but also provide power to large industrial companies that are heavy consumers of electricity, such as those dedicated to the production of aluminum.

Hydropower systems with capacities below 10 MW continue to increase in part because of subsidies and R&D programs, and because many places suitable for large hydropower plants are already occupied. In addition, the construction of large power plants, especially in developed countries, often face major environmental challenges and rejection by the local population. In developing countries, there exists a different context and these smaller plants are attractive because they require no large investments in dams and grids. Small hydropower systems are frequently placed in locations where a low cost hydro resource is available, in isolated areas where it is very expensive to bring the grid to the consumers or in countries where a national grid does not exist.

Mini hydropower plants are very often run-of-river systems. They avoid the potential environmental damage produced by dams and often have limited affect on flow. However, some run-of-river hydropower plants have storage and are said to have "pondage". They are very attractive for small rivers and tributary flows, especially in relatively mountainous areas.

Case Example 8.1

Designing a micro-dam reservoir in northern Ethiopia:

Reservoir leakage is one of the main problems facing micro-dam reservoirs (MDR) in northern Ethiopia. This part of the country is drained by intermittent rivers, which are dry 8–9 months of the year. Micro-dams are usually made of earth and stones, used mainly for irrigation and water supply to households. However, most of the MDRs are not functioning as planned because of siltation, water leakage, insufficient inflow and water logging. Geological, geophysical, hydrogeological and hydrochemical studies are needed to better understand and delineate leakage zones, path and mechanisms. Arato MDR is located about 35 km east of Mekelle City and entered in service in 1998. It was designed with a dam height of 20 m and a gross reservoir capacity of 2.59×10^6 m^3. The reservoir is almost abandoned due to leakage. Conventional geological fieldwork with the help of topographic maps, GPS, strategic test drilling, electrical resistivity measurements, water samples, and pre-planned traverses were used to characterize the MDR. The results show that the reservoir lies entirely within the Quaternary deposit underlain by the leaky limestome-shale-marl intercalation of the Agula Shale, although the southwest reservoir rims and abutment are composed of dolerite unit. Then, controlling the leakage through this unit may be difficult and expensive by an upstream or reservoir blanket. The installation of a cut-off wall at the upstream toe of the dam is considered a solution to minimize the leakage problem underneath the dam foundation and left abutment/flank [8].

The location for a hydropower plant site is characterized extensively prior to construction to reduce the environmental impact from flow modifications on the biodiversity of the ecosystem and to assure a safe, resilient and cost effective design. In the last two or three decades there has been numerous examples of adverse public response to large hydro projects in both developed countries and in emerging nations including South East Asia (Mekong river) and East Africa (Nile river). Thus, it is important to consider cumulative, large-scale impacts of multiple dams in a river as part of an integrated 'river systems' mindset [9]. The benefits of energy production from renewable sources must not be tainted by environmental concern. For this reason the support for mini-hydro power is also increasing, as the hydrological impacts of a run-of-river technology is usually much smaller.

Case Study 8.2

Hydropower conflicts and resettlements in Brazil:

The authors study the conflict between hydropower and resettlements in an effort to understand the frequency and motivations of those affected to participate politically in demonstrations against their relocation. Results show that "collective rural resettlement" is the main cause of mobilization. The primary motivation for demonstrations is human solidarity while other motivations are alignment with local movements, specific interests and retribution. Reasons for not mobilising are working duties (they do not want to get away from their new properties), accommodation, individualism, and alignment with the entrepreneurs [10].

In the case of reservoirs or dams, the water flow downstream cannot be drastically diminished or it will have impact on river fauna groundwater levels, and other critical characteristics. River volume serves as a representative and robust metric for assessing river systems across scales compared to more commonly used metrics such as river length or basin area [9]. Thus, downstream of a dam an ecological minimum flow must be maintained which can limit the water available for irrigation. Hydropower reservoir systems are also used to control the risk of flooding, ensuring a constant flow of water downriver for agricultural irrigation, and even support recreational activities. Other effects considered are the environmental impact caused by the process of constructing a large dam that can take many years. The disappearance of land with native flora and fauna, which is often replaced by lakes for fishing and leisure activities, must be evaluated. These ecological transformations must be evaluated before the beginning of the construction of dams, to avoid unwanted effects, as occurred after the construction of the Aswan Dam, where annual flooding ceased and, consequently, the soil was impoverished of nutrients.

Sometimes the environmental impact is partly diminished to ensure fish passage in general and the periodic migrations of salmon in particular, by building scales close to the dams. To prevent fish from passing through the turbines, some countries establish that the water inlet to the turbine should be protected by a mesh. Moreover, the water reaching the turbine contains very little sediment, leading to loss of sandbars downstream and erosion of the seabed at the river mouth, accelerated by intermittent openings of the turbines gates. Also variations in the oxygen content of water, before and after the hydropower plant, are detected. In addition, the water leaving the turbine is usually warmer than the stored water.

In tropical countries water reservoirs produce significant amounts of methane. Methane emissions from these reservoirs may become comparable to emissions from power plants producing electricity from fossil fuels, especially if the reservoirs are oversized with respect to the installed power, and if a cleaning of biomass was not considered prior to filling the reservoir [11]. It is also very important to consider the social impact when there is need to move people living in the affected territory. The construction of the Aswan Dam produced the displacement of nearly 100,000 inhabitants, a small number compared to the Three Gorges Dam on the River Yangtze, which required the displacement of more than 1,200,000 inhabitants.

8.2.3 *Wave Power*

Waves, created by the wind, can travel thousands of kilometers through the oceans with minimal energy losses until they are modified by other winds. The major drawback of this kind of energy resource is its variability, although some seasonal variation patterns can be identified [12]. The wave energy level is usually defined in terms of power per unit of transversal length. Adequate locations for the exploitation of this kind of energy are those showing values between 20 and 70 kW/m. These locations are generally situated in moderate to high latitudes, with mostly constant values throughout the year in the southern hemisphere. The wave power is proportional to the period of the waves and to the square of the wave height. Approximately, the energy flow in excess of 30 kW/m is located in a 2% of the 800,000 km of coastline on Earth. More specifically, the best sites to generate electricity from the waves are located in Southern America, Western Australia and west of the British Isles, where the wave power can be considered about 60–70 kW/m (Fig. 8.5).

Fig. 8.5 Map of wave power per unit length (kW/m) worldwide [13]

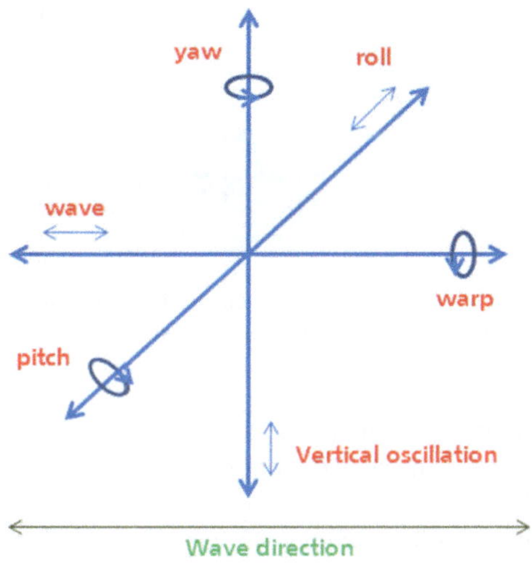

Fig. 8.6 Scheme of potential oscillating movements on a wave device

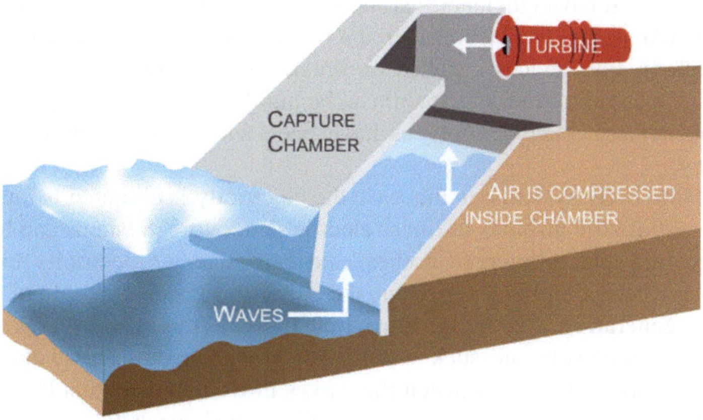

Fig. 8.7 Scheme of a shoreline oscillating water column device

Typically, devices that extract energy from waves have only one degree of freedom, although some may have many more degrees of freedom that improve the conversion process efficiency to mechanical energy. Six potential oscillating movements exist (Fig. 8.6) [13].

The devices used to obtain energy from waves can be classified according to the principle of conversion:

– Oscillating Water Column: In these devices, the wave motion of water induces a similar movement in the air column enclosed in the structure (Fig. 8.7). The electricity is obtained from a wind turbine moved by variations in surrounding air pressure. The movement of these turbines is independent of the flow

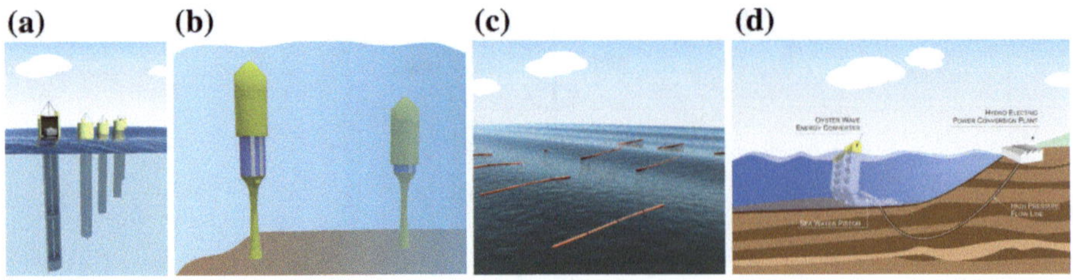

Fig. 8.8 **a** Floating system with translational movement, **b** submergible system with translational movement, **c** floating system with rotational movement, and **d** submersible system with rotational movement

direction (auto-rectifying). This is due to the symmetric rotor profile, producing an identical movement in each direction. These devices are divided between those with fixed structure (isolated or breakwaters) and floating structures. Among the isolated fixed structure, we can distinguish those installed on the shoreline and the ones anchored to the seabed. The LIMPET device, located in Scotland has a nominal ouput ranging from 2000 and 500 kW. Another device is installed at the island of Pico (Azores). These devices are connected directly to the grid, and have efficiencies around 30%. The device installed at the port of Kembla (Australia) can be considered a fixed structure, isolated and anchored to the seafloor. For devices installed on breakwaters, those located at the port of Sakata (Japan) and Mutriku (Spain) can be highlighted. Finally, in relation to the floating devices, many prototypes have been tested, as the Mighty Wale and Sperboy.

– Oscillating body systems: in this case the electricity is obtained from hydraulic motors, hydraulic turbines or linear electric generators. Systems can be divided into floating and submerged bodies, and can generate electricity from translation or rotation movements. Thus, the AquaBuoy is an example of a floating system generating electricity from translational movements (Fig. 8.8a), in which buoys elevate and suck water when waves arrive. This water is driven down by a hydraulic pump when the waves move away, so that the water flow moves an oil driven motor. Submerged systems generating electricity from translational movements (Fig. 8.8b) apply a similar method to generate electricity. Floating oscillating bodies use rotational movement to generate electricity. Multi-segment systems called attenuators, oriented parallel to the wave direction (Fig. 8.8c) are now being tested. The relative motion between floating bodies serve to pump high pressure oil through hydraulic motors, which move electricity generators. Finally, for submerged oscillating bodies using rotational movement to generate electricity, surge converter devices can be mentioned (Fig. 8.8d). These devices consist of platforms anchored to the seabed oscillating back and forth, in such a way that its kinetic energy is collected by a pump with a piston that produces electricity.

Case Example 8.3

Wave energy resource assessment in Uruguay:

Because Uruguay is located in a microtidal zone, the authors believe that wave energy is the most promising marine energy resource. A model was developed and validated with altimeter data and in situ measurements. The study zone includes the Río de la Plata (290 km length, 5–20 m shallow waters) and the Uruguayan waters in the Atlantic Ocean (taking 200 m depth as a reference, the width of the continental shelf varies between 140 and 180 km). The results show that although the wave energy potential is moderate, it is fairly steady throughout the year, and the extreme waves are relatively benign. The Atlantic region has values of 30 kW/m at 200 km offshore and 20 kW/m at 70 km offshore, and at 20 m depth the mean wave power varies between 8 and 14 kW/m. Inside the Río de la Plata the power is negligible. Seasonal variability related to wave direction was detected [14].

– overtopping converters: Electricity from these systems is obtained by means of low head hydraulic turbines that are are divided among floating systems and those with fixed structure (in shoreline or breakwater). All of these devices are quite similar to those used in tidal range dams, as they have reservoirs located at altitudes above the ocean but are filled with water from the waves. The water lifted by the waves, aided by the funnel effect of a channel, is stored in a reservoir. This water is then released through a pipe which is connected to a Kaplan turbine to generate electricity. The reservoir can store energy and the electricity can be produced when it is demanded. In addition, the wave power is irrelevant as long as it is capable of adding water to the reservoir. The most popular overtopping converter is the Wave Dragon, with a floating structure (Fig. 8.9a). In relation to fixed shoreline devices, the best known is the Tapchan (Fig. 8.9b), having been successfully installed on the Norwegian island of Bergen in 1985, with an output power of 350 kW and connected to the electricity grid. The locations where these overtopping converters are placed use to

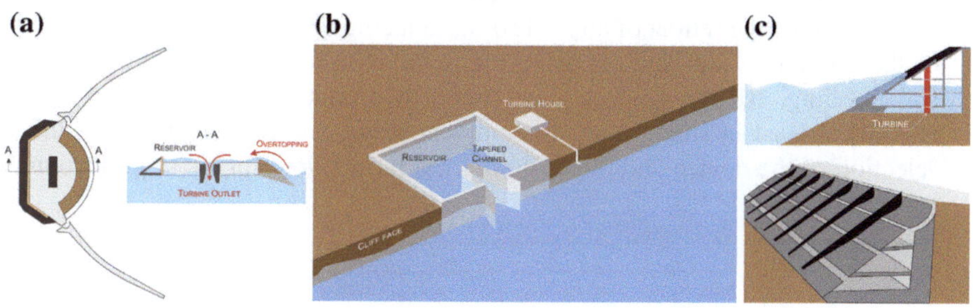

Fig. 8.9 **a** The Wave Dragon overtopping converter floating structure diagram; **b** diagram of the Tapchan device; and **c** diagram of the breakwater fixed-structure overtopping SSG

be near vertical cliffs that provide deep waters and, therefore, high-energy waves. Finally, as for breakwater fixed-structure overtopping converters, the most popular is the SSG (Seawave Slot-cone Generator), composed of several tanks at different levels to collect water from waves (Fig. 8.9c). The device, designed in Norway, has a special multi-stage vertical axis turbine to optimize the energy production from water stored in the different tanks.

The most common power-take-off systems installed on these devices are: (i) pneumatic turbines, (ii) hydraulic circuits with oil pressurizer and hydraulic motor, (iii) linear electric generators (less advanced), and (iv) hydraulic turbines (Kaplan or Pelton) as described earlier.

In terms of device location, the off-shore systems are more complex than those located near-shore or on the shoreline, mainly due to problems related to mooring, access to maintenance and the need for large seabed power transmission cables. Also, the waves have less energy near-shore, but this drawback can be compensated for by wave refraction and/or diffraction in these areas. For devices placed near-shore, the shallow seabed is advantageous for installing these devices submerged on the seabed.

Case Example 8.4

Beach response to wave energy converter farms acting as coastal defense in Mexico:

Las Glorias beach on the North Pacific coast of Mexico inside the Gulf of California was selected for a multi-purpose use of wave energy converters (WECs). Maintenance, coastal safety, energy supply and prevention against climate change effects were emphasized. The beach represents a case where the long-shore sediment transport is dominant, and the hydrodynamics induced by these devices (the Wave Dragon, the seabreath, the blow-jet and the dexa) is analyzed to understand the response of the coastline in the presence of WECs. The beach presents a short fetch with short wave periods and small wave heights, with the exception of waves coming from the South which are more energetic. The beach is limited to the North and South by rivers, and groynes built for navigation blocked sediment transport and caused a coastline retreat of about 150 m, reducing dramatically the resilience of the coastal system and exposing the population to a high risk of erosion and flooding. All the devices were found to produce a positive trend (accretion) at least in small areas. Recommendations are given to facilitate the selection of the device and the design of the farm layout for shore protection support [15].

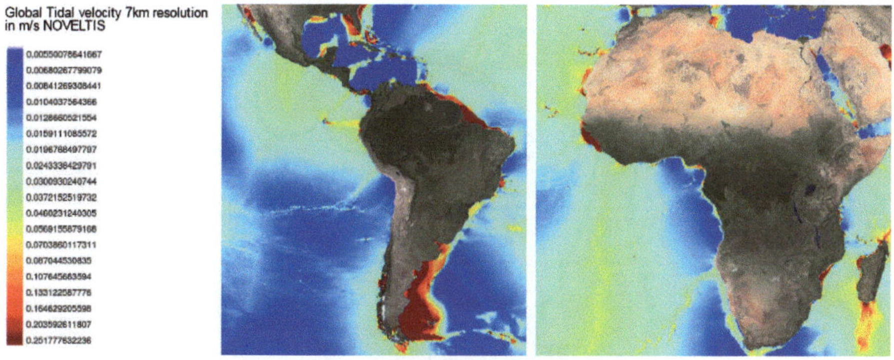

Fig. 8.10 Tidal velocity 7 km resolution. *Source* NOVELTIS [16]

8.2.4 Currents

Marine current technologies are focused primarily on the use of fast flows produced by tidal induced marine currents. These flows are greatly increased by coastal topography in certain areas of the planet and reach their maximum values at half tide periods. Generally, the power flow of the tides is greatest in shallow water, where funnel-effects occur, such as straits or bays. There are areas in Africa and Latin America where marine tidal currents provide a substantial tidal velocity (Fig. 8.10) suggesting opportunity areas for further characterization and the preliminary assessment of technical potentials. A more detailed description of long-term time series for tidal elevation and velocity, and tidal information such as current direction, maximum velocity and power density can be found elsewhere [16].

The technology uses concepts already developed for the production of electricity from wind resources, but the fluid is about 800 times more dense than air. It is estimated that if the marine current reaches an average speed of 2–2.5 m/s the installation can be economically viable. The technology used for the exploitation of tidal currents can be classified in any of the following: (i) horizontal axis turbines, whose operation is very similar to those used in wind energy and optionally covered to create flow concentration effects (Fig. 8.11); (ii) vertical axis turbines, including cross-flow turbines; (iii) oscillating hydrofoils, which move up and down in a plane perpendicular to the flow of the tide, creating pressure on a hydraulic fluid which drives a hydraulic motor, and by extension, an electric generator; and (iv) Venturi effect devices, where the flow of water is introduced into a tube that concentrates it and produces a pressure difference which is exploited by a second fluid to drive a turbine. On November 12, 2009 the first commercial turbine, 1 MW, was installed using a horizontal axis technology with some variations. The plant has been installed at the Minas Passage area in the Bay of Fundy, three kilometers from the coast (Canada), in one of the world's largest tidal range areas.

A similar technology is considered for obtaining energy from the ocean currents. These currents are mainly produced by wind and solar warming affecting oceans

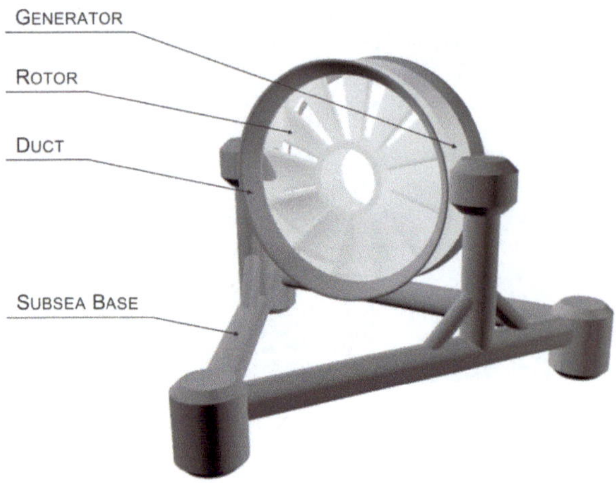

Fig. 8.11 Prototype of tidal current horizontal axis turbine

near the equator, although variations in the density and salinity of the ocean water can have an influence. These currents are relatively constant, and always flow in the same direction, in contrast to the currents produced by tides near the coast. Examples of ocean currents are the Gulf Stream, the Florida Straits Current and the California Current (Fig. 8.12) [17]. These currents tend to flow at the ocean surface and the potential global power has been estimated at 5 TW, with power densities

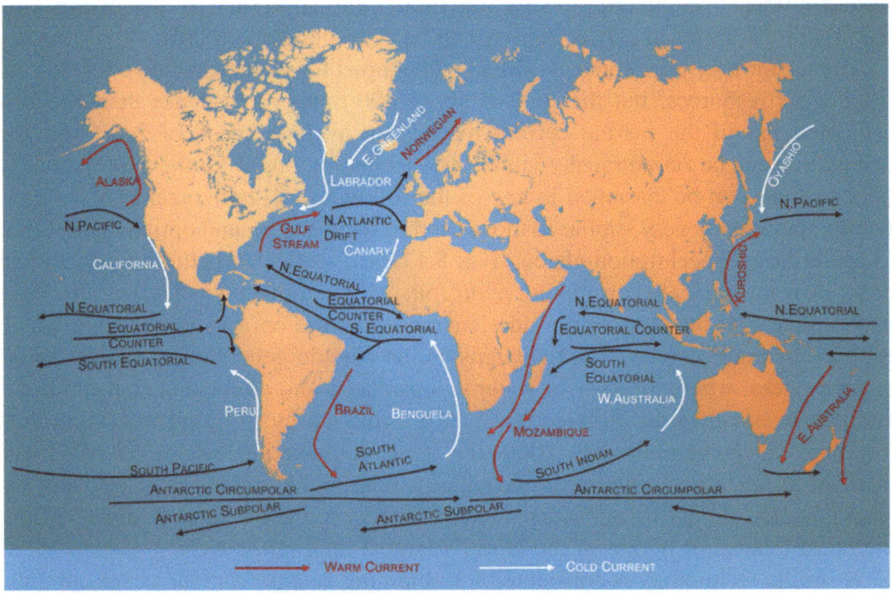

Fig. 8.12 Map showing major surface ocean currents on Earth [17]

reaching 5 kW/m^2 [18]. The tidal currents could provide more than 10 TWh/year (about 0.4 EJ/year) in the larger estuaries, where currents can reach values up to 10 m/s [17]. The best sites are located on the west coasts of America, Britain and Australia, with energy flows that can exceed 80 kW/m^2.

The average power generated by ocean currents per unit area perpendicular to the flow is proportional to the density of water and the cube of its speed so that, since the air density is about 800 times lower than that of water, ocean currents can carry much more energy at much lower speeds than the wind. Although electricity generation from ocean currents is based on relatively simple physical principles conceptually similar to those of wind generation, devices are not as developed as those for tidal currents.

8.2.5 *Tidal Range*

Tidal range is currently the most mature marine technology for energy production. The most common technology to extract energy from the tidal range is by installing a dam in a bay or estuary, acting similarly to a hydropower plant. Consequently, water stored during the high tides drives a turbine as it is released at low tides. Turbines can be powered in the period of high tide, but it does not increase power generation significantly and, in contrast, can increase costs and operating risks significantly. There are very few tidal dams in operation for commercial exploitation. The best known is the plant at La Rance, in Brittany (France), built in 1956 with a capacity of 240 MW.

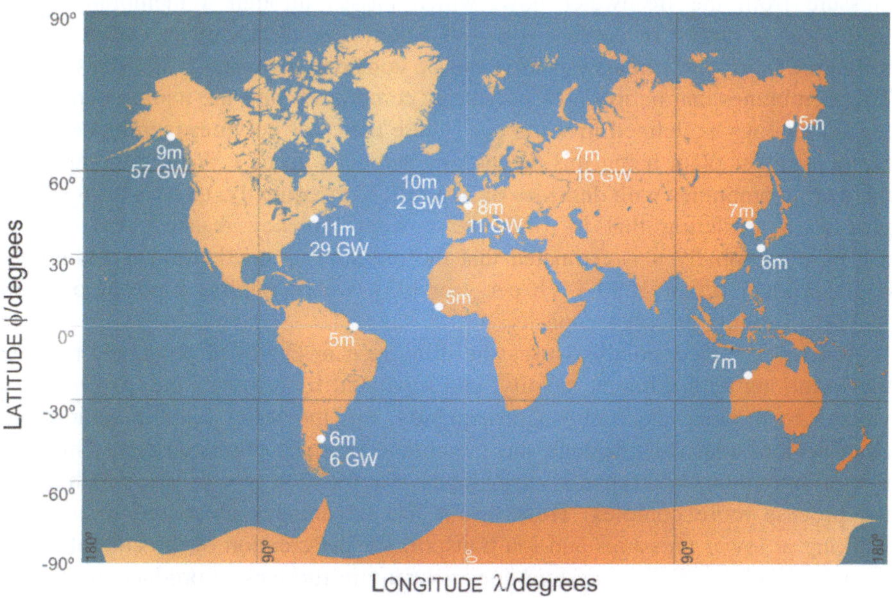

Fig. 8.13 Tidal ranges at different locations worldwide [20]

Other processes that exploit energy from the tides are called tidal reefs, lagoons and fences. Tidal reefs use fixed flow turbines operating with a constant head difference of two meters obtained by floating concrete blocks or moving locks. The tidal lagoons are independent structures built off-shore or in a semicircular shape connected to the shoreline, operating similarly to the dams, although there is no such system operating to date. Tidal fences are installed in shallow open waters where there are large water flows due to tidal periods, so that they can produce tidal energy in both upwards and downwards modes.

The potential of the tidal energy resource is about 800 TWh/year [19]. The most promising sites are indicated in Fig. 8.13 [20]. Some of the most favorable places are located in United Kingdom, where up to 25% of its total electricity could be generated using this technology. Other suitable locations are situated in France, Ireland, Canada, United States, Argentina, Chile and Australia.

8.2.6 Other Marine Technologies

Technologies based on salinity gradients are an alternative for producing energy by increasing the entropy of mixing water with different salinity [21]. The technology is under development, and some analyses consider that the theoretical global resources could reach 2000 TWh/year [19]. Two strategies for converting the energy produced by the increase of entropy are pressure-retarded osmosis and reverse electrodialysis.

Pressure retarded osmosis is based on the movement of a turbine by increasing pressure from the freshwater flow, which passes through a membrane, when freshwater is placed in contact with salt water. The electricity produced by this technique is very advantageous as it can be considered constant, 365 days a year. The membranes can be made of polymeric compounds and the main parameter of this technology is defined in watts per square meter of polymer membrane. Thus, the first power plant using this technology achieves up to 1 W/m^2 [22], although there are membranes with densities of 3 W/m^2. Consequently, there is a substantial progress considering that in 1999 the power efficiency reached was only of 0.1 W/m^2. Globally it is estimated that the pressure retarded osmosis technology could produce about 1600 TWh per year [23], which implies a power of about 183 GW running at 100% capacity.

The reverse electrodialysis uses the difference in chemical potentials between two solutions with different salinity, and separated by a membrane, to generate a voltage to produce electricity. Consequently, not the water but the sodium and chlorine ions that pass through the membrane produce the electricity. This configuration requires two types of membranes, one for each type of ion, which makes the process more complex. However, it has advantages compared to osmosis: working at lower pressures and ion transport for the creation of the current. Then, the reverse electrodialysis system does not require turbines to produce electricity.

Finally, ocean thermal energy conversion (OTEC) uses the temperature difference between deep and shallow waters to run a heat pump. For this reason, the highest efficiencies are achieved at the lowest latitudes (tropical and equatorial regions), where shallow water temperature is high. However, the conversion efficiencies are very low due to the very small temperature differences (1–3% efficiency, but a theoretical 6–7% is considered). The Rankine cycle is the only one that has been considered, commonly used as the operating cycle for steam power, and a closed loop (mainly ammonia and heat exchangers) or low-pressure open cycle water-turbine. The simplicity of the open loop system is counterbalanced by the increased demands for much larger low-pressure turbines and more complex compression systems [24], limiting the maximum power to 10 MW.

The resources for the temperature gradients technology are located in an area of over 100 million km^2 on the tropical regions of the oceans [24]. This is a technology that can deliver power with very high load capacity, so it can be part of the base load of any production system, estimating a theoretical power capacity from 3 to 10 TW [24] and an annual energy output of about 10,000 TWh [19]. Considering the temperature gradients demanded (about 20 °C), it is more appropriate to install these systems on floating structures than on the shoreline. Cold water flows are estimated from 2.5 to 3.0 m^3/s per MW of energy produced, and the required hot water flows are even larger [19]. The world's largest pipeline installed to date is a 1.4 m in diameter and 2.8 km in length installed on the west coast of Hawaii in 2001 [24].

The main projects in this area are based on applying temperature gradient systems to the desalination of seawater, mainly in India. Also, there are operating plants to supply power to heat pumps for heating and cooling [25]. Plants have achieved 1000 m^3/day desalinated water, converting about 1% of surface seawater pumped into vapor [24]. Another potential application in parallel is the use of large amounts of nutrients that can be found in deep water to be transferred to marine aquaculture.

8.2.7 *Costs*

8.2.7.1 Hydropower Costs

Hydropower is one of the most cost-effective technologies to generate renewable electricity (Table 8.4) [2]. The total installed cost for a large-scale hydropower

Table 8.4 Key costs for hydropower systems [2]

USD_{2010}	Up to 1 MW	1–10 MW	Larger than 10 MW
Investment (USD/kW)	3400–10,000 or more	1000–4000	1050–7650
O&M (USD/kW/year)	45–250 or more	40-50	45 (average)
Economic lifetime (year)	30	30	30
Interest rate (%)	10	10	10
Production (USD/MWh)	270 or more	20–100	20–190

project typically ranges from a low USD 1000/kW to around USD 3500/kW. The cost of pumped storage systems is about USD 2000–4000/kW [2]. However, exceptions abound if the site is far from the transmission line and without access and infrastructure. No significant cost reduction due to technology learning is expected in the future as hydropower is a mature technology.

Since the hydropower resource has no fuel cost, the investment expenditures are concentrated in the first years. Moreover, the average life of such plants is over 50 years and the lifetime of the different components of the plant is diverse. Investment plans and borrowing costs to meet the construction and operation phase of these plants must take into account this condition. Turbines have a life expectancy between 25 and 50 years, and external structures can reach a lifetime above 100 years. Annual O&M costs ranges 1–4% of the investment costs [2].

Case Example 8.5

Lack of maintenance and drought effects:

The Guri hydroelectric plan is the fourth largest in the world offering a capacity of 2.1 GW and supplying around two-fifth of Venezuela electricity. However, the water level reached less than two meters from its "catastrophic" point in May 2016, at which the generator must be switched off to avoid damaging its turbines [26]. Lack of water flow also affects other hydropower plants downstream. The result of this situation was rolling power supply restrictions all over Venezuela. The immediate cause was drought, increased by El Niño, but also other important reasons are that this country is very poorly interconnected, back-up thermal power plants are not a full option as they have been not properly maintained, and the lack of planning the repowering of Guri dam as electricity demand has been increasing for many years. Guri was built in 1969 with a combined installed capacity of 1750 MW and upgrades to 2065 MW in 1978. However the Venezuelan population has increased from 15.8 million inhabitants in 1981 to 30.7 in 2015. Also in Brazil, the prolonged drought put water supply and energy matrix into stress, exposing the market to costly generation from fossil fueled thermal plants. In the Brazilian wholesale market, where the impact of the energy crisis was felt more immediately, prices increased, and average spot prices in 2014 reached BRL 642/MWh (USD 273/MWh), up 146% compared to the average in 2013. Given the drought, and the necessity to add new capacity to the Brazilian's matrix, in 2014 the government contracted a total of 4 GW of biomass, solar, small hydro and wind projects through auctions. On the other hand, on August 2016 Brazil's Environment Ministry denied a licence for the 8 GW Sao Luiz do Tapajos Dam—conceived as the largest hydroelectric plant to be built in the Brazilian Amazon. Grounds for denial were that it would harm the environment and impact indigenous populations. Also, due to severe droughts from the first half of 2013 to the second half of 2014, Costa Rica has relied on the regional market (251 GWh) to help meet its domestic electricity needs, but in 2015, the country was able to meet most of its power demand needs with renewable resources, thanks to the hydro and wind plants installed in the country [27]. Similar drought problems and rely on the regional grid happened in Panama in the first half of 2014.

Official forecasts show that the investment costs are not expected to decrease from new operational insights as is evident iwth other renewable technologies (e.g., solar photovoltaics). Since hydropower investment costs are strongly associated with the prices of raw materials, fluctuations in these costs notably influence the cost of the technology. The main considerations to future costs reductions are related to the increase of rehabilitation projects for old hydropower systems, run-of-river systems, use of components produced in emerging countries and reduction of profit margins. The average cost of pumped storage systems strongly depends on the site configuration and the operational service provided. They are estimated to range between USD 2000–4000/kW [30].

Case Example 8.6

Potential use of water spilled for producing hydrogen in Ecuador:

The study is focused in the water spilled from hydroelectric plants of Ecuador and the potential it offers for hydrogen production via electrolysis. The hydrogen obtained is planned to be used as fuel for public transportation in the city of Cuenca (500,000 inhabitants), in Ecuador and the evaluation of the fuel cost is included. The hydropower plants (Mazar, Molino and Sopladora) with regulating reservoirs are considered and the information related to spills of the four reservoirs identified. The maximum operation level (limit height before dumping occurs) and the outflow level (height at which water is replenished to the river) of each plant is considered to estimate the head of water when there is energy excess. This energy excess (ideal case) is considered in monthly periods for the production of hydrogen. The cost of hydrogen is substantially lower than the cost of unsubsidized diesel. However, the work considered the mini-hydro capacity fully for H_2 production, and doesn't consider the extra cost of hydrogen infrastructure, a new fleet of hydrogen fuelled buses and some additional expertise requirements for O&M activities.

In terms of levelized cost of electricity (LCOE), the International Renewable Energy Agency (IRENA) establishes a range of USD 0.030–0.210/kWh in Africa with an average value of USD 0.100/kWh, USD 0.038–0.170/kWh in Central America with an average value of USD 0.060/kWh and USD 0.015–0.026/kWh in South America with an average value of USD 0.040/kWh (Fig. 8.14) [3]. Thus, on average and compared to other regions, production of electricity from hydropower is favourable in South America but not yet in Africa compared to other regions.

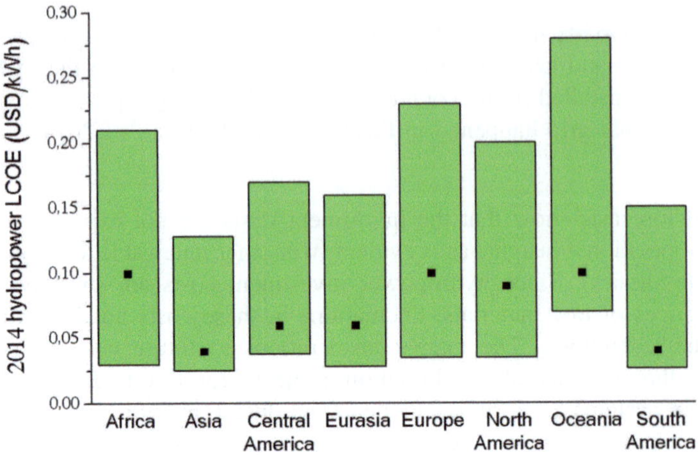

Fig. 8.14 Levelized cost of electricity from hydropower in the different global regions (2014) [3]

Regarding CO_2 emissions, no rigorous studies have been found to provide such data. It is recognized that this technology does not emit CO_2, however, this is an underestimation since, especially in the construction phase, CO_2 emissions are very significant, as in other renewable technologies. However, taking into account the long life of hydropower plants, the impact of CO_2 emissions per kWh produced is considered the smallest of all renewable technologies [28].

Also the emissions of CO_2 and methane from the reservoirs of the hydropower plants should be considered. Thus, there is some controversy about the impacts of hydropower on GHG emissions as a result of methane production. In this regard, rigorous methodologies to evaluate the carbon balance in freshwater reservoirs, which includes life cycle analysis and considers the potential impacts of floods are being defined.

8.2.7.2 Marine Costs

In general, it is somewhat difficult to analyze the costs per kWh for electricity generated from different marine energy technologies, since most of them are still in the demonstration and prototype phases. In addition, the unit cost of electricity reported by the research centres and companies developing different technologies are often not very reliable. In wave energy, production costs can be placed about three times higher than those associated with onshore wind energy and, consequently, prototypes are being installed on the shores of countries offering generous bonuses to the energy produced by such sources of energy [12].

Due to the limited commercial experience, the estimates for LCOE of wave energy technologies in 10 MW demonstration projects is in the range of EUR

Table 8.5 Current key cost elements for marine technologies [21, 28–32]

	Investment costs	LCOE
Wave energy	EUR_{2014} 5000–9000/kW	EUR_{2014} 330–630/MWh
Tidal current	–	EUR_{2014} 250–470/MWh
Tidal range	USD_{2012} 117–340/kW (site)	USD_{2012} 24–40/MWh (site)
Salinity gradient	–	USD_{2009} 740–1490/MWh
OTEC (large)	USD_{2014} 2500/kW	USD_{2014} 70–190/MWh

330–630/MWh (Table 8.5). However, there is considerable scope for economies of scale and learning to substantially lower this amount if deployment levels of more than 2 GW are achieved. The capital cost of farms is about EUR 5000–9000/kW and O&M cost EUR 48–97/kW/year [29].

Tidal current technologies are also in a demonstration stage. Then, cost values are also expected to decrease. Current demonstration projects suggest a LCOE in the range of EUR 250–470/MWh. The cost of tidal range power plants mainly depends on the location. The construction cost for La Rance was around USD 340/kW (2012 values) and the Sihwa barrage (South Korea) was constructed for USD 117/kW in 2011. Then, electricity production costs for La Rance and Shiwa are about USD 40/MWh and USD 24/MWh. Also, it is expected that both tidal range and tidal current technologies could fall by up to 40% in cases where they are combined and integrated in the design and construction [30].

For salinity gradients and plants based on the reverse osmosis technology, estimates of production costs of electricity are estimated in the USD 0.74–1.49/kWh range (2009 value) [32]. Small scale OTEC plants (<10 MW) have high overheads, and installation costs are about USD 16,400–35,400/kW (2010 value). Then, small OTEC needs to produce valuable by-products (e.g., fresh water or cooling) to be economically viable. For large scale OTEC the installation costs lie in USD 2500/kW (2010 value) and LCOE around USD 70–190/MWh [31].

There are very few studies evaluating the relation between the production of marine energy and CO_2 emissions. Also, depending on the technology used, CO_2 emissions resulting from marine energy production can be similar to other technologies. For example, tidal range is similar to hydropower energy, and tidal currents can be approximated to offshore wind energy although, especially in the latter case, the materials used in both systems can differ substantially. There is some research covering all renewable technologies which suggest that CO_2 emissions from ocean energy is about 7 g CO_2/kWh of electricity produced [33]. These values can fluctuate considerably due to the different technologies being included in marine energy. Other research suggests that up to 300 g of CO_2 could be avoided from fossil fuel emissions for each kWh generated by ocean energy [34].

Limited research [19] has examined in detail the different greenhouse gas emissions for the life cycle of marine energy technologies. This work focused on the Wave Dragon system (Fig. 8.9a), used to extract energy from waves, although it justifies its extension to other ocean technologies. 13.2 g CO_2/kWh is established as average emission rate, and considers that it is mainly due to the manufacturing process (12.5 g CO_2/kWh) and, to a lesser extent, to the operation process (0.51 g CO_2/kWh). Steel and concrete used in the construction phase are the main contributors to these emissions. Consequently, if a proper choice of building materials is considered, and reuse is included at the end of their life cycle, the emission rates can be substantially lowered.

8.3 Regional and National Perspectives on Technology

8.3.1 Hydropower

The production of electricity from hydropower in African countries depends mainly on the availability of the resource and the power demand (Fig. 8.15). Thus, several African countries obtain almost all electricity from hydropower. However the hydro output is low due to the low GDP in many countries. Also, it is evident that African countries situated in desert areas (Sahara Desert) have very low capabilities for producing electricity from hydropower resources. In terms of hydropower potential, DR Congo has the highest on the continent (Grand Inga Dam potential reaches 40 GW), and Cameroon (20 GW), of which 723 MW have been developed [27]. The Grand Inga Dam in DR Congo and the Grand Ethiopian Renaissance Dam (6 GW) are the main projects in this region [27]. In terms of technology development, Kenya offers expertise in small hydro project development, engineering and O&M [27]; and South Africa offers expertise and manufacturing capacity in small hydro project development, engineering, O&M, turbines and balance of plant [27].

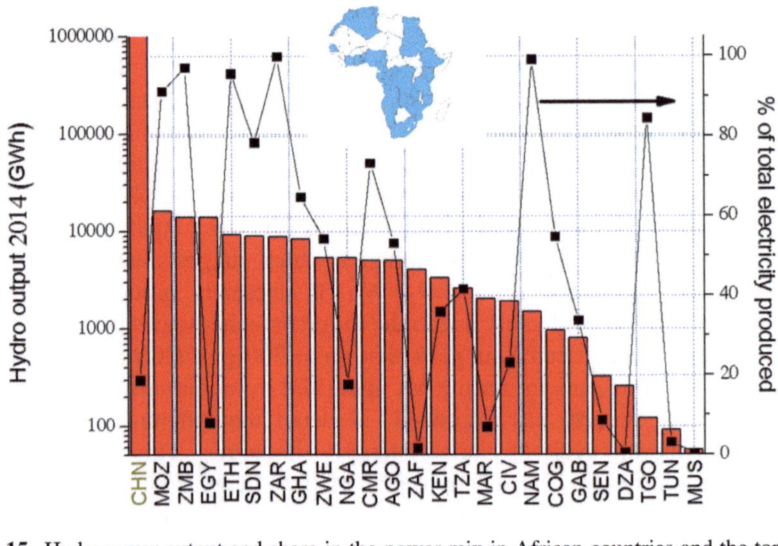

Fig. 8.15 Hydropower output and share in the power mix in African countries and the top world producer in 2014 [1]

Case Example 8.7

Grand Inga:

The Grand Inga Dam in the Congo river is a 40 GW hydroelectric project in the DRC that could fundamentally change the electricity market in sub-Saharan Africa much stronger than the Itaipú Dam between Paraguay and Brazil. If it were to happen, it would be the largest infrastructure project in history and could deliver more than 200 TWh per year, the equivalent of 13% of all of sub-Saharan Africa's total demand by 2040. It also can deliver 63 Mt of carbon savings by replacing many conventional power plants. The cost estimations comprise roughly USD 40 billion for generation and USD 40 billion for transmission. Multiple transmission routes have been planned, but the most likely is based on the massive expansion of the transmission line that runs from DRC, through Angola and Namibia, and into Botswana and South Africa. However, it could be more economical if flows north to Cameroon, Ghana and Nigeria. The expected generation cost is USD 25/MWh (LCOE), with transmission cost expected to be about USD 2–3 per MWh. But there are some important elements to be addressed: (i) ensuring the bankability of the project; (ii) obtaining sufficient funding to deliver the project; and (iii) securing effective stakeholder cooperation and collaboration. The greatest limitation has always been the need for a viable power buyer, or off-taker, to make the Inga project bankable. Potential options include BHP aluminum smelters in the DRC and the South African utility Eskom. It is important to remember that the power of the mighty Congo was first tapped in 1972 with the 385 MW Inga 1, which was followed in 1982 by the 1425 MW Inga 2, but utilization of those plants has fallen because of political instability, low demand, and poor maintenance [35]. The current phase of development, Inga 3, has a total potential of 7.8 GW, of that total 4.8 GW of capacity is under development, to be commissioned by 2023, and transmission lines totalling 1850 km will support Inga 3 exports of electricity [36]. However, the project suffered a severe setback when the World Bank suspended funding (July 2016) due to the government taking "the project in a different strategic direction to that agreed" with World Bank in 2014. Some 6% of the USD 73.1 million granted by the World Bank in March 2014 has been disbursed so far [37]. Current hydro installed capacity in DR Congo is 2420 MW, of which only 1281 MW is operational [27].

In Latin America the production of electricity from hydropower resources is larger than in Africa, as the resources and GDP are also larger. One country, Paraguay, produces 100% of its electricity from hydropower (Fig. 8.16). Paraguay is a very special case, as this country produces much more electricity from hydropower than consumed, exporting its surplus to Brazil and Argentina. In fact, Paraguay could act as the main storage system for a future Pan-American power grid if Latin American countries substantially increase their power interconnections. The Inambari

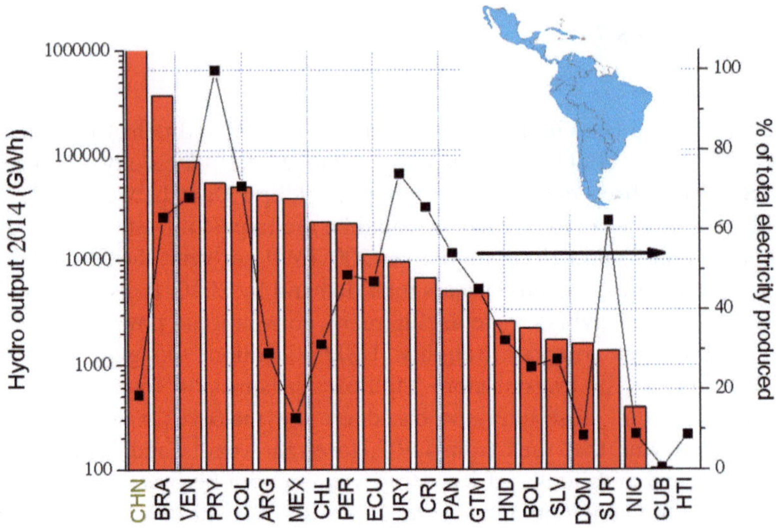

Fig. 8.16 Hydropower output and share in the power mix in African countries and the top world producer in 2014 [1]

hydropower complex (5 plants) between Perú and Brazil, adding a capacity of 20 GW, and the Cachuela Esperanza hydropower plant between Bolivia and Brazil, adding 0.8 GW capacity, are the main projects in this region [38]. In the region, Brazil and Mexico offer expertise and manufacturing capacity in all the small hydro chain: project development, engineering, O&M, turbines and balance of plant [27].

Case Example 8.8

La Esperanza run-of-river hydroelectric project in Honduras:

A hydroelectric power plant is being built in Intibucá, a province of Honduras, to exploit the hydropower potential of the Intibucá river in La Pozona, and expecting a contribution of about 50 GWh per year to Honduras national grid. The remains of a small electric power generation plant built by Germans immigrants right after the Second World War and operated until 1969 was located in the area. CISA (Consorcio de Inversiones, S.A.), a joint venture between a Canadian firm (85% of the ownership) and the Honduran firm SIMET, is the owner of the project, and expect at least a minimum return of 15% on their own funds invested. SIMET had operated for years primarily as a subcontractor installing electrical equipment and building electrical lines, connections and subsystems for ENEE (company that manages and supervises Honduras power system) and other private companies in Honduras. The SIMET 15% percent of CISA is a partial compensation for building the transmission line required to connect the hydropower plant with the local ENEE grid. La Esperanza was selected as the first project to be developed as a result of the use of detailed geographic maps, information from previous studies and visits to various sites. CISA requested a modification of a USD 250,000 loan approved by E+Co for funding only Stage 1A of the project instead of Stages 1A and 1B. E+Co is a firm specialized in providing funding services to renewable energy companies in less developed countries, and most of the funding comes from international development banks, donations from philanthropists, and NGOs. CISA has a 25-year concession (starting in 1999) from the local government to develop the project under certain conditions (create jobs for local inhabitants, plant about 20,000 trees, conduct land protection projects along the river banks and clean the reservoir area). In October 2001 CISE signed a PPA with ENEE to purchase all the energy from 12.7 MW (3 turbines in cascade) with a 15-year term, paying USD 0.057/kWh at the beginning and increasing a 1.5%/year for the first 12 years. In addition, the current incentive law would allow the project to receive a 10% premium. The catalytic loan was finally approved, and E+Co final investment in the project was USD 1.25 million [39].

The evolution of the hydropower electricity generation is mostly slightly upward (Fig. 8.17), however there are some substantial variabilities during the period where

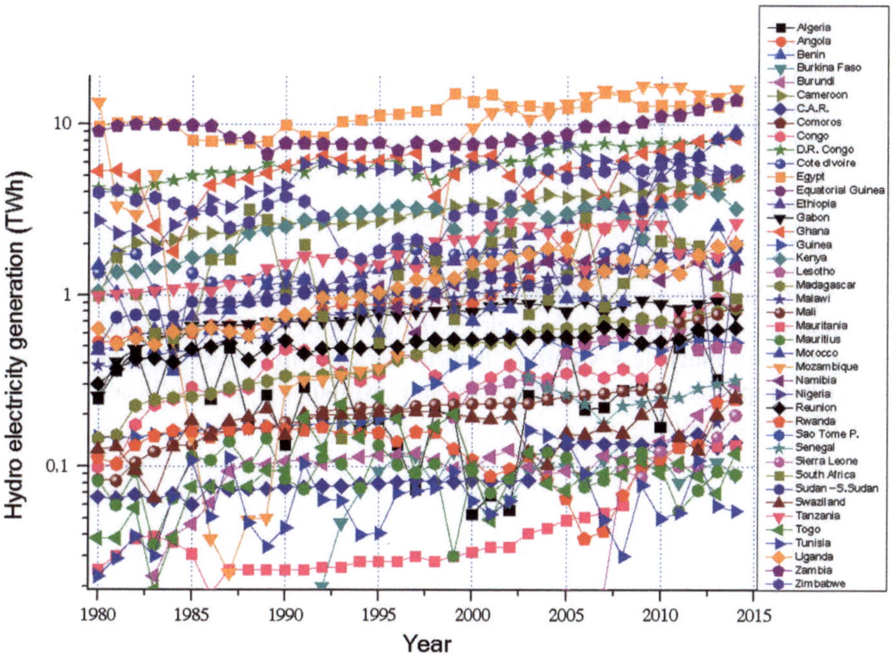

Fig. 8.17 Evolution of hydropower electricity generation in African countries in the period 1980–2014 [1]

data are available. This variability may be attributed to differences in water resources due to climatic conditions but also to internal difficulties and lack of O&M procedures.

Case Example 8.9

Hydropower planning in fragile and conflict states: South Sudan:

The authors expose that conventional methods for energy planning are likely to provide results that may be incorrect in fragile and conflict-prone countries because of the risks of violence and damage, or significant delays and cancellations. On the other hand, providing energy services to these countries helps to avoid youth joining gang or rebel movements because of unemployment and idleness. Thus, least-cost planning must explicitly address these inherent risks, very uncommon in the standard methodology. South Sudan is trapped in a cycle of violence because of ethnic and political issues. This country (as of June 2015) has less than 30 MW of installed diesel capacity serving pockets of loads in the towns, with no backbone transmission, but long-term demand is expected to reach 2500 MW in 2045. Recent reports plan the capacity expansion on developing a series of relatively large-scale hydro projects that would require more than USD 5 billion in investment. Building a large hydro plant may well be the least-cost option under certainty, perfect foresight, but if we include risks inherent in fragile and conflict states (i.e., security costs, contractual costs for regular O&M, delays, damage costs, premature termination and influence in the lifetime of the system), the risk-adjusted levelized cost may start rising quickly to a point where more distributed technologies as mini-grids can be cheaper. As conclusion, the authors consider the implementation of distributed generation technologies and systems to diminish these risks [40].

In Latin America also a slightly upward tendency is detected in relation to the evolution of the hydropower electricity generation (Fig. 8.18). The trends in these countries show much less volatility than in Africa. Also some relations between annual hydropower generation and economic evolution can be in observed some countries (e.g., Haiti).

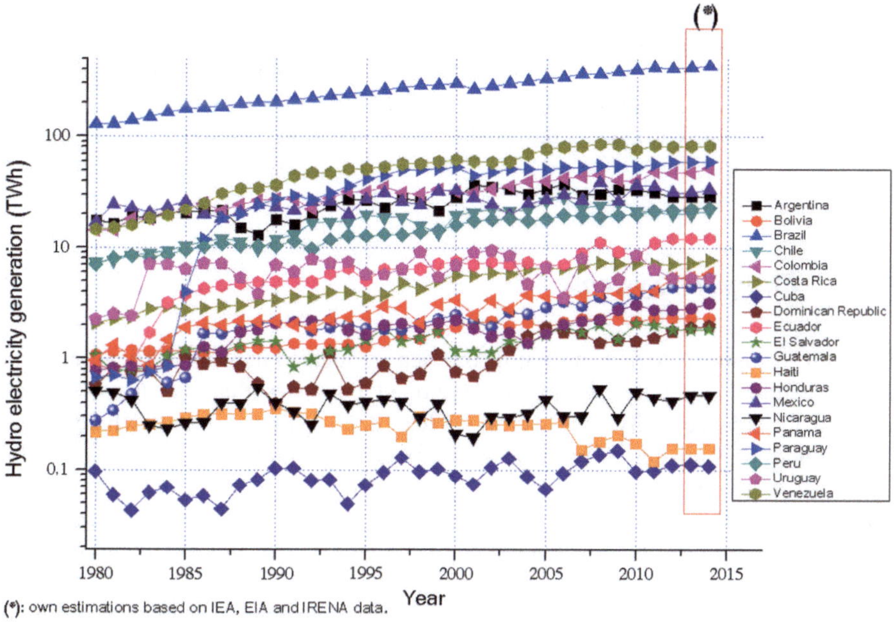

Fig. 8.18 Evolution of hydropower electricity generation in Latin American countries in the period 1980–2014 [1]

Case Example 8.10

Alternative use of the spilled water at Itaipu 14GW hydraulic plant in Paraguay:

Itaipu Dam has a nominal installed capacity of 14 GW, 20 generating units of 700 MW each: 10 operating in 50 Hz and 10 in 60 Hz. The spillway has the function of discharging all the water not utilized for electricity generation. The maximum discharge capacity of the Itaipu spillway is 62,000 m^3/s. A consistent amount of energy is spilled from the dam because electrical grids in Paraguay and Brazil do not require it at a particular time of the day, and because of the water demand of the Parana River. Average spilled values strongly depend on the season of the year: during the winter season spilled energy is nearly zero, from May to July it is between 100 and 200 MW, and for the remaining period of the year it is larger than 300 MW, with peaks of about 900 MW in the summer season. The average yearly value is about 340 MW, but in recent years it has reached a 264 MW low (2008) and a 1508 MW high (2010). The authors propose the production of hydrogen by water electrolysis employing time-dependent hydraulic energy related to the spilled water. Two different scenarios are considered: (i) classical hydrogen physical storage using tanks at 200 bar; and (ii) reduced size hybrid physical storage, where energy to feed electrolyzers is taken from Paraguay electrical grid when spilled water energy is not available. Results depend on the cost of electricity and hydrogen in the market, plant location and size [41].

8.3.2 Pumping Systems

Pumping systems in Africa and Latin America are located in South Africa, Morocco and Argentina, compared to many developed countries that use these systems initially to balance daily surplus of nuclear energy in the hours where power consumption diminishes. Currently, as renewables are increasing their share in the power grid, these systems are also very helpful to increase the penetration of non-dispatchable technologies. The share of hydro pumping capacity can vary widely depending on the characteristics of the power mix and interconnections to neighbouring countries. In many cases pumping capacity is small because it is only considered as a balancing option (Fig. 8.19), as in Argentina. However, there are countries as Luxembourg (highest worldwide share of pumping capacity on the

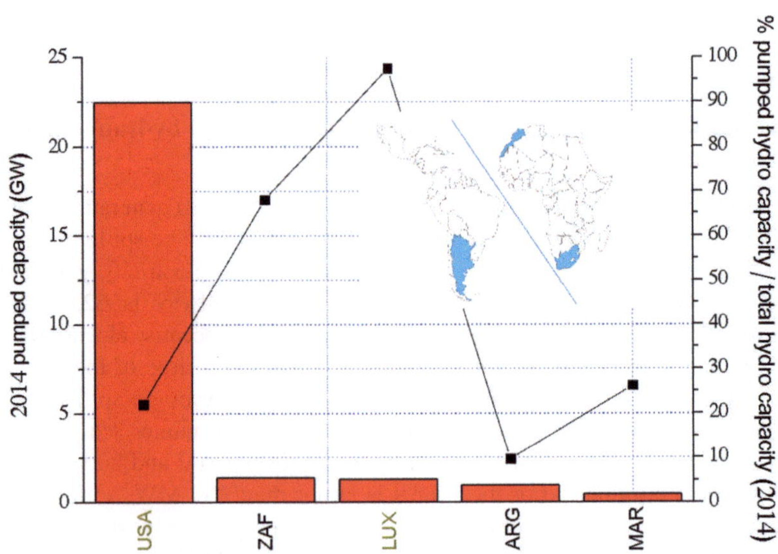

Fig. 8.19 Hydro pumped capacity of the top pumping country (USA), the country with the highest share of pumping hydro capacity (Luxembourg) and the countries with pumping systems in Africa and Latin America (2014) [1]

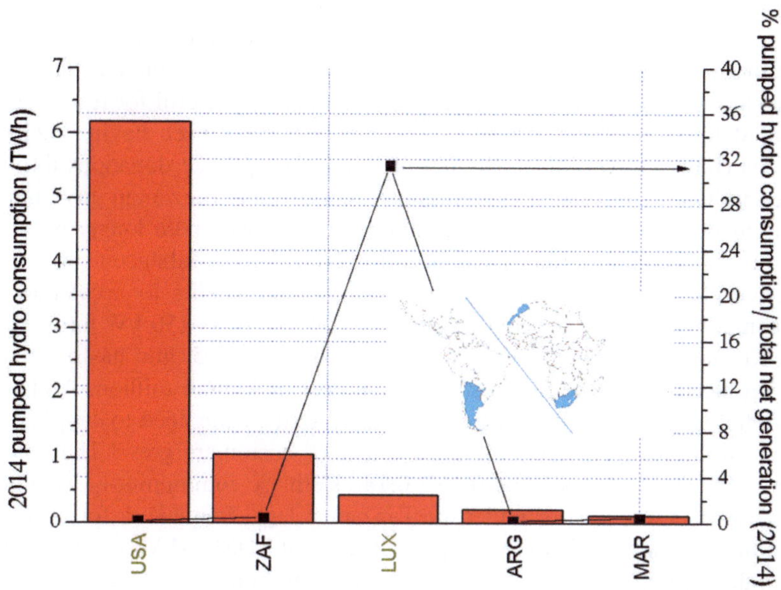

Fig. 8.20 Annual hydro pumping consumption and percentage of electricity consumed in pumping systems compared to total net generation for the top pumping consumer (USA), the country with the highest penetration of pumping activity (Luxembourg) and the countries with pumping systems in Africa and Latin America (2014) [1]

hydropower system) where the share is higher due to their location in a regional power grid. This role that could be played by similar countries in Africa and Latin America in future power grid extensions and upgradings.

The share of pumped hydro consumed by national power systems is normally very low (Fig. 8.20) because these systems are considered for storing electricity surpluses. However, in the case of Luxembourg the electricity consumed for pumping is high because the country is highly interconnected with other countries. Consequently the pumping systems of Luxembourg are not only used for self-regulating the national grid but also for trading electricity with neighbours. These trading activities can also play a role in Africa and Latin America in future as interconnections increase because of the large hydro capacity and hydro potential that some of the countries in these regions can offer.

Case Example 8.11

Economic changes after the NGO-based implementation of a small-scale off-grid hydropower system in Tanzania:

The authors explore a number of constraints an Italian NGO in partnership with a local church and funded by international and national donors have overcome in order to achieve sustainable electricity supply for rural electrification in Ludewa District, Tanzania. NGOs are often considered champions of local participation and empowerment, aspects very useful for reaching the goals defined in energy projects. Key strategies have been having multiple donors (Italian government, Region of Lombardy, private donations through Intervita Onlus, European Commission, World Bank Tanzanian government and District Council Ludewa), an integrated approach and a longer presence than usual. Another strategy has been the NGO active enhancement of productive electricity use, community services and benefits to accelerate the customer base. The hydropower system of 300 kW (2 × 150 kW turbines) in the Kisongo river basin started with the project in 2005, and has been supplying electricity since 2010, with a total cost of EUR 6 million. A change from flat tariffs to pre-paid meters was motivated by the need to simplify the administration of payments as the number of customers grew. The cost of electricity defined was EUR 0.067–0.148/kWh. A community-based utility (LUMAMA) was created under the Tanzanian legal framework for NGOs to own the electricity system. An important share of LUMAMA's income comes from the milling machines and businesses. Households use electricity mainly for lighting and for charging of the mobile phone, and some use much less than the expected average (as little as 5 kWh per month). Interviews in 2013 showed that the large majority of local people, including LUMAMA staff, didn't think the utility could be economically independent but, on the other hand, they also think that the utility makes a good job [42].

8.3.3 Marine Technologies

In relation to marine energy, only a 50 kW onshore wave energy converter installed at the port of Pecem in San Goncalo do Amarante (Brazil) is producing electricity from 2012 [43]. The converter cost was about USD 7.1 million and has two modules consisting of a float, arm, and pump that are fixed to a breakwater and linked to a turbine-generator and hyperbaric chamber. No other device or prototypes are detected in Latin America.

In Africa, there are only two countries showing some activity in marine energy. The Nigerian Institute for Oceanography and Marine Research is starting some research activities for defining OTEC facilities offshore the Nigerian Continental shelf. Moreover, the South African Stellenbosch University shows research activities on wave energy, the University of Cape Town on environmental and regulatory requirements, and the University of Witwatersrand related to the linear synchronous generator [44].

8.4 Conclusions and Future Perspectives

Hydropower plays an extraordinarily important role in the energy future of both Africa and Latin America. More than 20% of total global hydropower generation occurs in Africa and Latin America combined, and more than 90% of existing renewables in both regions is produced with hydropower. Also, the hydropower capacity in Africa and Latin America is expected to increase over time as several nations are expected to pursue new hydropower development or expansion of existing capacity. Moreover, it is estimated that 92% of the existing total hydropower capacity in Africa is untapped. DR Congo has the highest potential on the continent (nominally 40 GW) and Cameroon the third highest potential (nominally 20 GW). Conversely, hydropower production in some countries exceeds 85% of demand with more than 100% of electricity demand in Paraguay produced with hydropower. Consequently, in many countries energy-intensive industry users (e.g., production of aluminum) can be located as they can consume large quantities of low-cost power produced by hydropower.

The production of electricity from hydropower in African countries depends on the availability of the resource and the existing power demand. Several African countries obtain almost all electricity from hydropower whereas others have almost no capacity because of limited or no hydro-resource or a GDP that limits electricity demand. Small hydropower capabilities are likely to become more pervasive over time as environmental concerns, drought and climate change may limit efficiency of larger facilities. In particular, mini-hydro could provide many of the rural electricity. Moreover, climate change may drastically impact the hydropower and consequently, the system's capability to supply load. Indeed, two lessons can be extracted from recent Brazilian and Venezuelan hydro crisis: the need for diversification from hydropower and the importance of distributed generation.

Finally, marine energy plays a very marginal role in renewable electricity production worldwide (999 GWh global production in 2014) and virtually no role in Africa or Latin America. As technology evolves and costs decrease marine energy may play a role in select future locations however it is largely believed that more traditional renewable resources will be adopted much earlier in the energy development cycle.

References

1. International Energy Agency database (2015). http://data.iea.org/. Accessed 30 Nov 2015
2. IEA-ETSAP and IRENA (2015) Hydropower. Technology brief
3. IRENA (2015) Resource. http://resourceirena.irena.org/gateway/dashboard/. Accessed 30 Nov 2015
4. IEA-OES (2006) Policy report
5. IEA-OES (2015) Annual report

6. IRENA (2014) Tidal energy. Technology brief. http://www.irena.org/documentdownloads/publications/tidal_energy_v4_web.pdf. Accessed 30 Nov 2015
7. Ramage J (2004) Hydropowerity. Renewable energy. Oxford University Press
8. Berhane G, Kebede S, Gebreyohannes T, Martens K, Van Camp M, Walraevens K (1996) An integrated approach for detection and delineation of leakage path from Micro-dam reservoir (MDR): a case study from Arato MDR, Northern Ethiopia. Bull Eng Geol Environ 75:193–210
9. Grill G, Lehner B, Lumsdon AE, MacDonald GK, Zarfl C, Liermann CR (2015) An index-based framework for assessing patterns and trends in river fragmentation and flow regulation by global dams at multiple scales. Environ Res Lett 10:015001
10. Pase HL, Da Rocha HJ, Dos Santos ER, Patella APD (2016) The sociopolitical conflict in hydroelectric enterprises. Ambiente & Sociedade 19:45–66
11. World Commission on Dams Final Report (2009)
12. de Falcão AFO (2008) The development of wave energy utilisation. Annual report 2008. IEA Implementing Agreement on Ocean Energy Systems
13. Centre for Renewable Energy Sources (2006) Coordinated action on off-shore energy: ocean energy conversion in Europe recent advancements and prospects. Greece 2006
14. Alonso R, Solari S, Teixeira L (2015) Wave energy resource assessment in Uruguay. Energy 93:683–696
15. Mendoza E, Silva R, Zanuttigh B, Angelelli E, Andersen TL, Martinelli L, Nørgaard JQH, Ruol P (2014) Beach response to wave energy converter farms acting as coastal defence. Coastal Eng 87:97–111
16. NOVELTIS' Global marine tidal currents. http://tips.noveltis.com/tips-metadata.html. Accessed 30 Nov 201517
17. Schiermeier Q et al (2008) Electricity without carbon. Nature 454:816–823
18. U.S. Department of Interior (2006) Technology white paper on Ocean current energy potential on the U.S. Outer Continental Shelf. Minerals Management Service
19. NEEDS (2008) Final report on technical specifications of reference technologies (wave and tidal power plant). New Energy Externalities Developments for Sustainability (NEEDS)
20. Sorensen B (2000) Renewable energy. Academia Press, London
21. IEA-ETSAP and IRENA (2014) Salinity Gradient Energy
22. The Economist (2009) No pinch of salt. Science and Technology. Available 3 Dec 2009
23. Skråmestø ØS, Skilhagen SE (2008) Status of technologies for harnessing salinity power and the current osmotic power activities. Annual report 2008. IEA Implementing Agreement on Ocean Energy Systems
24. Nihous GC (2008) Ocean thermal energy conversion (OTEC) and derivative technologies: status of development and prospects. Annual report 2008. IEA Implementing Agreement on Ocean Energy Systems
25. IEA (2008) Energy technology perspectives
26. The Economist (2016) Venezuela. Lights out. Available 7 May 2016
27. Climatescope (2015) The clean energy country competitiveness index. The Multilateral Investment Fund, UK Department of International Development, Power Africa and Bloomberg New Energy Finance
28. ExternE-Pol (2007) Extension of accounting framework and policy applications
29. IEA-ETSAP and IRENA (2014) Wave energy. Technology brief
30. IEA-ETSAP and IRENA (2014) Tidal energy. Technology brief
31. IEA-ETSAP and IRENA (2014) Ocean thermal energy conversion
32. The Economist (2009) No pinch of salt. Science and Technology. Available 3 Dec 2009
33. Krewitt W (2009) External costs of energy technologies. NEEDS Meeting. Brussels, 16–17 Feb 2009
34. EU-OEA (2010) Oceans of energy—European ocean energy roadmap 2010–2050. European Ocean Energy Association (EU-OEA)
35. McKinsey & Company (2015) Brighter Africa. Electric power & natural gas

36. World Bank (2014) International development association project appraisal document on a proposed grant in the amount of SDR 47.7 Million to the Democratic Republic of Congo for an Inga 3 Basse Chute and Mid-size Hydropower Development Technical Assistance Project
37. Bloomberg New Energy Finance (2016) Week-in-review. Available 1 Aug 2016
38. Banco de Desarrollo de América Latina (2012) Nuevas Oportunidades de Interconexión Eléctrica en América Latina. CAF
39. Jenkins M, Guevara LM (2014) Financing renewable energy: La Esperanza hydroelectric project. Manag Decis 52:1724–1749
40. Bazilian M, Chattopadhyay D (2016) Considering power system planning in fragile and conflict states. Energy Sustain Dev 32:110–120
41. Rivarolo M, Bogarin J, Magistri L, Massardo AF (2012) Time-dependent optimization of a large size hydrogen generation plant using "spilled" water at Itaipu 14 GW hydraulic plant. Int J Hydrogen Energy 37:5434–5443
42. Ahlborg H, Sjöstedt M (2015) Small-scale hydropower in Africa: socio-technical designs for renewable energy in Tanzanian villages. Energy Res Soc Sci 5:20–33
43. Laboratório de Tecnologia Submarina. Universidade Federal do Rio de Janeiro. http://www.hydroworld.com/articles/2012/07/brazil-wave-power-prototype-generates-first-energy.html. Accessed 30 Nov 2015
44. OES-IEA (2015) Annual report 2014. Ocean energy systems

Chapter 9
Geothermal Energy

Ricardo Guerrero-Lemus and Les E. Shephard

Abstract This chapter shows that geothermal electricity provides a small amount of electricity in Africa and Latin America. That is consistent with global trends, and that capacity is being developed in many countries in East Africa and Central America. All of these countries are located close to the edges of tectonic plates where high temperature geothermal resources are available near the surface. Geothermal power plants mainly use flash or binary configurations because enhanced geothermal configurations are still under development. Kenya is currently developing its own geothermal plants for increasing the availability of the resource and for exporting the technology and associated services. Heat energy obtained from shallow underground areas at nearly constant temperatures of about 60–70 °C and injected to heat pumps is not used in Africa and Latin America. However, there are options to use this geothermal heat to locate specific industries and spas in hotels in the future. Again, Kenya is leading the way to achieve success in these areas. The need to find the necessary demand and connection to the national power grids are observed to be the main impediments to reach a long term growing trend in geothermal energy.

Keywords Energy · Africa · Latin America · Electricity · Geothermal energy · Heat pumps · Enthalpy · Permeability · Binary systems · Flash systems · Tectonic plates · Hydrothermal

9.1 Introduction

The temperature gradient in the earth's crust is typically 30 °C/km, although there are areas where it is above 150 °C/km. When the gradient is low, the geothermal region is qualified as a normal or low enthalpy region, while for high gradients of temperature it is designated a high enthalpy region. Consequently, geothermal energy is defined as the energy available as heat contained in or from the earth's crust. The main sources for geothermal energy are the heat flow from the earth's core and mantle ($\sim$40%) and the heat flow generated by the gradual decay of

R. Guerrero-Lemus and L.E. Shephard, *Low-Carbon Energy in Africa and Latin America*, Lecture Notes in Energy 38,
DOI 10.1007/978-3-319-52311-8_9

radioactive isotopes in the earth's continental crust (~60%) [1]. Exploitable geothermal resources are hydrothermal systems containing water in pores and fractures with sufficient permeability to produce fluids in adequate volume. Traditionally, the geothermal energy has been extracted close to the edges of tectonic plates where high temperature geothermal resources are available near the surface (Fig. 9.1). Thus, many locations in East Africa and Central America are suitable for geothermal energy production and, in fact, several countries in these areas are producing electricity based on this resource. Also heat energy obtained from underground areas at nearly constant temperature of about 60–70 °C, and injected into heat pumps is used for heating buildings and homes, aquaculture, horticulture and industrial processes.

Direct heating from renewable sources has grown slowly over recent years. Bioenergy, solar thermal and geothermal energy accounted for 0.55 EJ devoted to produce heat in 2014, compared to 0.19 EJ in 1990 with geothermal energy representing 5.4% in 2014 and a 7.9% in 1990 [2]. However, geothermal heat capacity has grown slightly ahead of the average trend for renewable technologies (+5.9% annual) [3]. In the building sector in particular, district heating has gained importance and heating from renewable technologies currently provides 7% of the final energy used for heat. However, no geothermal energy has been accounted for producing heat in Africa and Latin America [2].

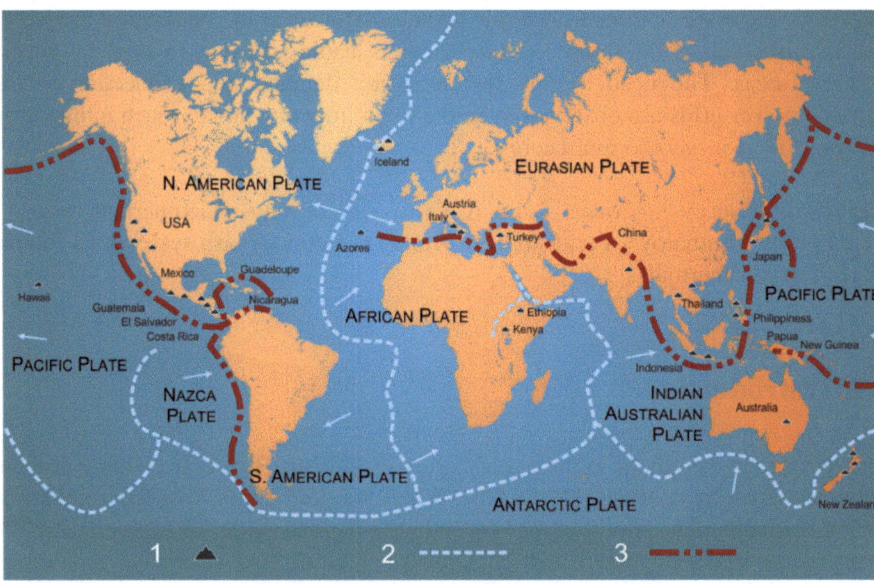

Fig. 9.1 World pattern of tectonic plates, oceanic ridges, oceanic trenches, subduction zones, and geothermal fields. Arrows show the direction of movement of the plates towards the subduction zones. **(1)** Geothermal fields producing electricity; **(2)** mid-oceanic ridges crossed by transform faults (long transversal fractures); **(3)** subduction zones, where the subducting plate bends downward into the asthenosphere

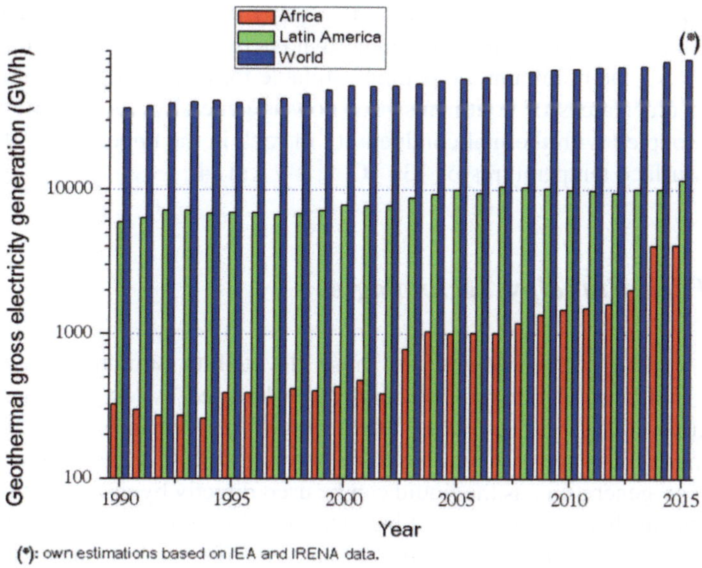

Fig. 9.2 Evolution of the percentage of geothermal gross electricity generation worldwide, in Africa and Latin America in the period 1990–2015 [2]

Geothermal electricity represents 0.32% of all electricity produced worldwide in 2014 (last data available), and geothermal heat represents 0.22% of all heat produced [2]. However, in Africa these values are 0.55 and 0% and in Latin America are 0.63 and 0%. If we consider geothermal generation in regional values, the trend is upward worldwide with sharp increases in Africa and Latin America in recent years due to new capacity added in some specific countries (Fig. 9.2). Thus, in 2015 it is estimated 4.1 TWh geothermal generation in Africa, 11.6 TWh in Latin America and 80.2 TWh worldwide.

Geothermal energy has many positive aspects, which can be summarized as follows: (i) ability to provide base load power; (ii) stable to seasonal variations; (iii) stable against effects of weather and climate change; (iv) integrated in centralized and decentralized production plants; and (v) availability of resources in all regions of the Earth.

Also geothermal energy has some negative aspects: (i) high investment costs; (ii) high risk to obtain the geothermal resource at an affordable cost; (iii) lack of knowledge about exact resource availability; and (iv) environmental effects.

9.2 Technology State of the Art

The geothermal technology is mainly mature but there are areas where the technology can further improve, mostly to obtain electricity from dry rocks and/or low temperature resources. For describing the state-of-the-art of the geothermal energy it is adequate to distinguish between high enthalpy and low enthalpy technologies. Although direct use applications of geothermal energy can use temperatures as low as about 35 °C, the minimum temperature suitable for electrical generation is about 135 °C. In older areas of continents heat flow is generally 40–60 mW·m^{-2}. This heat flow coupled with thermal conductivity of rock in the upper 4 km of the crust yields subsurface temperatures of 135 °C at 5–7 km [4].

9.2.1 High Enthalpy Technologies

The high enthalpy technologies are those devoted to simultaneously produce electricity and heat. Geothermal electricity can be used as low cost base load in power grids, as the resource offers high performance and reliability. High-temperature water or steam-based resources (>180 °C) are the most efficient for electricity generation, as the liquid can be used directly by dropping the pressure to create steam that can drive a turbine. The initial exploration (e.g., surface geophysics) is used to map the underground and identify geothermal reservoirs, followed by exploratory drilling, which provides additional information on underground conditions and defines the extent of the reservoir and its characteristics (e.g., pressures, temperature, flow rates, etc.). The geothermal field characteristics will have a significant influence on what type of power plant can be used, and on well productivity. The operational life of a geothermal power plant is considerably high, reaching in some cases more than 50 years.

Geothermal formations are generally characterized as hot, abrasive, highly fractured, under-pressured, and containing corrosive fluids. These conditions mean that drilling is usually difficult (rate of penetration and bit life is usually low). All commercial hydrothermal resources exist as dual porosity systems which have both matrix and fracture permeability. When high productivity fractures are coupled to a low permeability and porosity matrix, there can be steam in the fractures and water in the matrix, resulting in "dry steam" to create power generation economically. However, in these systems highly productive fractures can introduce circulation losses (loss of drilling fluid to pores or fractures in the rock formations being drilled), requiring treatment and, thus, increasing well costs [5].

The geothermal power plants mainly use flash, binary or enhanced configurations, as exposed below. The thermal efficiency of the geothermal electric plants is low (10–23%), as the geothermal fluids obtained do not reach the high temperatures of steam applied to conventional power boilers. However, as the geothermal resource is almost constant over time (unlike other renewable energy sources), the capacity factor of geothermal systems to produce electricity can be large (up to 96%). A minimum 200–300 m distance between boreholes is needed to avoid any interference between them.

Flash geothermal is the technology most widely used. In this case, the well tank collects the high-pressure hot water and reduces this pressure in the separator, using

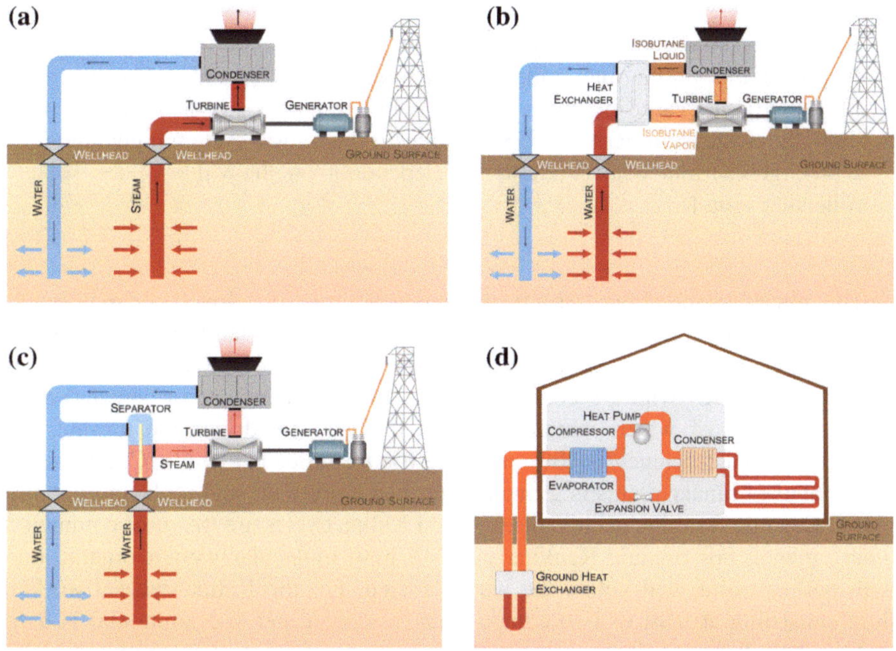

Fig. 9.3 Scheme of the **(a)** flash technology; **(b)** binary technology; and **(c)** enhanced geothermal technology for electricity heat production from geothermal resources. Also, the scheme of a **(d)** geothermal heat pump is included

the generated steam to drive turbines for electricity production (Fig. 9.3a). The steam is condensed after crossing the turbine and re-injected. The required fluid temperature must be above 180 °C.

Case Example 9.1

Study of a solar-geothermal hybrid power plant in northern Chile:

This work evaluates the performance of single and double-flash geothermal power plants assisted by a parabolic trough solar concentrating collector field for the superheating or evaporating (geothermal brine) processes. The reduction of the brine flow rate allows extending the geothermal reservoir life. The goal is to achieve maximum efficiency and to reduce the geothermal resource consumption for a constant power production in a location where daily average direct normal irradiation (NDI) is higher than 9 kWh/m^2 and there are 6 already explored geothermal areas. Results show that the hybrid single-flash power plant is able to produce at least 20% additional power output and reduces the consumption of the geothermal fluid up to a 16% (and 19% for the hybrid double-flash), depending on the characteristics of the geothermal resource. For the single-flash hybrid system the superheating process generates additional 0.23 kW_e/kW_{th}, while supplying solar heat to evaporate the geothermal brine only delivers 0.16 kW_e/kW_{th}. The energy efficiency is increased by at least a 3%. A LCOE analysis is also included to show the decrease in terms of USD/MWh for the hybrid system compared to a pure geothermal configuration, but depending on the value of the solar collection area [6].

The binary plant (Fig. 9.3b) is the most recently developed technology for electricity production from geothermal resources. Binary plants typically use low temperature fluids (<180 °C) in the exchanger (having accepted fluids with temperatures of 57 °C) with a secondary fluid boiling point (butane or pentane) much lower than that of water, which generates the gas that drives the turbine. The average thermal efficiency is 10%, reaching 0% for fluids below 60 °C. At present, the number of binary plants is growing very fast worldwide and replacing the flash plants. Organic Rankine cycles (mainly) and Kalina cycles are the most common in binary plants. For organic Rankine cycles a high molecular weight organic fluid with boiling point well below the water is used. For the Kalina cycle a working fluid consisting at least of two components (usually water and ammonia) is used, and the ratio between them varies in different parts of the system to obtain higher reversibility and, therefore, to increase the total thermodynamic efficiency of the system.

The EGS (enhanced geothermal systems) technology, formerly called hot dry rock, creates or enhances artificial geothermal reservoirs underground, circulating water (closed loop) from the surface through wells (Fig. 9.3c) to fracture the hot, dry rock and artificially enhance the permeability of the reservoir rock by fracturing. The injected water is heated and extracted through other wells and sent to binary plants to generate electricity. This system can be extended to geothermal areas that are not tectonically active. Some of these plants operate in a cascade mode, as the water used to produce power in the geothermal plant is subsequently derived for district heating before being re-injected into the geothermal reservoir.

There are several methods to solve the environmental problems caused by the small amounts of chemical pollutants (boron and arsenic) and gases (H_2S and CO_2) contained in the extracted water. Mainly, these techniques are performed as the water is re-injected in the reservoir, the gas is condensed, and some chemical treatments are applied for the extraction of minerals. The costs of these processes are estimated at about 1–2% of the total costs [7]. As a drawback, these systems require large amounts of water for recirculation. Thus, a small plant of 5 MW may require about 350 m^3 of water per hour to operate, and a commercial plant would require 10 times that amount [8].

Case Example 9.2

Geothermal power plants impact on surrounding plants and soil in Kenya:

This work tested the hypothesis that emitted elements from two geothermal power plants Olkaria I (operated since 1981) and Olkaria II (operated since 2003) would have an impact in close plant tissue and soil with consequences for plant health. The concentrations of sulphur, arsenic, boron, antimony and mercury in the soil and leaves of the dominant shrub were essayed and associated foliar injury and growth traits assessed at variable distances and directions. The study detects elevated sulphur concentrations close to the power plants and decreased with increasing distances. Similar trends were not detected in soil and other elements. The study didn't support the observed higher degree of leaf injury or growth traits close to the power plants. Overall, the levels of sulphur, arsenic, boron and antimony in leaves, and sulphur and boron in soil around the power plants were higher than the reference site well out of the range of element deposition from the power plants. Follow-up experimental studies and studies on more sensitive ecosystem components are recommended to advise mitigation measures against chronic exposure from the emitted gases and associated impacts [9].

Depending on market penetration, the generation of energy from geothermal resources is led by the flash technology, as it is the most mature. Binary technology is becoming more cost competitive, with significant ongoing R&D activities, mainly in organic Rankine and Kalina cycle technologies. Furthermore, EGS systems (advanced systems) require still great technological efforts to be competitive.

9.2.2 Geothermal Heat Pumps

This technology uses underground areas located at a constant temperature at variable depths obtaining heat through a closed water circuit. The resource is considered to provide thermal stability for space and water heating. This technology works in two different configurations: (a) vertical well heat exchangers (Fig. 9.3d); and (b) horizontal subsurface loops. The former are more expensive, but have improved performance and occupy less space. This technology is used for multiple applications: spas and swimming pools, snow melting, radiators, heating of water, radiant panels, air conditioning, cereal forage, breeding barn, greenhouse conditioning, vegetable drying, food processing, fisheries, metal leaching, sludge digestion, curing of concrete blocks, dry cloth, pulp and paper, etc. Geothermal heat pump technology is mature, although some degree of innovation is still expected in the future.

Case Example 9.3

Evaluation of earth-air heat exchanger for cooling and heating a poultry house in Morocco:

One of the main obstacles facing the poultry industry in Morocco is climate conditions including heat stress waves in summer and cold waves in winter, causing mortality and drop in poultry performance (weight). Often a high level of coventional energy is required to address this problem. An alternative approach is to consider geothermal heat pumps to thermo-stabilize the area. The depth of the buried tubes are about 2 m to avoid unnecessary cost. The tube is nonmetallic material (polypropylene and PVC). The heat exchange is considered in a steady-state, constant ground temperature and physical properties. The total pressure drop is calculated by choosing the proper length, diameter of the tube and the blower (fan) to compensate linear and singular losses (tees, restrictions and elbows). The indoor building description is given by the American Society of Heating, Refrigeration and Air Conditioning Engineers (ASHRAE): 30–33 °C, 50–80% relative humidity, 0.1 L/s ventilation rate per kg live mass during winter and 1–2 L/s in summer. Ground temperatures of six representative climatic regions in Morocco are chosen and results show that the air temperature inside poultry houses could be lowered to below 26 °C in hot periods and heated up to above 14 °C in cold periods. No consideration about costs are included [10].

Case Example 9.4

Self-powered desalination of geothermal saline groundwater in Tunisia:

The authors considered groundwater sources at <100 °C and the Organic Rankine Cycle (ORC) for driving Reverse Osmosis (RO) for four different locations, one of them in the Kebili geothermal field, Tunisia. This field lies above the intercalaire aquifer which covers an area of 600,000 km^2 extending into Tunisia, Algeria and Libya, and typically the water emerges at 70 °C. The aquifer is already exploited for heating of greenhouses, bathing, and for irrigation of crops. However, water salinity is in the range 2500–5000 mg/kg, making it unsuitable for human consumption and only marginally suitable for agriculture. The authors conclude that the overall recovery ratio strongly depends on the geothermal source temperature, but it offers attractive recovery even for temperatures below 70 °C, and it can be as high as 91.5% for a hot spring emerging at 96 °C with a salinity of 1,830 mg/kg. Great care is needed in the design of the plant to minimize losses: (i) the use of a multi-stage continuous flow RO design; (ii) specific valves; (iii) specific membrane area; (iv) large evaporator and condenser surface area [11]. This option could be a consideration in many locations. Geothermal resources can be depleted if aquifer recharging is not considered in parallel to resource exploitation.

9.2.3 Costs

The cost of geothermal power generation depends on the resource pressure and temperature, depth and permeability of its location, the chemical composition of the fluid, the location, drilling market characteristics, plant size and type of plant. Costs also are affected by the price of raw materials (oil, steel and concrete). Also, geothermal power presents more dynamic questions than for some other renewable technologies and projects must be carefully managed in order to optimise the resource (manage a program of reservoir monitoring, well testing and maintenance and drilling) to reduce uncertainties and costs.

The total installed costs of conventional flash geothermal power generation projects grew to USD 1900–3800/kW in 2009, and the more expensive binary power plants saw installed costs for typical projects increase to USD 2250–5500/kW [12]. Cost ranges for small-scale, low-temperature resource binary plants are likely to be higher than those for excellent geothermal reservoirs and resources (Fig. 9.4), and are typically in the range USD 5000–10,000/kW [13].

The costs of electricity generation from geothermal energy can be known only after the project is operational, but as an orientative approximation, flash technology ranges from USD 52 to 80/MWh for high temperature resources, for lower temperature binary technology costs range between USD 59 and 107/MWh [15], and for EGS from USD 100 to 300/MWh [15].

In terms of levelized cost of electricity (LCOE), the International Renewable Energy Agency (IRENA) establishes a range of USD 0.045–0.080 kWh in Africa with an average value of USD 0.080 kWh and a range of USD 0.045–0.245 kWh in

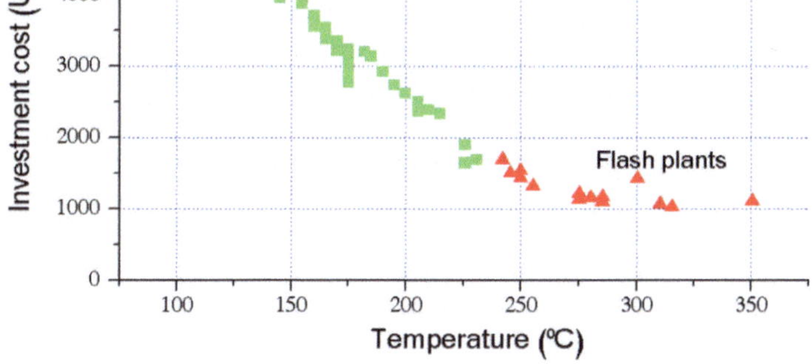

Fig. 9.4 Indicative power plant only costs for geothermal projects by reservoir temperature in the USA [14]

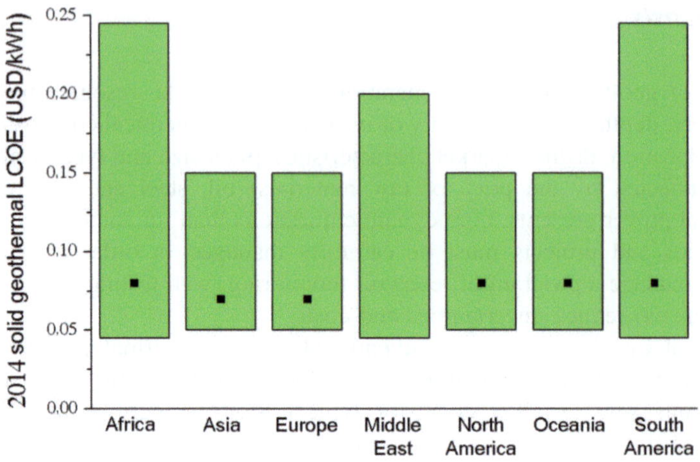

Fig. 9.5 Levelized cost of electricity from geothermal energy in the different global regions (2014) [16]

South America with an average value of USD 0.080 kWh (Fig. 9.5) [16]. Thus, on average and compared to other regions, the production of electricity from geothermal energy can be favourable in both regions.

Investment costs for geothermal district heating can be approximated to USD 570–1570/kW$_{th}$ [15]. The district heating costs oscillates between USD 45–85/MWh$_{th}$ [15]. The O&M costs range from USD 9 to 25/MWh$_{th}$ [17], and it is a small percentage of the total costs due to the lack of fuel expenses. These costs are dependent on the power capacity, characteristics of the location, type and number of plants and the use of remote control systems. Such costs would experience significant increases if the well needs to be replaced. The estimated investment cost for the energy supplied by geothermal heat pumps is highly variable depending on the country being analyzed, ranging from USD 439 to 600/kW$_{th}$ for China-India to USD 1,170–2,267/kW$_{th}$ for Europe. The estimated average heating costs for this technology are about USD 80/MWh$_{th}$ [2].

The key parameters to reduce costs are higher temperatures and flows. Consequently, the drilling characteristics are critical to make a geothermal resource profitable. Because drilling costs increase substantially as the required drilling depth increases (Fig. 9.6), the EGS technology nearly doubles in cost per kWh compared to the costs corresponding to more conventional systems [15]. Geothermal drilling is more expensive (in cost per depth) than on-shore oil and gas drilling for three principal reasons: (i) special tools and techniques are required for the harsh downhole conditions; (ii) large diameters are required because the produced fluid (hot water or steam) is of intrinsically low value; and (iii) geothermal wells, even in the same field, are more different among each other than oil and gas wells in the same field, so the learning curve from experience is less useful. Also, an indirect cost is attributed to the fact that almost all produced fluids must be re-injected, thus requiring additional

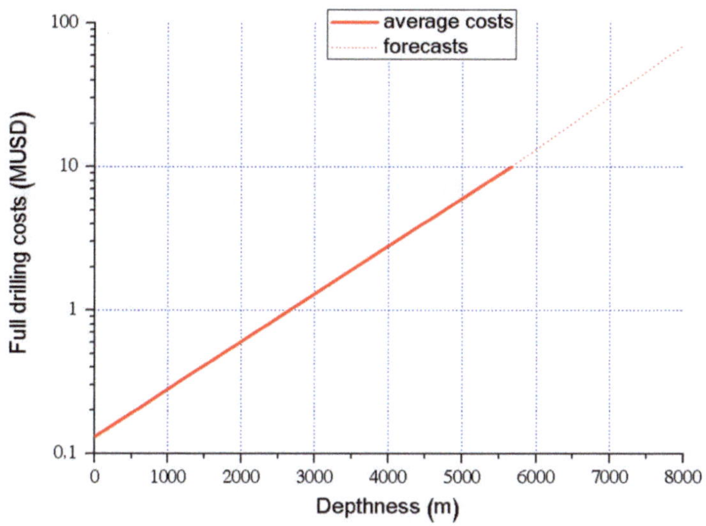

Fig. 9.6 Drilling cost for a typical geothermal plant [16]

wells. Added to all of these factors, the cost of drilling an injection well field can rise to 50% of the total project cost for a geothermal power plant [5].

On the other hand, CO_2 emissions from low-temperature geothermal resources are negligible (0–1 g/kWh [8]) and most binary systems are operating in closed loop. Consequently, their emissions are negligible. In addition, CO_2 emissions from geothermal heat pumps are reduced by at least 50% compared to oil boilers, depending on the source considered for electricity production [8].

9.3 Regional and National Perspectives on Technology

9.3.1 Evolution on Electricity Produced from Geothermal Energy

Geothermal electricity output is only produced in 2 African and 5 Central American countries (Fig. 9.7) but is an important share of the power mix in the smallest Central American countries. In Africa, Tanzania is developing geothermal projects, Djibouti is aiming for projects to come on-stream in 2020 [18] and some activity is detected in Uganda, Rwanda, Zambia and the Comoros Islands. All of these countries are located close to the edges of tectonic plates where high temperature geothermal resources are available near the surface, as it has been exposed above (Fig. 9.1). In absolute terms, Mexico is leading the geothermal power sector in both regions under study, but in terms of power mix generation, geothermal resources accounted for more than half of Kenya's electricity output in 2014 [18]. Ethiopia has

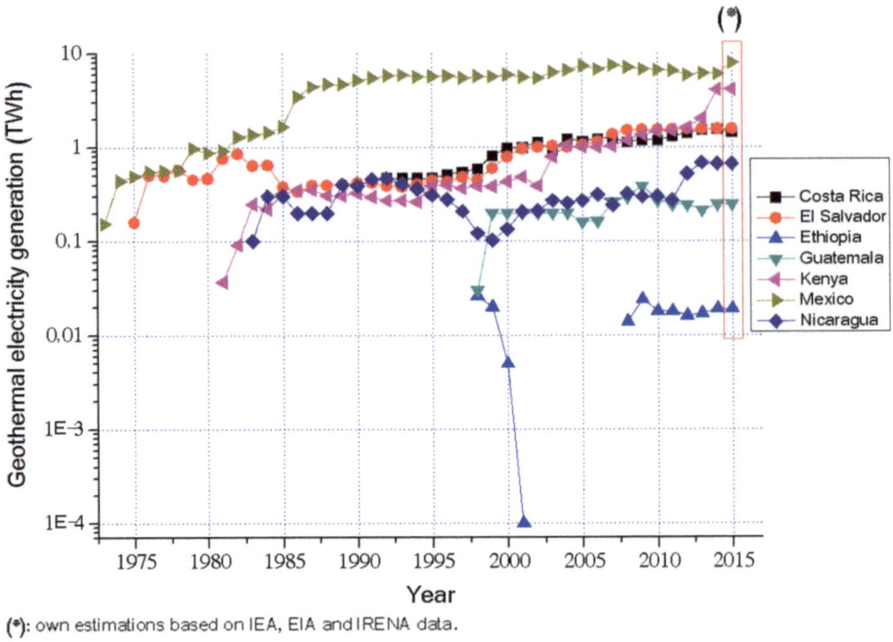

Fig. 9.7 Geothermal electricity output in African and Latin American countries in the period 1973–2015 [2]

vast geothermal resources but the technology remains untapped with only 7 MW installed and an additional 20 MW to be added in the near future from the Corbetti project (the goal is to expand this project to 1 GW) [19].

Case Example 9.5

The Corbetti Geothermal Power Plant in Ethiopia:

This power plant project is the result of an agreement between the Government of Ethiopia and the Ethiopian Electric Power Corporation (EEPCO) with Reykjavik Geothermal (Iceland) to establish the first independent power producer project in Ethiopia's history. The plan involves the construction of a geothermal power project of 1000 + MW_e in two phases, with the first 500 MW_e of capacity producing electricity in 2018 and the remaining in 2021. Reykjavik Geothermal has acquired geothermal exploration licenses covering an area of more than 6500 km^2, and the scientists have pinpointed an area of 200 km^2 in which high temperatures up to 350 °C have been identified. Heads of Terms for a 25-years + PPA is negotiated with EPPCO. The total cost of the first phase is estimated at USD 2 billion [4].

Electricity production from geothermal energy for the Latin American countries started in Mexico in 1959, with the commissioning of the first commercial plant in Pathé field (central Mexico) that was in operation until 1973 [21]. In Africa, Kenya was the first country introducing geothermal power generation (Fig. 9.7). An upward trend is observed in most countries but some instabilities are observed in Nicaragua and a large interruption and very low values evident in Ethiopia. In Mexico the new Geothermal Energy Law, passed in August 2014, should facilitate private exploration and development of new geothermal areas, and regulations issued in October 2014 detail the process to be followed [21].

Case Example 9.6

San Jacinto-Tizate geothermal power plant in Nicaragua:

The largest geothermal power plant in Nicaragua (72 MW) is owned and operated by Polaris Energy Nicaragua, S.A., a subsidiary of Ram Power, Inc., of the U.S. It is based on two condensing flash units, each of 36 MW capacity, with a dessert course of a binary bottoming cycle of 10 MW. In 2013 it supplied the country with 12% of its total electricity supplies. Spot market prices for suitable fuels over a three-year period have ranged from USD 80–160/MWh, a level at which geothermal power can be competitive. The geothermal area (40 km^2 concession) has a resource capacity estimated at over 200 MW. An initial 10 MW power plant project was commissioned in 2005, and the 72 MW net capacity became fully operational in December 2012. It required the drilling of two injection and production wells to a depth of 2500 m. The drilling programme encountered some technical well damage problems but solved the issue with acid-stimulation. The total cost of the project was USD 309 million and power produced is under a long term PPA with local utilities [22].

9.3.2 Electricity Share from Geothermal Energy

Kenya plays a significant role in geothermal energy in terms of capacity and share in the power mix (Fig. 9.8). This country is planning to reach 5.5 GW geothermal capacity in 2030, has expertise in project development, engineering, O&M and resource development, and the Kenyan Geothermal Development Company (GDC) has become somewhat of a regional center for geothermal, assisting neighboring countries with exploration of their resources and training their workers [19]. Thus, GDC is promoting the Bogoria-Silali Geothermal Project Phase I. Only Mexico is a country in Latin America that offers expertise and manufacturing capacity in all the geothermal production chain: project development, engineering, O&M, resource development, turbines and balance of plant [19].

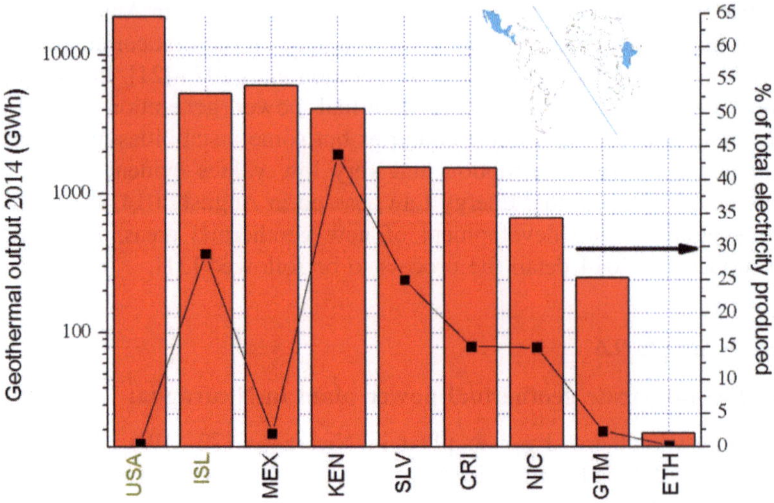

Fig. 9.8 Geothermal electricity output and share of geothermal energy in the top geothermal producer countries in absolute terms and as the share of the power mix and in African and Latin American countries (2014) [2]

Case Example 9.7

KenGen and geothermal policy support in Kenya:

Kenya Electricity Generating Company Limited (KenGen) is the leading electric power generation company in Kenya, producing about a 75% of electricity capacity installed in the country. It has assisted neighboring countries, including Rwanda, Eritrea and Zambia in the assessment and development of their geothermal resources. The main lessons learned in Kenya are: (i) long term focus from the government, the donor community (i.e., World Bank, European Investment Bank, and KfW from Germany), the private and public sector (the government showed that was willing to bear part of the risk of developing resources); and (ii) specialization and training of geothermal experts who currently integrate the core of KenGen specialized unit on geothermal power. KenGen is the largest geothermal producer in Africa (508.8 MW, plants availability of 89.17% in 2014) and completed the single largest geothermal power plant in the world (Olkaria, 280 MW) in 2015. The company is promoting geothermal wellhead projects to move quickly to produce power from a single well while the whole geothermal plant is being built, and also expects that some industries will consume geothermal steam directly, promoting a Industrial Park in Olkaria for this purpose. Other activities are the construction and management of a geothermal health spa in Olkaria and the provision of commercial drilling services to external companies [20].

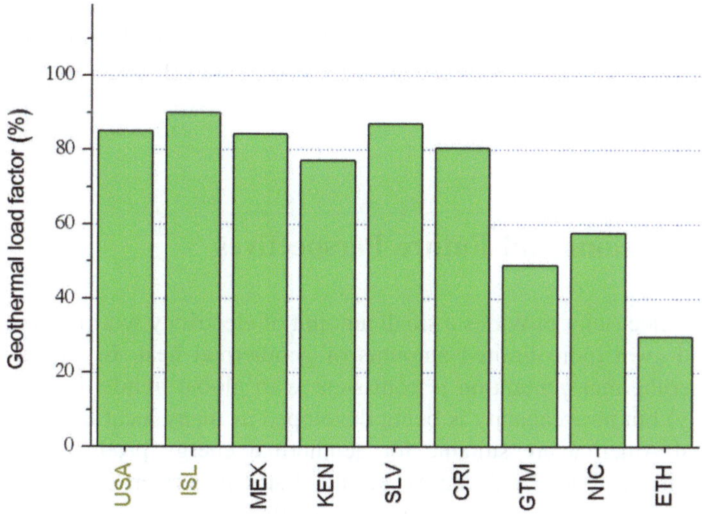

Fig. 9.9 Geothermal load factor in African and Latin American countries in 2012 [2, 16]

Finally, is it interesting to show the load factor of the African and Latin American countries producing electricity from geothermal resources because there is as substantial variability in the range (Fig. 9.9) [2, 15]. Most countries show a capacity factor very close to 80%, except Guatemala, Nicaragua and Ethiopia. These results mean that the operation of geothermal power plants in many of these countries is similar in terms of efficiency to the most advanced countries in the world in the area of geothermal energy (Iceland and USA).

Case Example 9.8

Geothermal energy in El Salvador:

El Salvador is the Latin American country with the largest share of geothermal energy integrated in the power grid (204.4 MW). Only one company (LaGeo) is devoted to the development of geothermal energy. The technical potential of geothermal energy in El Salvador is estimated at 791 MW. There are 12 high enthalpy geothermal areas. The capacity values (for 25 years) have been estimated in 1988 by means of the USGS (United States Geological Survey) method. The average temperature varies between 153 and 233 °C. LaGeo has made some estimates about a 30 MW geothermal project and considers 8 years for the commision of the plant and investments costs about USD 150–200 million. There exists also the possibility to exploit low enthalpy geothermal resources (118–141 °C) using the binary technology. The main barriers for the development of geothermal energy in El Salvador are: (i) a large amount of administrative procedures and time to complete (at least 2 years); (ii) difficulties for the acquisition of the land;

(iii) high population density in the geothermal areas; (iv) difficulties to obtain water for the drilling procedure; and (v) the success ratio of exploration drilling is about a 25% [23].

9.4 Conclusions and Future Perspectives

Geothermal electricity provides a small amount of electricity worldwide (nominally 0.31%) and even more limited amounts of geothermal heat. In Africa and Latin America geothermal generation is consistent with global trends (0.27 and 0.64%, respectively) but new capacity is being developed as many locations in East Africa and Central America are suitable for geothermal energy production. Electricity production from geothermal energy for the Latin American countries started in Mexico in 1959 while in Africa, Kenya was the first country introducing geothermal power generation. Two African and five Latin American countries currently produce electricity using geothermal energy but development is ongoing in several small Latin American countries and in Tanzania, Djibouti, Uganda, Rwanda, Zambia and the Comoros Islands. All of these countries are located close to the edges of tectonic plates where high temperature geothermal resources available near the surface.

Geothermal power plants mainly use flash or binary configurations because enhanced geothermal configurations are still under development. The geothermal plants typically have low thermal efficiency and high capacity factors (up to 96%). Indeed, it is remarkable how Kenya is developing its own geothermal plants for increasing the availability of the resource and for exporting the technology and associated services.

Heat energy obtained from underground areas at nearly constant temperatures of about 60–70 °C, and injected to heat pumps is used for heating buildings and homes, aquaculture, horticulture and industrial processes, but not in Africa and Latin America. However, there are options to use this geothermal heat to locate specific industries and spas in hotels. Again, Kenya is leading the way to achieve success in these areas.

In conclusion, geothermal power and heat is growing in Africa and Latin America because of endogenous resources in some very well defined locations. The need to find the necessary demand and connection to the national power grids are the main barriers to attain a long term growing trend.

References

1. IEA-GIA (2011) Geothermal energy. International efforts to promote global sustainable geothermal development. 13th Annual Report—2009
2. International Energy Agency database (2015). http://data.iea.org/. Accessed 30 Nov 2015

3. Renewables (2016). Global Status Report. Renewable Energy Policy Network for the 21st Century. REN21 Secretariat, Paris
4. Reykjavik Geothermal. Corbetti geothermal power. http://www.rg.is/static/files/about-us/rg-corbettigeothermalpower.pdf. Accessed 12 Apr 2016
5. Finger J, Blankenship D (2010). In: Handbook for best practices for geothermal drilling. Sandia report (SAND2010-6048), Sandia National Laboratories
6. Cardemil JM, Cortés F, Díaz A, Escobar R (2016) Thermodynamic evaluation of solar-geothermal hybrid power plants in Northern Chile. Energy Convers Manag 123:348–361
7. IEA (2010) Renewable energy essentials: geothermal
8. IEA (2008) Energy technology perspectives
9. Mutia TM, Fridriksson T, Jónsdóttir IS (2016) Concentrations of sulphur and trace elements in semi-arid soils and plants in relation to geothermal power plants at Olkaria, Kenya. Geothermics 61:149–159
10. Laknizi A, ElMaakoul A, Abdellah AB, Bouya M, Dhimdi S, Said S (2015) Evaluation of earth-air heat exchanger for cooling and heating a poultry house: case study in Morocco. In: Proceedings of the 2015 IEEE international renewable and sustainable energy conference, IRSEC 2015, 7455027
11. Davies PA, Orfi J (2014) Self-powered desalination of geothermal saline groundwater: technical feasibility. Water 6:3409–3432
12. IPCC (2011) IPCC special report on renewable energy sources and climate change mitigation. IPCC, Geneva
13. IRENA (2015) Renewable power generation costs in 2014
14. NREL (2012) Renewable electricity futures study, vol 2
15. International Energy Agency (IEA) (2011) Technology roadmap. Geothermal heat and power
16. IRENA (2015) Resource. http://resourceirena.irena.org/gateway/dashboard/. Accessed 30 Nov 2015
17. EIA (2011) Short-term energy and winter fuels outlook, Oct 2011
18. IRENA (2015) Off-grid renewable energy systems: status and methodological issues
19. Climatescope (2015) The clean energy country competitiveness index. The Multilateral Investment Fund, UK Department of International Development, Power Africa and Bloomberg New Energy Finance
20. KenGen (2016) 2015 Integrated annual report and financial statements. http://www.kengen.co.ke/sites/default/files/financial-reports/2015%20KenGen%20Annual%20Report.pdf. Accessed 12 Aug 2016
21. IEA (2015) Annual report 2014. IEA Geothermal implementing agreement
22. Chin C, Wallace K, Harvey W, Dalin B (2013) Long M (2013) Big iron in Nicaragua: a muscular new geothermal plant! The San Jacinto-Tizate geothermal project. GRC Transactions 37:687–692
23. Consejo Nacional de Energía (2012) Plan Maestro para el Desarrollo de la Energía Renovable en El Salvador

Chapter 10
Wind Energy

Ricardo Guerrero-Lemus and Les E. Shephard

Abstract This chapter shows that wind energy is an established technology for producing low cost electricity in Africa and Latin America. However, these regions contribute less than 5% of the total global wind production and have no offshore production installed or planned. In Latin America wind resources are being developed in 12 nations and in a few countries it has become a significant contributor to total power production. As importantly, the manufacturing of wind turbine blades, towers, and balance of plant components has become a key contributor to local economies in some Latin American countries and is contributing to a nascent industry throughout the region. Future expansion of wind resources in both regions will likely require significant investments in transmission as the best wind resources are located distant from existing urban demand centers. Also, the usually strong public support in both regions should be accompanied with a defined strategy to avoid a growing resistance in some communities opposing large-scale wind production. Sharing benefits from wind energy with the local communities and avoiding environmental damages are the best recipes to increase wind energy in Africa and Latin America.

Keywords Energy · Africa · Latin America · Electricity · Wind energy · Wind turbines · Off-shore wind · Wind generation

10.1 Introduction

The production of energy from wind is an established technology today that can produce electricity at lower costs than that obtained through conventional sources in many areas of Africa and Latin America. This is especially true in places where there are appropriate wind speeds for extended periods of time or areas where electricity costs are high because they rely on small systems. Globally, the latest data available (2014) estimates a 3.1% wind energy share of the world electricity production [1]. If we consider the wind energy generation deconvoluted into

R. Guerrero-Lemus and L.E. Shephard, *Low-Carbon Energy in Africa and Latin America*, Lecture Notes in Energy 38,
DOI 10.1007/978-3-319-52311-8_10

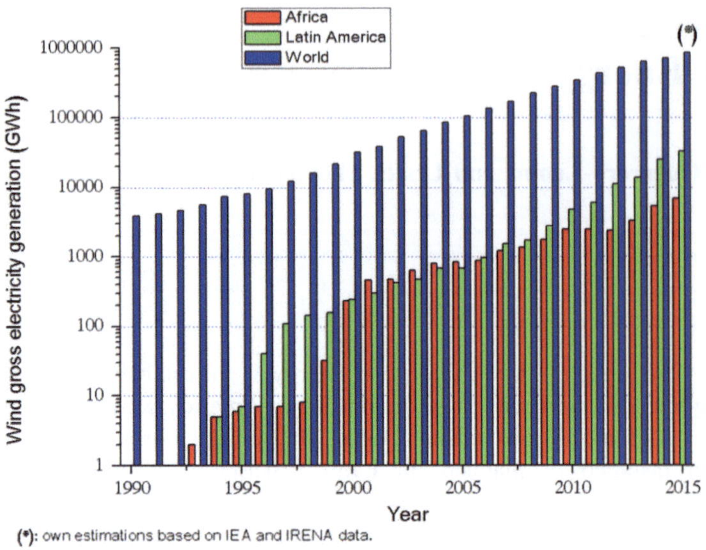

Fig. 10.1 Evolution of the annual production of electricity from wind systems in the period 1990 to 2015 [2, 3]

regional values, non negligible values are obtained in Africa from 1993 and Latin America from 1994. Wind generation in Latin America has surpassed Africa from 1994 to present (Fig. 10.1). In 2015 there was an estimated 6.9 TWh wind generation in Africa, 32.1 TWh in Latin America and 864.2 TWh worldwide.

In the future we anticipate similiar expansion of wind energy in Africa and Latin America as both regions have substantial wind resources (Fig. 10.2). However, in both regions the best wind resources are located away from major demnad centers and will rely on a regional power grid to transport the electricity produced to large urban areas, as it is being done in the USA. Consideration of wind capacity factor values using annual peak demand hours (capacity factors estimated during the peak demand hours within a year) substantially increases the geographic distribution and abundance of favorable wind zones, compared to only considering annual average capacity factors [4].

Globally there is significant general public support for wind energy but increasing resistance by local communities to specific wind farm projects. The social conflict can be related to multiple factors: (i) the perception of damages to natural and cultural heritage; (ii) the perception of negative strong effects that the wind farm development would have on local productive activities; (iii) local community distrust toward the company, local authorities and environmental regulators; and (iv) the threat to life projects of the local people [5]. This is also true in emerging countries.

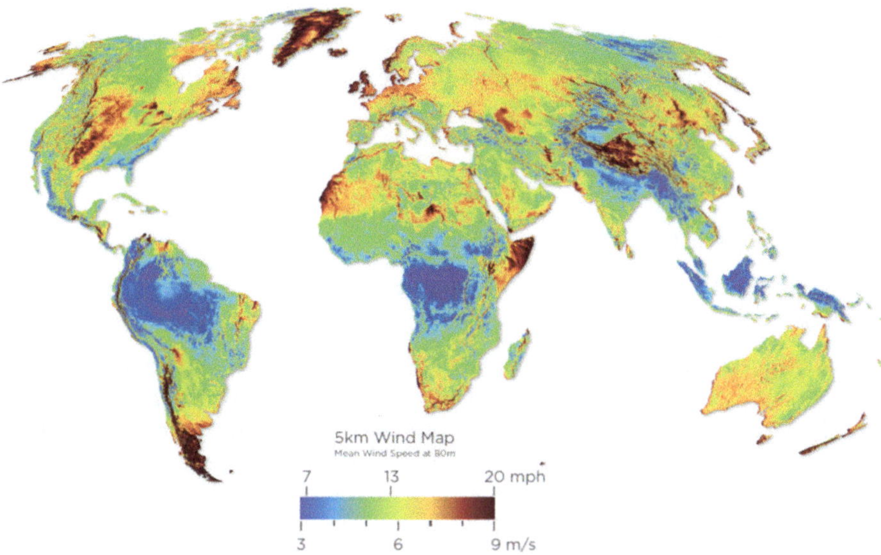

Fig. 10.2 Global Wind Dataset 5 km onshore wind speed at 80 m height units in m/s. *source* Vaisala Inc. 2015 ©

Case Example 10.1

Social response to the installation of a wind farm in Chile:

A 1000 ha wind farm (100.8 MW, USD 250 million investment) was projected for the Isla Grande de Chiloé (9181 km^2). 66.9% of the island is covered by native forest (extinct in many Chilean areas) and 27.4% by grassland. The island has 39,946 inhabitants (31.7% rural) and is only accessible by ferry. A wind farm (42 wind mills) was promoted by Ecopower, a Chilean-Swedish company, that also planned roads, a substation, underground power lines and a port in a rural area (Mar Brava) with about 2712 inhabitants including three indian communities (huichiches) and productive activities devoted to agriculture, livestock, fishing and tourism. The Chilean legislation asks for an environmental evaluation (EE) when such projects are close to populated or protected areas. The first controversy emerged when the company defined the area as non populated and non protected and the environmental qualification of the project was approved by the Government (August 2011). Subsequently, the local community considered the resolution illegal and arbitrary, and filed appeals and protection initiatives, firstly rejected by the Court but later admitted. In parallel, different political initiatives started in the Congress and appeals to the President of Chile. Finally, a new EE was restarted in November 2013 and the consultation to the native communities was included. As a result, the project design was modified, the local workforce was considered for the construction phase and a Foundation was established to manage a percentage of the earnings of the wind farm to fund projects of the local communities [5]. The EE was finally approved on September 2015.

Case Example 10.2

Determinants of community acceptance of a wind energy project in Tunisia:

The Sidi Daoud wind farm project that comprises 26 wind turbines of 1320 kW each, with a total installed capacity of 34.32 MW was analyzed [6]. A 22.5 km high voltage transmission line (90 kV) links the wind farm to the nearest central transformation station of Menzel Temime and allows for exporting the total quantity of power generated by the project to the Tunisian interconnected grid. Empirical data were collected from both primary and secondary sources: documents, archival records, in-depth interviews and direct observation. Participants made it clear that deriving private or collective gain (e.g., revenue sharing, job creation, preferential electricity tariffs, investment in local roads and public services) boosts the scope of resentment and protestation. Public attitudes toward wind energy are related to familiarity with the technology, education, financial situation, the visual impacts, land occupied and noise levels. Also, because local people are not familiar with the technology, acceptance may mostly depend on trust in the individuals that are responsible for the project and communication with them. This work recommends to policy makers the need to capture the full significance of community acceptance in order to ensure a smooth implementation of the investment. Project managers should also try to adapt the project to its context; the community approach to renewable energy initiatives can also contribute to the support of the project, as local people could take the initiative in developing their own wind farms [6].

10.2 Technology State of the Art

Wind turbines convert the wind's kinetic energy collected by the rotor into electrical energy through a generator. Usually the turbine rotor is connected to a gearbox and an induction or constant-speed generator. However, gearless or direct-drive turbines are increasing in recent years [7]. Wind power is obtained mainly using horizontal-axis 3-blade wind turbines because they have reduced mechanical wear and the lowest visual impact. The power extracted from the wind is proportional to the circular area covered by the blades when they rotate, the air density and the cube of the wind speed. Consequently, it is very important to correctly evaluate the wind resource in every potential location because small variations in average wind speed largely influence the electricity produced over the years.

10.2.1 Wind Turbines

Most wind turbines are installed on-shore but efforts are increasing to locate wind turbines off-shore, in shallow waters (<50 m depth) or as floating turbines in deeper waters (hundreds of meters) largely because wind speed at a given height from the sea surface is usually higher than at the same height on the ground. In addition, offshore wind is generally more stable, predictable and has less turbulence. Moreover, the best on-shore locations in many developed countries have already been occupied. However, the technologies related to offshore wind turbines are less mature than those installed on-shore because of the greater technological challenges related to installation and O&M.

The maximum rotational speed of the turbine blades is limited by the turbulence generated by the wind at their tips. To limit the turbulence the linear speed of the blade tip is limited to about 100 m/s (360 km/h), which corresponds to wind speeds up to about 17 m/s (60 km/h) in current large turbines [8]. Wind turbines generate electricity from wind speeds ranging from 3 m/s or 4 m/s to 25 m/s (even 34 m/s with storm control). If the wind speed exceeds these maximum limits then the turbine stops to avoid damage [7]. This stop is achieved by modifying the angle of the blade with respect to the wind direction, or automatically, as wind turbulences increase at specific blade edges above a given wind speed.

Each turbine is characterized by a specific power curve that is a function of the wind speed, which depends on the characteristics of the turbine. The nominal power of the turbine is its maximum in the power curve, and the average power is the power provided in terms of the average winds in the specific location.

The maximum wind power available for exploitation is limited, according to Betz law, to 16/27 (approximately 60%) of the instantaneous wind power. This limitation is due to the fact that a 100% wind power exploitation would stop completely the wind, preventing its evacuation after moving the wind blades and, consequently, being an obstacle for new wind reaching the turbine. In addition, the mechanical power obtained from wind should be further converted into electricity. Then, in practice, the maximum conversion efficiency of wind energy into electricity is usually around 40%. With respect to offshore wind farms, growth has been also very significant, mainly in Europe, growing from 67 MW (0.4% of total wind capacity) in 2000 to 8761 MW (2.4% of total wind capacity) in 2014. No offshore wind farms are located or expected in the near term in Africa and Latin America.

As mentioned above, the standard wind turbine has three blades, a horizontal axis and is connected to the grid. The maximum diameter of the circle covered by the blades has increased from 20 m in the 80's, to 50 m in the 90's and to about 164 m at present (Vestas V164). In the long term, this maximum diameter is expected to grow more moderately, mainly due to increasing logistic and planning difficulties.

Generally, the power of a generator is expressed for a given wind speed, which is usually 12 m/s. Currently, the installed turbines typically have a power range between 2 and 8 MW. The efficiency for electricity generation is measured in terms of electricity produced per unit area swept by the blades (kWh/m^2), and it has been growing over 2% annually in recent years [9]. For example, a 5 kW small turbine can produce 200 W/m^2, which is almost half the value obtained for a 1 MW turbine.

The capacity factor is about 0.3 for current turbines. In some locations with intense and almost constant winds, the capacity factor can reach 0.4, and there are locations where the capacity factor can exceed 0.5. Capacity factors have increased, but can be impacted by curtailment and inter-year wind resource variability [10]. Onshore turbines can guarantee 97% time availability for the production of energy over many years [9] and offshore power plants availabilities are also over a 90% [7]. The vast majority of wind turbines now installed have fault ride-through capabilities and offer active and reactive power control, thanks to power electronics, to increase wind integration in the grid.

10.2.2 Small Wind Turbines (SWT)

For the small-scale wind sector, there are different definitions of 'small' wind turbines, but the third revision of the international small wind design standard IEC 61400-2 defines them as wind turbines with a rotor-swept area less than 200 m^2. This definition corresponds to approximately 50 kW of nominal generation capacity or less. The last annual report of the World Wind Energy Association (2016) notes that at the end of 2013, a cumulative total of at least 945,000 SWTs (830 MW) were installed worldwide, increasing by 8.3% (+10.9% in capacity) compared to the previous year [11]. Most of this growth is in China, the United States and the United Kingdom but also some growth is observed in Africa (Morocco) and Latin America (Cuba). Globally, the average size of small wind turbines is growing: 0.66 kW in 2010 and 0.87 kW in 2014. 74% of the commercialised one-piece small wind manufacturers invested in the horizontal axis orientation, while only 18% have adopted the vertical axis design and 6% use both technologies. Over 80% of the manufacturers produce stand alone applications [11].

The SWT have typically been applied to off-grid supply and battery charging, but currently there is an increase in application for small distributed power grids. The SWT uses a permanent magnet alternator (direct-drive) configuration that produces "wild" AC or variable frequency and voltage electricity converted through an inverter into a stable signal. Although higher voltages have more efficient electrical conversion, lower-voltage systems are more typical of global off-grid applications. The system for electricity generation in microturbines is usually simpler than for large turbines, since it supplies mono-phase current instead of three-phase current. Moreover, despite using a variable speed rotor, it does not affect the frequency of the signal, since the electricity is usually stored in batteries previously to being injected in a local grid. Consequently, the concept of urban wind power is growing in relation to the introduction of small wind turbines on buildings in cities and isolated residential areas. Also small wind turbines are being introduced in medium-sized public areas and parks. For their proper exploitation, very detailed aerodynamic models on the behaviour of wind around buildings, streets, river banks, canals, etc., have been created. The industry faces a continuous and highly damaging influx of products with inflated performance claims and/or misleading product labels; therefore, the major SWT manufacturers strongly support standards and certification requirements [12].

Case Example 10.3

Wind pumps for greenhouse micro-irrigation in Ciego de Ávila, Cuba:

A methodology based on daily estimation balance between water needs and water availability was used to evaluate the feasibility of the most economic windmill irrigation system. The windmill is a 6 m-high conical tower with a triangular section, 2.2 m rotor diameter, eight blades and 2.77 m/s minimum wind speed. Different wind speeds, elevation heights and evapotranspiration as a function of the crop (tomato) planting date were considered. Monthly volumes of water required for irrigation, water tank capacity and monthly irrigated area were estimated. The results show the optimum period of wind-pump driven irrigation, the tank capacity and the cultivated area for a defined system operating at an established height elevation [13].

10.2.3 *Offshore Wind Turbines*

Offshore wind turbine technology (Fig. 10.3) emphasizes the development of large wind turbines to offset costs associated with construction and operation in the marine environment. Unlike onshore turbines, off-shore turbines often include power generators that supply power at high voltage to eliminate the need for transformers, and modifications to the nacelles to incorporate dehumidification technology and other advances to reduce maintenance costs.

Most offshore wind turbines have been built in shallow waters (<25 m depth) and relatively close to the shoreline (<20 km). The water depth limit for offshore foundations is established at present at to about 50 m [9], as the foundation costs increase very rapidly with depth.

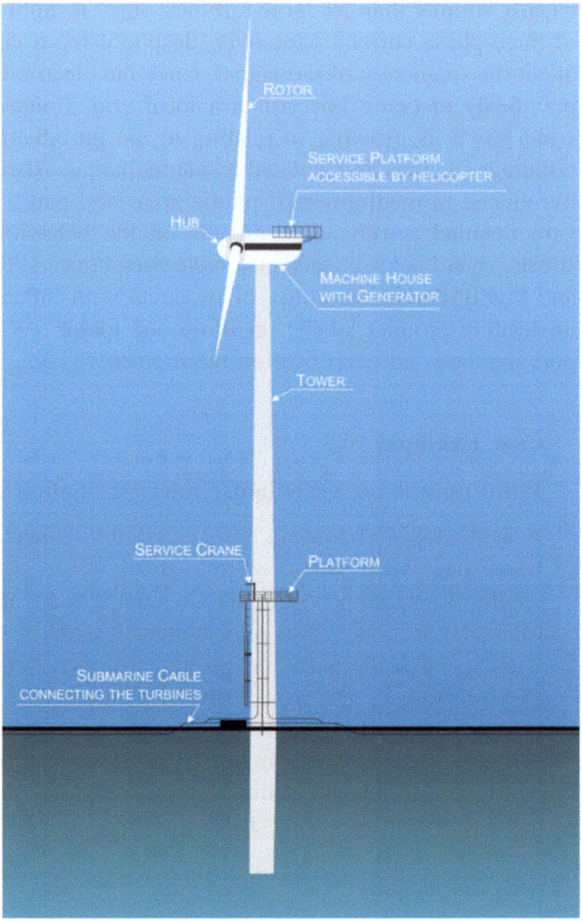

Fig. 10.3 Scheme of a typical offshore wind turbine [11]

A significant challenge for offshore wind turbines as they move into deeper water with more hostile weather environments is related to the design and construction of foundations or moorings to the seabed. Figure 10.4 shows the three most common underwater foundations used at present [9]. The monopile consists of a large pile inserted into a deep boring with the seafloor. This technique is usually applied for shallow waters (<15 m). For deeper depths the gravity foundation is very common, which consists of a base, usually of concrete, placed on the seabed. For depths, up to about 25 m, a tripod-like foundation is also used but it is not as common as the previous two types of foundations. The tripod can be composed of three (Fig. 10.4) or more piles. There is also the option of floating wind turbines, which are gaining attention because of the expected decrease in cost and their use in deeper waters anchored to the seabed.

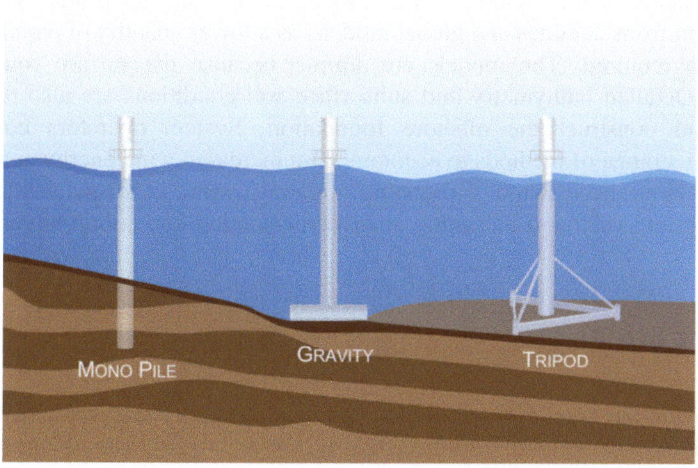

Fig. 10.4 Common foundations for offshore wind turbines: monopile, gravity and tripod

Commonly, transportation and installation for offshore wind turbines has been based on conventional engineering. However, as wind farms are installed in more remote and difficult locations, an alternative strategy is based on building the offshore turbines in port and transporting them in one piece to their location similar to practices empoyed in the offshore oil industry. Another key challenge is the complexity of the regulatory environment in many potential areas for the location of offshore wind turbines.

10.2.4 Wind Resources

The correct evaluation of wind resources is of great importance, both before the wind farm is built in order to determine the best location and after construction to more accurately predict the amount of the electricity production of the wind farm. This second objective is gaining in importance as non-manageable renewable energies increase their share to the grid and the grid operators require knowing in advance the amount of wind energy that should be delivered to the grid.

Commonly the wind resource analyses is made using an array of meteorological stations equipped with anemometers, and pressure and temperature gauges, properly located to ensure a prediction that should match the future production of the wind turbine. In contrast, more sophisticated predictions combine information from satellite data, global forecasts, radar, surface observations and other variables, which, combined in a suitable computer model, allow predictions decreasing in accuracy as the temporal distance between the measurement and prediction is extended.

In the analysis for offshore wind resources, the data are primarily based on information from satellites and global models, as a lower quantity of parameters and values are required. The models are simpler because the surface roughness is minimal. Detailed bathymetry and subsurface soil conditions are also required to design and construct the offshore foundation. System operators continue to implement a range of methods to accommodate increased wind energy penetrations, including centralized wind forecasting, treating wind as dispatchable, shorter scheduling intervals, and balancing areas consolidation and coordination.

10.2.5 Costs

Onshore wind has been able to compete in many electricity markets where the wind resourse is steady without subsidies for several years [14]. This is attributed to a combination of low wind turbine prices and improved turbine technology. Thus, wind turbine prices have continued to decrease and remain well below levels seen several years ago. For example, in Brazil, wind projects have reached grid-parity with conventional sources and have become one of the main sources of new capacity [15].

Case Example 10.4

Grid parity for wind energy in Brazil:

In Brazil, wind projects have reached grid-parity with conventional sources and have become one of the main sources of new capacity [15]. Reaching grid parity in Brazil is due to a combination of several characteristics: large wind resources in several areas, manufacturing capacity of wind turbines, expertise and the use of auctions to define PPAs. Thus, in the last auction there have been 18 tenders in which renewables have competed, contracting a total of almost 17 GW, mostly in the form of wind energy (12 GW). As the country is immersed in a macroeconomic crisis, the renewable energy sector has started to feel the impact, especially in 2015 and 2016, Brazil has decided to cancel the only auction for this year while costs have increased and project financing has become more difficult [16].

The total investment costs for onshore turbines in 2015 ranged from USD 850–1250/kW depending on the manufacturer and the turbine model [17]. It means an average costs 20% below the peak in 2008 [18]. The capacity-weighted average installed project cost at USD 1690/kW [17], down USD 690/kW from the apparent peak of 2009–2010. Installed project costs exhibit some economies of scale, at least at the lower end of the project and turbine size range. Wind PPA prices have reached all-time lows in the USA to around USD 20/MWh, including tax credits [17]. Studies in the USA also show that wind energy integration costs are almost always below USD 12/MWh-and often below USD 5/MWh-for wind power capacity penetrations of up to or even exceeding 40% of the peak load of the system in which the wind power is delivered. Recently, an unsubsidized USD 30/MWh bid for an onshore location in Morocco by Enel Green Power was signed in January 2016 [19].

There are endogenous influences (labour costs, warranty provisions, turbine manufacture profitability and turbine design) and exogenous influences (raw materials prices, energy prices and foreign exchanges) that impact wind turbine costs, but there is no a single dominant factor. Turbine up-scaling is, by a significant margin, the largest single driver, as capital costs increase with size. But the advantages in terms of balance of plant costs and higher levels of energy production typically outweigh those turbine price increases with size. It is also important to consider that profit margins are falling from 2008 due to the global financial crisis and the resulting softness in turbine sales [20]. Raw materials prices have the largest exogenous influence, with five materials (steel, fiberglass/resin/plastic, iron, copper and aluminium) accounting for more than 98% of the total. The price of steel (66–81% of total turbine mass) has by far the largest impact on turbine prices. Also, there is a natural temporal lag between turbine purchases and approving the project of the corresponding installation. For example, the wind power plant installed in any year often relies on turbines purchased two years earlier.

Case Example 10.5

Cabeolica wind farm:

Cabo Verde's Cabeolica is the first commercial-scale privately financed, public-private partnership wind farm developed in sub-Saharan Africa. It has 25.5 MW capacity and was established in 2009. The partnership is integrated by the Government of Cabo Verde; Electra, Cabo Verde's government owned national utility; and InfraCo, a private company backed by European Governments and the World Bank, who served as the project's lead developer. The USD 78 million project was financed through debt and equity financing provided by public and private entities: approximately USD 60 million, debt financing provided by the European Investment Bank and the African Development Bank accounted for roughly three-fourths of the financing requirements, with equity financing coming from the African Finance Cooperation, Finnfund, and InfraCo to cover the remaining portion of projects capital costs. Additionally, the project benefits from its participation in the Clean Development Mechanism (CDM), providing an additional stream of financial support through the allocation of over 70,000 Certified Emission Reduction (CER) credits [21].

It is also interesting to note the differences in costs between the facilities onshore and offshore, the causes having been already discussed in the previous section. IEA estimates that, in 2013, offshore investment cost ranged from USD 3600/kW to USD 5600/kW [20], and NREL estimates that the levelized cost of offshore wind energy was USD 215/MWh. Recent data suggest that costs have stabilized (after rising from 2008 to 2013), and many expect decreases through 2020 (as evidence, recent winning bids for competitive subsidies in Europe). In fact, in the Netherlands, Dong Energy broke through its EUR 100/MWh (USD110/MWh) aspiration for offshore wind in 2015, winning an auction on 5 July 2016 with a record-low bid of EUR 72.70/MWh for a 700 MW wind farm close to the shore and the Dutch authorities covering developing expenses and transmission costs. This is the lowest ever price for offshore wind and, on the face of it, is competitive with onshore wind in most of Europe [22]. Water depth, distance from shoreline and project size influence the final investment cost for specific offshore wind farms.

The investment cost for the world's largest floating turbine located in Norway (2.3 MW), has been evaluated at about USD 30,435/kW [18]. These prices make the technology far above the values exposed for offshore foundations-based wind technology. However, a new study considers floating offshore wind LCOE about USD 180/MWh in 2020 and at about USD 110/MWh in 2050, very similar to the LCOE obtained for fixed-bottom offshore wind energy [23].

For wind power plants, information about O&M costs is scarce, of limited content and variable in quality. The O&M costs are very low compared to conventional fuel plants. Therefore, wind power plants can be considered capital intensive compared to conventional ones, for which fuel costs play a major role. Typical annual values of O&M costs for wind farms are between 2 and 4% of the total costs of the facility but represent an important component—15–25%—in the cost of wind power [20], and they have remained almost constant in recent years [11]. Sample analysis in the USA shows that average O&M costs of wind systems constructed in the 1980s equal USD 34/MWh, dropping to USD 24/MWh for projects installed in the 1990s, USD 10/MWh for projects installed in the 2000s and USD 9/MWh for projects installed since 2010 [10]. Offshore O&M costs can be two or three times higher than those of land based systems [9], ranging from USD 20/MWh to USD 48/MWh in 2013 [20].

For SWT in the USA, the installed cost estimates in 2014 ranged between USD 8200/kW (<2.5 kW), USD 7200/kW (2.5–10 kW) and USD 6000/kW (11–100 kW). The Chinese small wind industry yielded, in comparison, a significantly lower average turnover of RMB 12,000/kW (1900 USD) in 2011 [11].

In terms of levelized cost of electricity (LCOE), the International Renewable Energy Agency (IRENA) establishes a range of USD 59–133 MWh in Africa with an average value of USD 90/MWh, USD 62–129/MWh in Central America with an average value of USD 80/MWh, and USD 32–138/kWh in South America with an average value of USD 90/MWh (Fig. 10.5) [3]. Thus, in average and compared to other regions, production of electricity from wind is favourable in South America but less favourable in Africa and Central America.

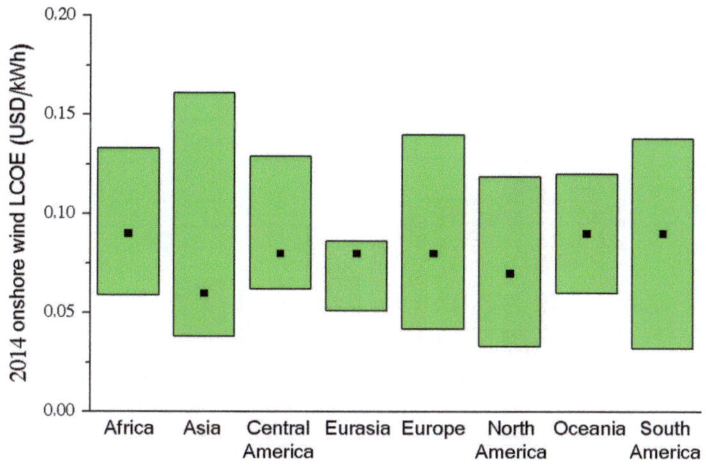

Fig. 10.5 Levelized cost of electricity from wind in the different global regions (2014) [13]

10.3 Regional and National Perspectives on Technology

10.3.1 Electricity Share from Wind Energy

Wind energy is underdeveloped in Africa. Only 6 countries show significant electricity production in 2014 (Fig. 10.6) from wind which is attributed to the high wind potential in the areas where these countries are located: coastal areas in the Maghreb, East and South Africa. These values are compared with the countries where the largest quantity of electricity is produced from wind energy (China) and the country where wind energy adds the largest share to the power mix. In this respect, the share of wind energy in the power mix in these African countries is very small. Also wind energy is observed in 2014 in Cape Verde, Mauritius, Reunion [24]. Kenya offers expertise in wind project development, engineering, O&M and balance of plant [15]. South Africa offers expertise and manufacturing capacity in project development, engineering, O&M, turbines, blades, towers and balance of plant [15].

Case Example 10.6

310 MW Lake Turkana wind farm:

The single largest project under development currently in Africa is a 310 MW wind farm near Lake Turkana, in northwest Kenya. This is a truly flagship project having taken nearly 10 years and a dozen investors to reach financial close in 2014. The developers broke ground in July 2015, with an estimated commissioning date of 2017. A forthcoming 428 km overhead transmission line—funded by the government—will connect the project to Suswa, some 100 km from Nairobi [15].

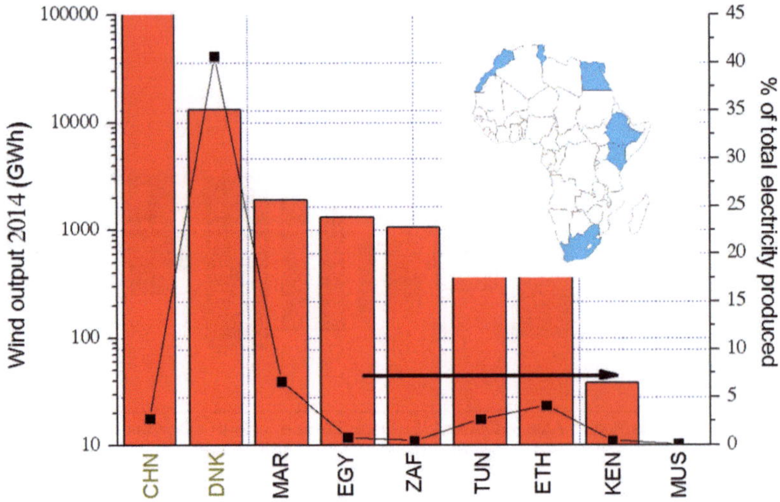

Fig. 10.6 Wind electricity generation per country produced in Africa in the last year available (2014) [2]

In Latin America, wind energy is more developed than in Africa, and wind electricity output is observed in 12 countries (Fig. 10.7). In some Central American countries the wind energy share in the power grid is appreciable. It is important to mention that Bolivia's first wind farm was commissioned in 2014, a 3 MW plant

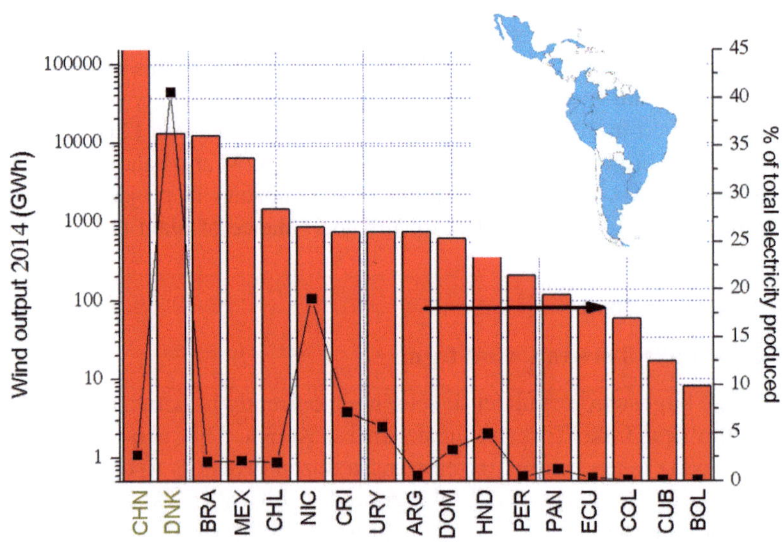

Fig. 10.7 Wind electricity generation per country produced in Latin America in the last year available (2014) [2]

located in the department of Cochabamba. Moreover, Latin America has begun developing a substantial wind power industry to complement its rich hydro and biomass resources. Thus, Brazil counts seven wind turbine manufacturers producing locally from towers to blades [15]. Global wind turbine manufacturers as Vestas, GE and Gamesa, have also opened factories in Brazil in recent years. Moreover, Mexico offers expertise and manufacturing capacity in project development, engineering, O&M, turbines and balance of plant.

10.3.2 Evolution on Electricity Produced from Wind Energy

Wind energy has been steadily growing since the first wind turbines were connected to the grid in Egypt and Cabo Verde in 1993 (Fig. 10.8). Thus, wind energy started to produce appreciable amounts of electricity in Africa earlier than in Latin America. Traditionally, the top wind energy producer has been Egypt but very recently it has been surpassed by South Africa.

In Latin America wind energy started in Mexico (Fig. 10.9) and it is growing fast, but Brazil leads the upward trend. Other countries with large increases include Argentina, Chile and Uruguay in recent years. Wind generation in Cuba and Colombia show stagnant trends.

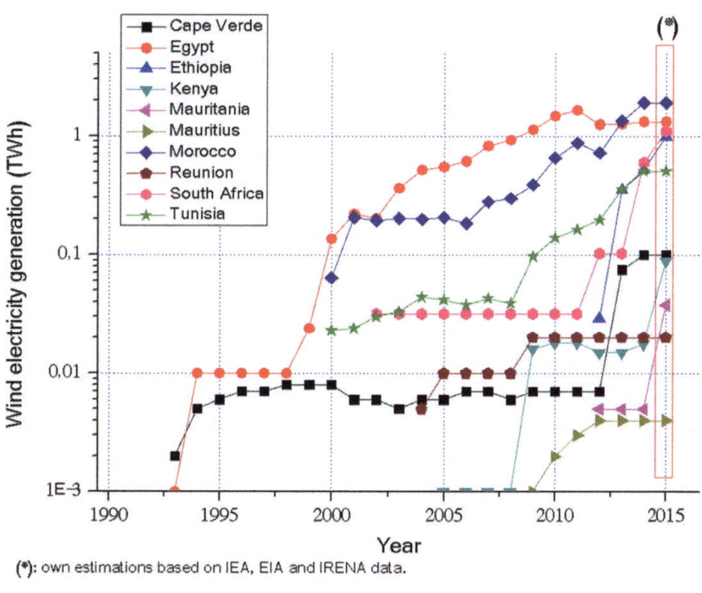

Fig. 10.8 Wind electricity generation per country produced in Africa in the period 1990-2015 [24]

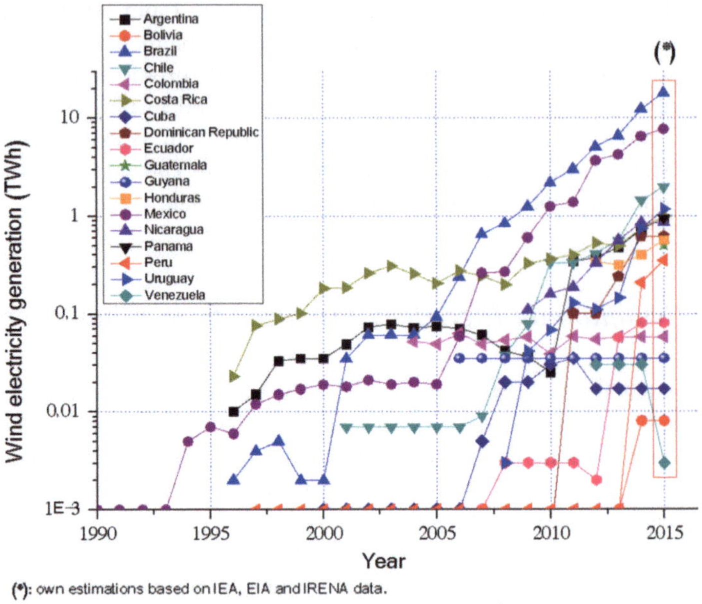

Fig. 10.9 Wind electricity generation per country produced in Latin America in the period 1990-2015 [24]

As it has been mentioned above, no offshore wind energy capacity is presently installed in Africa or Latin America through 2015 [3]. Worldwide there is 12,107 MW (2015) of offshore wind generation with the largest capacity in Great Britain (5,066.5 MW). SWT capacity is only detected in Argentina (5 MW; 8500 units) and Morocco (700 kW; 200), and also Cuba (150 units) [11].

Case Example 10.7

The largest wind farm in Central America (Panama):

In April 2016 the largest wind farm in Central America was initiated. It is placed in Penonomé (150 km west of Panama City) and it has been called Laudato Si in honour to the Pope Francisco's Second Encyclical, devoted to the environment. The wind farm is composed of 106 wind mills reaching a nominal capacity of 270 MW and that will supply about a 9% of the electricity demand of Panama. The wind farm is expected to reach 337.5 MW nominal capacity in a second phase [25].

10.4 Conclusions and Future Perspectives

Wind energy is an established technology for producing low cost electricity in many regions around the world, particularly in those areas with abundant wind resources. Wind generation produces more than 3% of all electricity globally and is growing, with much of the production occurring on land but with a trend toward offshore production in the future. Africa and Latin America contribute less than 5% of the total global wind production and have no offshore production installed or planned. In Latin America wind resources are being developed in 12 nations and in a few countries wind has become a significant contributor to their total power production (e.g., Nicaragua and Panama with ~8–9% wind). The largest wind farm (~270 MW and expanding) is located in Panama. As importantly, the manufacturing of wind turbine blades, towers, and balance of plant components has become a key contributor to local economies in Brazil and Mexico and is contributing to a nascent industry throughout the region. Future expansion of wind resources in both regions will likely require significant investments in transmission as the best wind resources are located distant from existing urban demand centers. In Africa, most wind production occurs in South Africa, Morocco and Egypt with limited development elsewhere, mostly along coastal regions.

In conclusion, wind energy will continue growing in most African and Latin American countries with available wind resources. However, the usually strong public support in both regions should be accompanied with a defined strategy to avoid a growing resistance in some communities opposing large-scale wind production. This resistance emerges for several reasons: (i) the perception of damages to natural and cultural heritage; (ii) the perception of negative strong effects that the wind farm development would have on local productive activities; (iii) local community distrust toward the company, local authorities and environmental regulators; and (iv) the threat to life projects of the local people. Sharing benefits from wind energy with the local communities and avoiding environmental damages are the best recipes to increase wind energy in Africa and Latin America.

References

1. Ren 21 (2015) Renewables 2015. Global status report. Key findings
2. International Energy Agency database (2015). http://data.iea.org/. Accessed 30 Nov 2015
3. IRENA (2016) Resource. http://resourceirena.irena.org/gateway/dashboard/. Accessed 30 Nov 2016
4. IRENA and LBNL (2015) Renewable energy zones for the Africa clean energy corridor. Multi-criteria analysis for planning renewable energy
5. Garrido J, Rodríguez I, Vallejo A (2015) Social responses to wind farm development: the case of the mar brava in the great island of Chiloé, Chile. Revista de Sociología 100:547–575
6. Hammami SM, Chtourou S, Triki A (2016) Identifying the determinants of community acceptance of renewable energy technologies: the case study of a wind energy project from Tunisia. Renew Sustain Energy Rev 54:151–160

7. OCDE/IEA (2013) Wind energy technology roadmap
8. Taylor D (2004) Wind energy (Chap. 7). Renewable energy. Oxford University Press
9. Guerrero-Lemus R, Martínez-Duart JM (2013) Renewable energies and CO_2. Cost analysis, environmental impacts and technological trends. Lecture notes in energy vol 3. ed. Springer. ISBN 978-1-4471-4385-7
10. Wiser R, Bolinger M (2015) 2014 Wind technologies market report. Energy efficiency and renewable energy. US Department of Energy
11. World Wind Energy Association (2016) Small wind world report 2016
12. IRENA (2015) Small wind turbines. Quality infrastructure for renewable energy technologies
13. Peillón M, Sánchez R, Tarquis AM, García-Fernández JL (2013) The use of wind pumps for greenhouse microirrigation: a case study for tomato in Cuba. Agric Water Manag 120: 107–114
14. IEA (2012) Energy technology perspectives
15. Climatescope (2015) The clean energy country competitiveness index. The Multilateral Investment Fund, UK Department of International Development, Power Africa and Bloomberg New Energy Finance
16. IRENA (2015) Off-grid renewable energy systems: status and methodological issues
17. Wiser R, Bollinger M (2016) 2015 wind technologies market report. US DoE
18. Mc Kenna P (2011) A new twist on floating wind power. Technology Review. MIT (4 Feb 2011)
19. Liebreich M (2016) In search of the miraculous. Bloomberg New Energy Finance Summit. 5 Apr 2016
20. Bolinger M, Wiser R (2011) Understanding trends in wind turbine prices over the last decade. Lawrence Berkeley National Laboratory (LBNL-5119E)
21. REN21 (2014) ECOWAS Renewable energy and energy efficiency. Status report 2014
22. Bloomberg New Energy Finance (2016) Week in review. Available 7 July 2016
23. Wiser R, Jenni K, Seel J, Baker E, Hand M, Lantz E, Smith A (2016) Forecasting wind energy costs and cost drivers. LBNL—1005717
24. Energy Information Administration (2016) US Department of Energy
25. Revista Eólica y del Vehículo Eléctrico (2016) http://www.evwind.com/2016/04/13/eolica-en-panama-inauguraran-el-mayor-parque-eolico-de-centroamerica/. Accessed 13 Apr 2016

Chapter 11
Biofuels

Ricardo Guerrero-Lemus and Les E. Shephard

Abstract This chapter shows that liquid biofuels are used in the transportation sector in Africa and Latin America and as an alternative cooking fuel option in Africa. In the transportation sector, biofuel production is limited largely to Brazil and to a lesser extent Colombia and Argentina with almost no production ongoing in Africa. Bioethanol can be directly used in some specific internal combustion engines (i.e., flex vehicles), with E85 serving as a standard for Brazilian vehicles. Biodiesel may be burned directly in diesel engines with additives or blended with diesel to promote improved combustion. In both Africa and Latin America, diesel engines are preferred over gasoline, although Brazil is the second largest producer of bioethanol in the world. Efforts to produce ethanol in many locations in these regions have been impacted severely by delays, water limitations, opposition from local communities, variable governmental policies, subsidies to fossil fuels and falling gasoline prices globally. Ethanol and ethanol gels are emerging as cooking fuel options in Africa because mixing ethanol with cellulose produces a clean-burning fuel that does not spill and it is denatured to prevent accidental ingestion. Thus, several ethanol gel stoves are now available. Then, stable policies, a balance between subsidies to fossil fuels and biofuels, and sharing profits from biofuels cultivation and transformation with local communities are key elements to increase the share of biofuels in the transportation sector and as an alternative cooking fuel option.

Keywords Energy · Africa · Latin America · Electricity · Biofuels · Biodiesel · Bioethanol · Biomass · Gasoline · Feedstock · Lignocellulosic fuels · Hydrogen · Algae · Crops · Water · Jatropha

11.1 Introduction

Liquid biofuels (mainly bioethanol and biodiesel), simply called biofuels, are produced from biomass through chemical and biological processes, and are chiefly used in transportation and in the chemical industry. Biofuels can also offer a lower

R. Guerrero-Lemus and L.E. Shephard, *Low-Carbon Energy in Africa and Latin America*, Lecture Notes in Energy 38,
DOI 10.1007/978-3-319-52311-8_11

carbon emission alternative to electric vehicles if these are recharged with electricity mostly generated by fossil fuels. Due to the importance of biofuels as a low carbon alternative to fossil fuels in Africa and Latin America, mainly in the transportation sector, but also as a cooking fuel option, a full chapter is devoted to them in this book.

The IEA estimates that 97.6% of biofuels are consumed in the transport sector (2014) [1]. The global final consumption of biofuels in the transport sector was a 3.64% (95,644 kt) in 2014, but for Africa was a 0.05% (51 kt) and 11.00% (24,481 kt) in Latin America [1]. Biofuels production and consumption in Latin America is mainly focused in Brazil and to a lesser extent in Argentina and Columbia. There is virtually no biofuels production in Africa (Fig. 11.1).

According to the IEA, the world bioethanol and biodiesel production grew steadily from 1990 until the economic crises in 2008 and now is starting to increase again (Fig. 11.1) [2]. For biodiesel in Africa and Latin America the production rates are currently stagnant but from 2005 to 2010 there was a significant increase in biodiesel production in Latin America that was comparable to bioethanol.

Bioethanol can be belended with fossil fuels to produce "gasohol" that is used directly with internal combustion engines and "flex vehicles". For example, E10 gasohol means that the proportion of bioethanol is 10% with respect to gasoline. The blended percentage currently is in the range of 10 to 15%, except in some countries, mainly Brazil, where E85 is a commercial standard for flex vehicles. Although the energy per unit volume of ethanol is only 67% of that provided by gasoline, the enhanced burning properties of bioethanol, together with its low proportions in the gasohol, result in near similar liters spent per kilometer of gasohol and unleaded gasoline in standard vehicles [3]. Biodiesel may also be burned directly in diesel engines, but its continued use may cause engine damage because of its incomplete

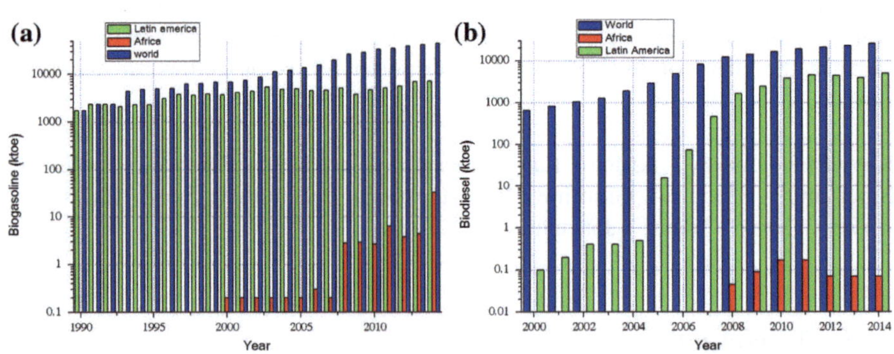

Fig. 11.1 Evolution of the world, Africa and Latin America biofuels (**a**) biogasoline in the period 1990–2014; and (**b**) biodiesel production in the period 2000–2014 [1]

combustion and high viscosity. Then, diesel blended with biodiesel at different percentages (e.g., B20) can be supplied to avoid engine damage.

11.2 Technology State of the Art

Following the recommendations of the IEA, biofuels can be classified in the following categories:

- First generation biofuels: those fuels that have reached a stage of commercial production. In general, they are obtained from crops grown following similar techniques to food crops and, consequently, compete with them for agricultural land.
- Second generation biofuels: these do not compete for agricultural land, since they are obtained from lignocellulosic biomass such as straw, grass, stems, stalks, roots, wood, shells, etc.
- Third generation biofuels: these fuels mainly consist of oils from algae and hydrogen from biomass. They are still at an early stage of development and far from large-scale production. Therefore, they are not expected to reach large production in the short term.

11.2.1 Bioethanol

First generation bioethanol (C_2H_5OH) can be produced naturally by fermentation of sugars by certain micro-organisms under acidic conditions (pH between 4 and 5). The most common microorganism used is yeast (*Saccaharomyces cerevisiae*) although yeast has the disadvantage of contaminating the ethanol in concentrations above 10%. Consequently, to obtain near pure ethanol concentrations ($\sim$95%) a series of distillations and fractionations of chemical compounds are required. Generally, bioethanol is produced from sugarcane molasses, being with the conversion of saccharose into ethanol the most important chemical reaction, with an approximate 80% efficiency [3]. The energy contents of the distillate is not very high, about 30 GJ/t [4] (oil is 42 GJ/t). For this reason, the overall balance of ethanol production is negative, unless the use of bagasse is considered for heating during the production process. In addition, the residue of the fermentation can be washed and dried to add value to the process, commercializing it as fertilizer or feed.

Case Example 11.1

Case study of a small scale ethanol production in Brazil:

The potential of small-scale ethanol production (up to 5000 l/day) can help to meet the demand for liquid fuel in rural areas, contribute byproducts to animal feed and reduce transportation costs in remoteregions in many parts of Brazil. However, in Brazil the sale of ethanol by the producer directly to consumers is not allowed and the yields for small-scale producers is much lower than for large units due to the simplification of the process. A study [5] aims to assess the economic feasibility and environmental impacts of a 30 l/h hydrous ethanol fuel (HEF) distillery. Sugarcane harvest seasons lasts for 180 days resulting in a high period of industrial inactivity, unless stored sugarcane molasses or sweet sorghum is used as substitute. For different raw material prices, the cost of HEF reaches USD 0.68, 0.92 and 1.16 per litre, where the raw materials reach up to 69% of the total cost. It is important to note that in Brazil ethanol is produced from sugarcane at autonomous distilleries or at distilleries adjacent to sugar production, making the HEF price susceptible to the sugar prices of the international markets (for the same amount of sugarcane 1 unit of ethanol or 1.47 units of sugar can be produced). The production has a positive energy balance (1.97) considering bagasse silage as byproduct or replacing firewood by bagasse to generate process heat (7.39). Byproducts in this process can constitute a valuable source of raw material for other products, such as livestock (between 25 and 30% of the fermentable sugars remains in the bagasse), biogas, and liquid fertilizer. The study also demonstrates that without the support of policies that provide economic incentives, the small-scale production will have only a secondary role in supplying HEF to meet the demand of liquid fuels in Brazil.

The IEA [6] has estimated the production yield for first generation bioethanol in units of liters of gasoline equivalent (lge) per hectare. The results are presented in Table 11.1 as average values of the yields for various geographic regions.

The objective of second-generation (lignocellulosic) bioethanol is to extract the raw material (sugars) from cellulosic biomass instead of from food-oriented agriculture that may occur with first generation bioethanol. Thus, biomass production is more sustainable, abundant, cheap and does not directly compete with food crops.

Table 11.1 First generation bioethanol average liters of gasoline equivalent (lge) production yield for various biomass species [6]

Feedstock	Region	lge/ha
Sugar cane	Brazil	4490
Sugar cane	World average	3630
Sugar beet	Europe	3300
Corn	North America	1980
Wheat	Europe	1650

lge litres of gasoline-equivalent

Examples of cellulosic biomass suitable for first-generation bioethanol would be stalks of many plants, herbs, wood, roots, straw, etc. Cellulosic biomass is mainly composed of cellulose and hemicellulose molecules consisting of several tightly bound sugars and lignin, a phenyl propene compound that acts as adhesive material between the sugar chains. The conversion of lignocellulosic biomass into bioethanol requires the following steps [7]: (i) obtaining lignocellulosic biomass (crops/waste); (ii) pretreatment to separate the biomass into cellulose, hemicellulose and lignin; (iii) hydrolysis of cellulose and hemicellulose to produce sugars; (iv) fermentation of sugars to produce alcohol; and (v) separation of ethanol from the co-products of fermentation.

11.2.2 Biodiesel

First generation biodiesel can be obtained from vegetable oils from oil seed plants, like a variety of soybean, rapeseed, sunflower, palm, etc. The energy content is about 36–40 GJ/t, 90% of the energy contents of diesel (42 GJ/t), and above that of bioethanol [6]. As discussed above, biodiesel may be burned directly in diesel engines, but its continued use may cause engine damage because of its incomplete combustion and high viscosity. To lower their viscosity, oils are submitted to hydrolysis, producing fatty acids and glycerin. Subsequently, the fatty acids are subjected to transesterification with methanol, which yields the less viscous methyl esters, with improved combustion properties in the engines.

For the production of 1 MJ of biodiesel from soybeans and methanol, 0.3 MJ of fossil fuels are currently required, mainly for the production of the methanol. Therefore, it is convenient to adapt the production process to obtain methanol from the biomass to reduce carbon emissions from biodiesel production. Table 11.2 presents the data for biodiesel yield in units of litres diesel-equivalent (lde) per hectare, corresponding to various geographic regions and average values. For comparison data from second generation biodiesel, including lignocellulosic biomass [6] and lipids from marine algae, are also presented [8].

Second-generation biodiesel is also known as synthetic biodiesel or advanced biodiesel. It is a liquid biofuel obtained from lignocellulosic biomass through various thermochemical processes. Biomass to liquids (BtL) is the most widely used of these processes. In the first stage of the BtL process the biomass is gasified.

Table 11.2 Yields in the production of biodiesel average liters of diesel-equivalent (lde) from various biomass species and geographic regions [6]

Feedstock	Region	lde/ha
Rapeseed (FAME)	Europe	1080
Soybean/oilseed rape (FAME)	US	720
Oilseed rape (FAME)	Brazil	630
Lignocellulosic	World	3000
Microalgae lipids	World	11,863 (lge)

lde litres of diesel-equivalent

Then, a gas purification stage is performed in order to remove tars, particles, and small amounts of pollutant gases producing a syngas with appropriate hydrogen and carbon monoxide proportions. The main catalytic procedure for syngas conversion into liquid biofuel is the Fischer-Tropsch process. In this process, the mixture of hydrogen and carbon monoxide is carried out at high pressures and temperatures and converted into various types of liquid and gaseous hydrocarbons. The gases produced in the reaction can be recycled or used for heating, while the liquids are refined to produce biodiesel suitable for use in vehicles. For BtL biodiesel synthesis the entire plants are used, not just the seeds, thereby reducing the amount of land required to produce BtL compared to first generation biodiesel. It is therefore expected that when large-scale commercialization is achieved, both BtL production costs and GHG emissions will be substantially reduced.

Case Example 11.2

Failure of many large-scale jatropha plantations in Ethiopia:

It has been argued that jatropha can be commercially grown well on marginal land without irrigation. The Ethiopian government has proposed about 23.2 million ha of marginal land be converted for biofuel feedstock production, mainly jatropha to secure its national energy supply by producing biofuels. More than 80 companies were licensed to invest in biofuels and had acquired more than 700,000 ha of land only for jatropha and castor bean production. Most of these projects have been delayed for years while many others that were initiated have failed. Moisture stress is observed as the key factor in the failure of many large-scale jatropha plantations in Ethiopia. Emami Biotech's large-scale jatropha plantation was the main focus of the research, but additional fields were visited in other districts. Other important factors that contributed to the failure of this effort included the use of untested planting material, conflict with local communities over the land, a lack of soil tests and adaptation trials, under-capitalization of the companies, failing crude oil prices and the world economic downturn [9]. Similar conflicts have been reported in Kenya where conflict management and research tools are proposed to avoid shortcomings in the technical feasibility studies and to enrich the participatory planning processes [10].

11.2.3 Third Generation Biofuels

Algae (macroalgae and microalgae) serves as feedstock for third generation biofuels. Macroalgae is used to provide a series of products including food, vitamins, pharmaceuticals, etc. Biofuels could be obtained from macroalgae but the low margins associated with biofuels compared to other products limits interest and production in using macroalgae for energy. Microalgae has a high concentration of lipids that makes it a better source for the production of biodiesel and other biofuels. An important advantage of algae cultivation is that does not directly compete for purely agricultural land or for large freshwater resources, because many algae species can grow in seawater, high salinity water and even wastewater.

Microalgae are very attractive as raw materials for energy production because of their much higher growth rates than agricultural crops. This is due to several factors including easy adaption to different water types containing various dissolved nutrients, CO_2 capture and access to sun radiation. Microalgae farming has been also proposed as a novel technology for carbon capture and sequestration since adding CO_2 to microalgae greatly improves the growth rate. Microalgae lipid concentrations can reach up to 60% and the potential yield per hectare can be up to 10–20 times higher than conventional crops [11]. In this respect, studies show typical yields for microalgae lipids reaching 11,863 l/ha·year [8]. These yield values are much higher than those projected for biofuels obtained from agriculture crops (Tables 11.1 and 11.2) due to the fact that microalgae crops can be harvested every few days.

Microalgae is grown in either open or closed systems. Open systems are usually ponds, which have the disadvantage that the more oily algae, which are not the quickest growing, may be invaded or contaminated by other algae species. Alternatively, closed systems or photo-bioreactors provide a controlled environment for the cultivation of microalgae. In photo-bioreactors the strict control of species cultivated and the controlled conditions for growth substantially increase crop yields. The use of photo-bioreactors generally imply higher production costs. These costs may be reduced if the photo-bioreactors are placed near industries that emit large amounts of CO_2 or if it is integrated into the carbon sequestration process.

In addition, production of ethanol through algae photosynthesis is being tested. In this process enzymes are introduced to directly synthesize this biofuel. Finally, hydrogen obtained from biomass can be considered as a third generation biofuel and his topic has been described in the gasification and fermentation sections in Chap. 5.

At this stage, it is also appropriate to provide information on the water resource footprint (WF) for the production of biofuels [12]. For this purpose, green water (rainwater) and blue water (surface water or ground irrigation) can be distinguished. The WF depends largely on the country analyzed and the production process employed. The WF related to electricity production would be lower than that in biofuel production if the residual heat is integrated into the process. Moreover, the results show that the WF values are lower for bioethanol production than for biodiesel (Fig. 11.2).

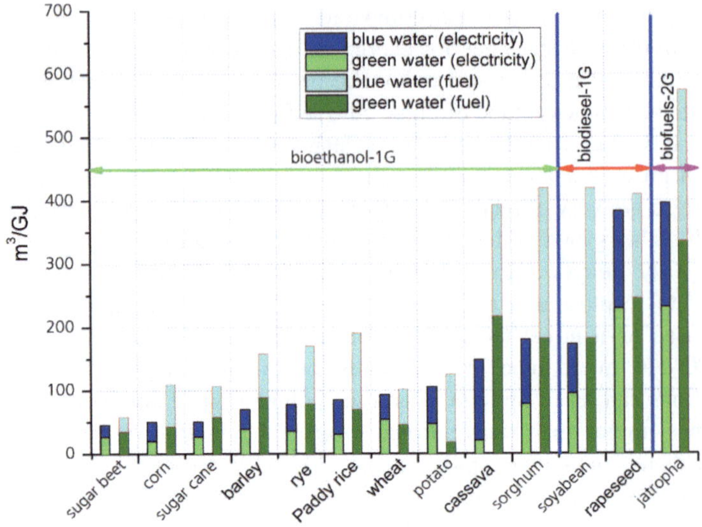

Fig. 11.2 Water footprint for various crop species for the production of electricity and biofuels

11.2.4 Ethanol as a Cooking Fuel Option

Ethanol and ethanol gels are emerging as cooking fuel options in Africa. In Mozambique, the first supply chain to distribute ethanol and efficient ethanol stoves is operational. About 1 million litres of gel fuel are produced in South Africa annually [13].

As a domestic fuel, ethanol gel has the advantages of safety over liquid ethanol and kerosene. Ethanol gel is manufactured by mixing ethanol with cellulose to produce a clean-burning fuel that does not spill. It is also denatured to prevent accidental ingestion. There are several ethanol gel stoves available that operate silently, incorporate single or double burners, which accommodate custom pots and allow heat control [14].

Ethanol stoves can utilise ethanol with a water content as high as 50%. This solution, given its low concentration of ethanol, can be easily produced locally. The efficiency of these cookstoves is typically in the range of 55–65% [15].

11.2.5 Costs

Biofuel production costs vary considerably in relation to the feedstock used, the efficiency and complexity of the production plant, the oil price, as well as the number of hours the plant operates per year. In the long term, substantial cost reductions are expected, mainly for second generation biofuels. For first generation biofuels, the rise and volatility in feedstock prices (Fig. 11.3) is a matter of concern, as they greatly influence the final production costs of biofuels. The instabilities in biofuel prices, typical of fossil fuels, are undesired.

We can check fuel parities between bioethanol/gasoline and biodiesel/diesel by analyzing the evolution of biofuel prices in the futures market and comparing them with the competing fossil fuel prices. The comparison considers the different calorific values of both. The comparison shows that when oil prices reach record highs (Fig. 11.4a), the price of ethanol and gasoline in the futures market overlap. For biodiesel (Fig. 11.4b), available data show that the price differential with diesel (gasoil) remains higher than the average differential for bioethanol/gasoline and far from overlap at any time.

Thus, production costs for the different technologies that commercialize biofuels are close to market introduction. However, if we study the production costs of lignocellulosic biofuels and hydrogen obtained from biomass, considering that they are the principle focus for future biofuel production, they are much higher (Fig. 11.5). The investment cost for a lignocellulosic biofuel plant with a production range of 50–150 Ml/year is about USD 131–262 million, up to ten times higher than the cost of a first generation biodiesel production plant [17]. For conventional biofuels, the main cost factor is feedstock, which accounts for 45–70% of total production costs. It must be recognized that biomass feedstock prices vary widely between countries. For advanced biofuels the main factor is related to capital investment costs (35–50%), followed by feedstock (25–40%) [18].

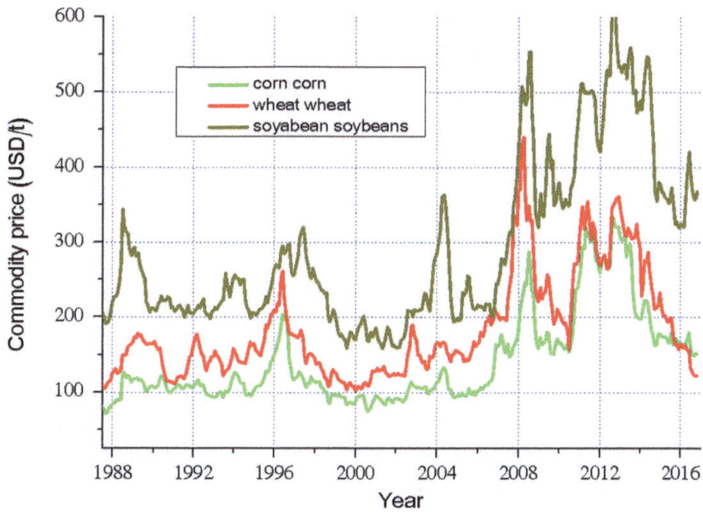

Fig. 11.3 Monthly commodity price for typical feedstock demand (corn, wheat and soybean) for first generation biofuel production [12]

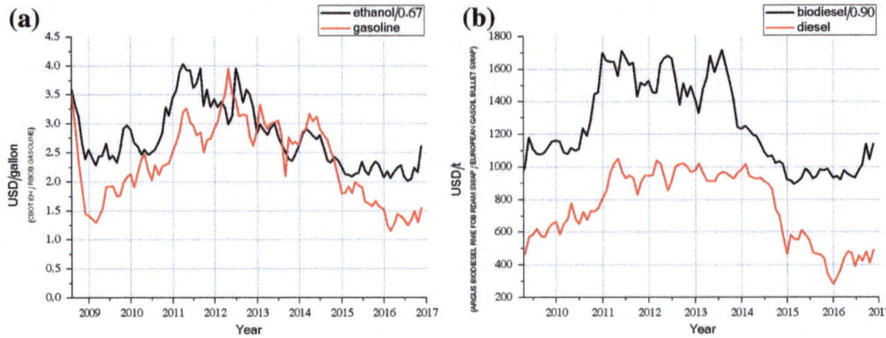

Fig. 11.4 Price evolution for different fuels in the futures market (balancing the difference in calorific values): **(a)** ethanol compared to gasoline, and **(b)** biodiesel compared to diesel (gasoil) [16]

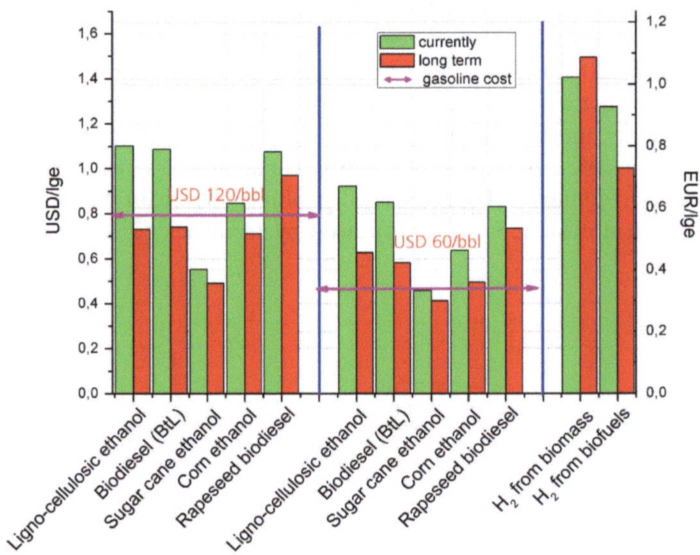

Fig. 11.5 Evolution of the production costs of biofuels and hydrogen from biomass considering different species of biomass and processes [19, 20]

For algae biofuels, official data show that the algal lipid upgrading design the state of technology cost is USD 5.5/l [21] and is expected to sharply decline in future. Initial studies [22] indicate that at present these costs are very high, considering that current technology would not be competitive with oil bbl prices even in the range of 100–200 USD. Production cost are mainly in the feedstock, where the cost decrease is expected to be more pronounced [21]. For this reason, different

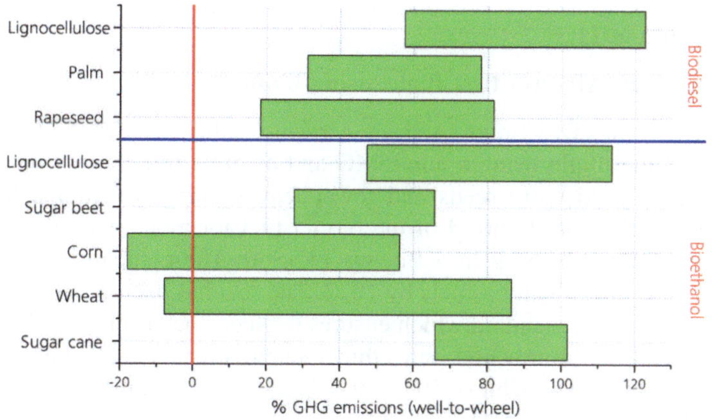

Fig. 11.6 Percentage of CO_2 well-to-wheel emissions avoided for different feedstock species for the production of bioethanol and biodiesel [11]

strategies are on course to reduce microalgae-based biofuel costs, mainly by adding valuable co-products (animal feed, chemicals, pigments, etc.). However, the current approximate values for these co-products may diminish strongly if the production of algae for biofuels increases substantially.

There are not specific cost data available in the literature on ethanol and ethanol gels for cooking However, cooking with ethanol is generally considered cheaper than cooking with ethanol gel. Moreover, ethanol has been found to be cost competitive against charcoal and wood in Ethiopia, Malawi, Mozambique, Senegal and South Africa in 2004, while ethanol gel was approximately three times the price of ethanol/water blends [14].

In relation to the emission of greenhouse gases as a function of the energy contents of the biofuels, Fig. 11.6 shows a summary of several studies by different authors and organizations [19], for first and second-generation biofuels. Thus, the majority of biofuels currently produced do not involve large reductions in greenhouse gas emissions. Even in some cases a slight increase in biofuel GHG emissions occurs when compared to competing fossil fuels. However, second generation biofuels are able to avoid large GHG emissions, although these studies have been carried out in pilot plants and need to be confirmed in commercial plants. The GHG emission reductions over 100% in Fig. 11.6 are attributed to arise from the benefits of co-products obtained during the processes of biofuel production and to also involving CCS in the process.

Case Example 11.3

South African Airways first flight using biofuels:

On Friday 14th 2016 South African Airways launched Africa's first commercial biofuel flight using a non-GMO and nicotine-free rich tobacco plant crossbred to yield more seeds and fewer leaves than a traditional tobacco plant. The airline used 6300 l of the bio jet fuel for a one-way trip to Cape Town from Johannesburg in a Boeing 737-800. The company plans to fly 50% of its airliners with biofuels by 2022. The fuel has been certified under a voluntary eco-label standard which ensures the feedstock is produced with the proper social, environmental and ethical safeguards. The certification also requires the biofuel to achieve 50% lower lifecycle GHG emissions compared to fossil jet fuels. This activity may provide a new financial market for millions of stakeholders around the world who depend on tobacco cultivation but risk losing their livelihoods as the demand for cigarettes declines [5].

For second generation biofuels CO_2 emissions are largely reduced because they utilize the entire plant residue from agriculture and forestry residue and do not use fertilizers. A distinction should be made between secondary lignocellulosic residues (rice husks, sugarcane bagasse, etc.), often considered as waste, and primary lignocellulosic residues (corn stalks, rice straw, leaves trees, etc.). These primary residues are often left in the soil to nourish it. Consequently, as a balance of nutrients in the soil is required, it will be necessary to consider an additional cost that should be adequately assessed.

In addition, we note that there is still no consensus on how to account for emissions of greenhouse gases, although some governments are defining official values for the different feedstock species [19]. The debate is particularly intense in relation to the indirect effects due to a strong increase in the production of biofuels, especially those arising from what is known as the indirect land use change (iLUC factor). This factor refers to the indirect destruction of forest to be devoted to agricultural production, as increasing agriculture land is required for feedstock production. These indirect effects are difficult to identify and evaluate but obviously attenuate the impact on reducing the GHG emissions listed in Fig. 11.6.

From a fundamental research perspective, studies based on partial equilibrium models indicate that the effects accounted for by the iLUC factor (e.g., corn produced in the U.S.) are substantial. Other studies indicate that although the iLUC factor is important in relation to carbon emissions, the expected increase in fertilizer use will cause nitrous oxide emissions to be more important than carbon in terms of global warming and therefore recommend improving the practices in the use of nitrogen fertilizers [23]. There are also more specific studies [19, 24] that consider for example, that if "cerrado" vegetation is cleared ("cerrado" is a name for a type of vegetation that occupies large areas of Brazil) and replaced by sugar cane it would produce CO_2 emissions equivalent to 165 t/ha and take roughly 17 years to compensate for the amount of CO_2 previously emitted. However, other studies [19, 25, 26] describe these approaches as based on exaggerated or incorrect methodologies, which implies the need to reach an agreement on the methodology that can be accepted by institutions and governments.

Case Example 11.4

The sustainability of sugarcane-ethanol systems in Guatemala:

Since 2010, Guatemala is exporting ethanol, principally to European markets, that they certify as sustainable. Based on more than on more than eighty qualitative, in-depth interviews a recent study [27] argues that the current framework runs the risk of exacerbating the plight of Guatemala's already marginalised rural communities. In Guatemala, 2% of agricultural producers, with an average of 194 ha, controlled 57% of arable land, while 87% of producers, with an average of 1.2 ha, occupied just 16%. In addition, Guatemala's sugar mills are owned by the major landowning families who constitute some of the country's political and economic elites. Many of the sugar mills are also connected to biorefineries. On the Pacific Coast, where more ethanol is produced, communities had traditionally relied on leasing land from landowners to produce basic grains; however, increasing competition from sugarcane had effectively forced subsistence and small-scale farmers out of the land rental market. This has affected food security, changes in diets, and the needs of monetized incomes for subsistence.

No detailed studies have been found that assess CO_2 emissions related to microalgae. Although in a recent scientific meeting, it was estimated a reduction in CO_2 emissions reaching 25–75% compared to oil [22].

11.3 Regional and National Perspectives on Technology

11.3.1 Supporting Policies

The difference between supporting policies governing gasoline or diesel engines in the transportation sector is significant. Traditional diesel is slightly cheaper to produce than gasoline because it is heavier and less volatile. New ultra-low-sulphur diesel (ULSD) increases diesel production costs because it requires more refining [28]. These trends provide an indicator about the potential future penetration of bioethanol and biodiesel in these countries. As observed in Fig. 11.7, most African countries have policies in place that promote diesel engines instead of gasoline. The promotion of diesel engines is more substantial in those countries where it is produced and in those countries where the supporting policies are

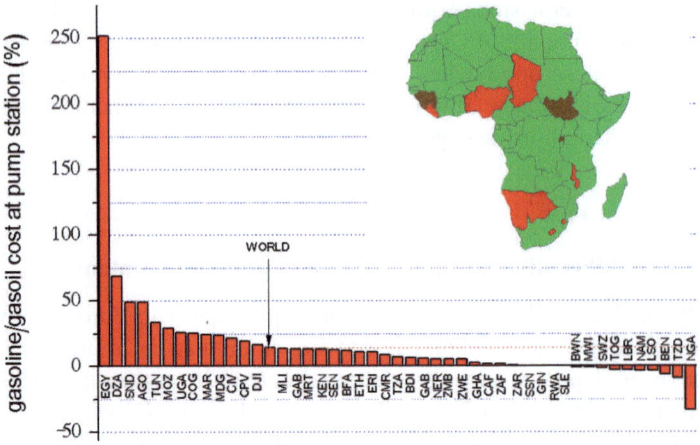

Fig. 11.7 Increase in pump gasoline prices (percentage) compared to diesel fuels in 2014 for African countries [29, 30]

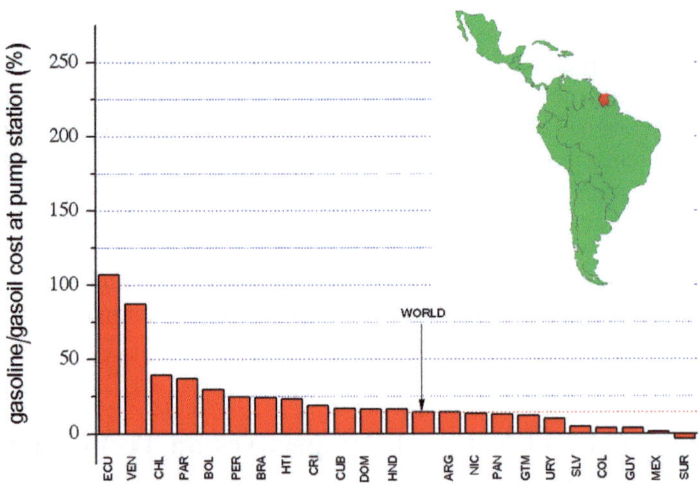

Fig. 11.8 Increase in gasoline pump prices in 2014 compared to diesel fuels in Latin American countries [29, 30]

oriented toward diesel. Many oil producing countries have similiar, but often more acute policies, favoring diesel.

Policies favoring diesel engines is even more acute in Latin America (Fig. 11.8) where again important top oil producers encourage diesel engines. Thus, in only one country (Suriname) is the gasoline pump price lower than the diesel price.

11.3.2 Biofuel Supply

Brazil is the largest supplier of biogasoline (mainly bioethanol) in Africa and Latin America (Fig. 11.9). Brazil is also the second largest ethanol producer worldwide, although the sector has struggled in past years due to gasoline price controls and weak crops [31]. Brazil is the largest consumer of these fuels worldwide, in part becasue they have a large fleet of flexible vehicles that can support combustion with high ethanol content. Available statistics show that biogasoline production in Africa is only present in Ethiopia which has had a 20% national biofuel blending mandate in place since 2008. The 20% mandate was eliminated in early 2015 due to delays in the commissioning of ethanol plants [31]. In terms of technology, South Africa offers expertise and manufacturing capacity in biofuels production, engineering, O&M, distribution and blending [31]. Also, Brazil offers expertise and manufacturing capacity in all the biofuels production chain: production, engineering, O&M, equipment manufacturing, distribution and blending [31].

For biodiesel, the top supplier countries are Brazil, the top producer in 2012 (Fig. 11.10) and Argentina. Argentina is a top global producer of raw materials (mainly soybeans) for biodiesel production. No significant biodiesel production is detected in the African countries. In the past, Argentina exported most of its biodiesel production (2bn litres out of the 2.9bn produced in 2011) mostly to European Union (EU) countries. However, in August 2012, the EU initiated an anti-dumping investigation on Argentine and Indonesian biodiesel, which resulted in a 10.6%

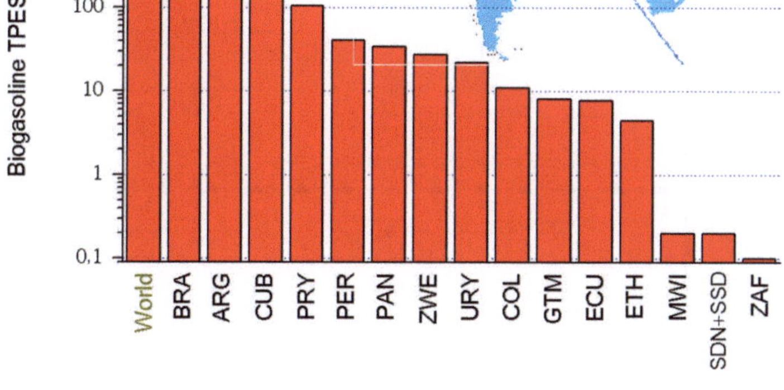

Fig. 11.9 Biogasoline total primary energy supply (TPES) in African and Latin American countries and the top global supplier in 2014 [1]

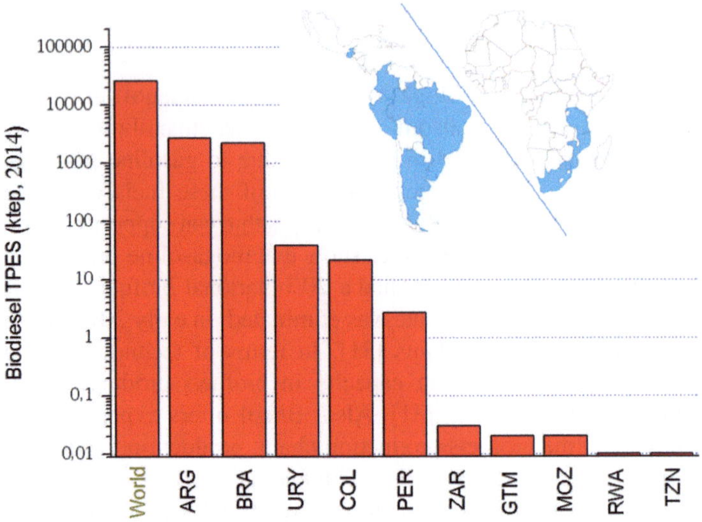

Fig. 11.10 Biodiesel total primary energy supply (TPES) in African and Latin American countries and the top global supplier in 2012 [1]

antidumping duty implemented in May 2013. In order to support local production, Argentina has increased its biodiesel blending mandate, mandate price and reduced export tax [31].

As discussed above, the production of bioethanol in Africa is very small. If we study the evolution of bioethanol production, only six countries show some limited production (Fig. 11.11) that has only been started recently.

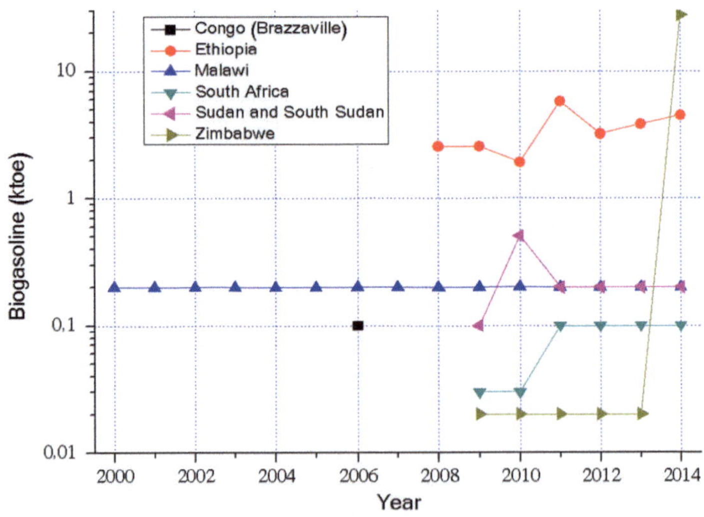

Fig. 11.11 ktoe of bioethanol and country produced in Africa in the period 2000–2014 [2]

Case Example 11.5

Evolution of the biofuel policy in Zimbabwe:

The production and use of biofuels is described by the authors as an “on-off relationship” from the early 1970’s. Indeed, Jatropha plant was introduced into the country in the 1940’s for its exploitation in soap and lubricant production, but changed to biofuels production during the 1970’s and 1980’s. In 1980, Zimbabwe started producing bioethanol using molasses as the feedstock, locating the first facility of its kind in Africa (Triangle Ethanol Plant: 120 m^3/day). Operations ceased in 1992 due to severe drought conditions and lower costs of fossil fuels. In 2005, the country once again started promoting jatropha and in 2007 a processing plant (300 m^3/day) was set up in Mount Hampden, Harare. The fuel crisis crippling the economy due to lack of foreign currency in the country was the major driving force to operate this plant. However, the government stopped funding the project in April 2010 because the plant has operated at under 5% capacity largely due to a severe shortage of feedstock. Other reasons for the Mount Hampden plant closure are attributed to lack of coordination between the Ministries involved, failure to pay an attractive price to farmers for jatropha, and the allowance to companies to import fuel independently. Thereafter, a 700 m^3/day bioethanol plant at Chisumbanje was setup in a joint venture partnership between the government and a private investor in 2011. Mandatory blending of 5% minimum anhydrous ethanol with unleadeded petroleum seems to have stabilized the sector. The lack of a long-term national biofuel policy is identified as the major factor impeding the biofuel sector. An implementation plan is proposed to allocate farming land and address the challenges associated with choice of feedstock, water scarcity, logistics, import tariffs and duties, biofuel standards, incentives, R&D, and certificates/licences [32].

In Latin America, Brazil maintains a dominant role in the production of bioethanol (Fig. 11.12) and its production rate has increased regularly since official data has been collected. Bioethanol production in the other 11 Latin American started in 2000 or later but also show increasing production trends.

The production of biodiesel in Africa (Fig. 11.13) is also extremely low, occuring in only 5 countries. As in most Latin American countries, production started very late (in 2008) compared to Brazil and the production rates are largely constant in recent years with the exception of Zimbabwe, where production vanishes in 2012.

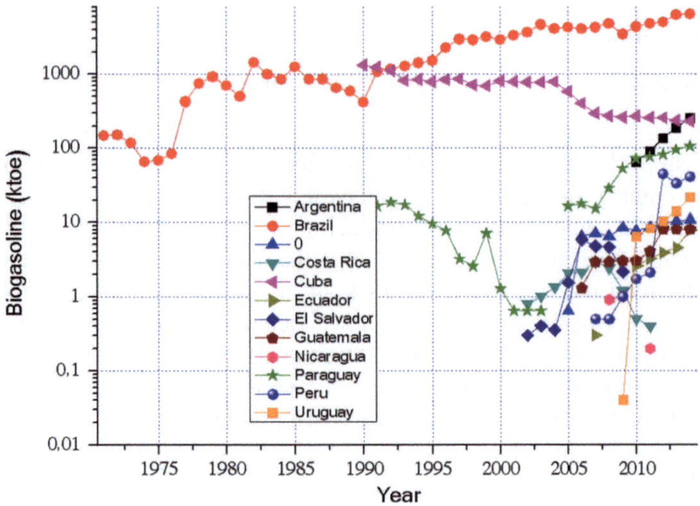

Fig. 11.12 ktoe of bioethanol barrels and country produced in Latin America in the period 1970–2014 [2]

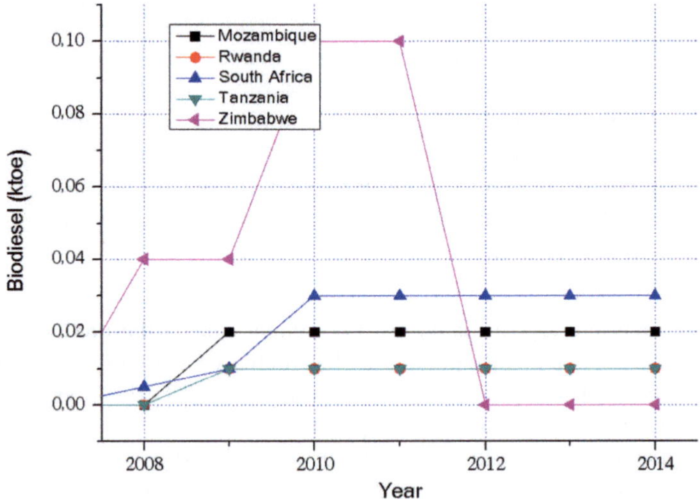

Fig. 11.13 ktoe of biodiesel and country produced in Africa in the period 2000–2012 [2]

Case Example 11.6

Impact of biofuel projects in food security in Mozambique:

Between 2005 and 2012, a total of 36 biofuel investment proposals had been submitted to the Mozambican government requesting 411,000 ha and projecting 148,225 jobs, but only 10 such projects have been approved totaling 8512 ha and 853 permanent jobs. Four biofuel projects have been visited and analysed between December 2008 and July 2012 [33]. For low resource endowed households, the feasibility of cultivating biofuel crops was questionable as they need the majority of their land to achieve food self-sufficiency and already face labour constraints. For the high resource endowed households, cultivating biofuel crops is a question of prioritizing labour allocation to different crops and activities when considering their relative revenues and benefits. Mozambique was one of the first Sub Saharan African countries to approve a national policy for biofuels (2009) for developing the agro-industrial biofuel sector and also smallholders farmers and communities. Agro-industrial biofuel projects need a license from the government and need to comply with the food and security criteria that assure food security at the local level. Social unrest due to agro-industrial projects forcing farmer activities to move to new and hardly used land is occurring. Smallholder biofuel producers are far less monitored, especially when they focus on local markets and use. Drops in the price of jatropha paid to the farmers forced them to produce higher value products from the jatropha oil such as soap. The case study raise questions about the extent to which detailed and operational sustainability frameworks and certification schemes developed at the global level (in terms of thresholds, indicators and/or methods and nature) are able to capture the complex dynamics and heterogeneity between and within different types of biofuel production systems.

The production of biodiesel in Latin America is dominated by Brazil and Argentina. Colombian biodiesel production also grew rapidly until recent years (Fig. 11.14). Biodiesel production is constant or slightly decreasing in most countries since 2012 in part because of lower demand for exports.

Finally, it is also important to remark that South Africa is also one of the very few countries in the world where coal-to-liquid accounts for a significant part of the transportion fuel [34], accounting for 2.74 Mtoe in 1973 (first data available) and 0.01 Mtoe in 2013 [1]. Then, it is expected that this country could produce second generation biodiesel using the Fischer-Tropsch processes required also for coal-to-liquid technologies.

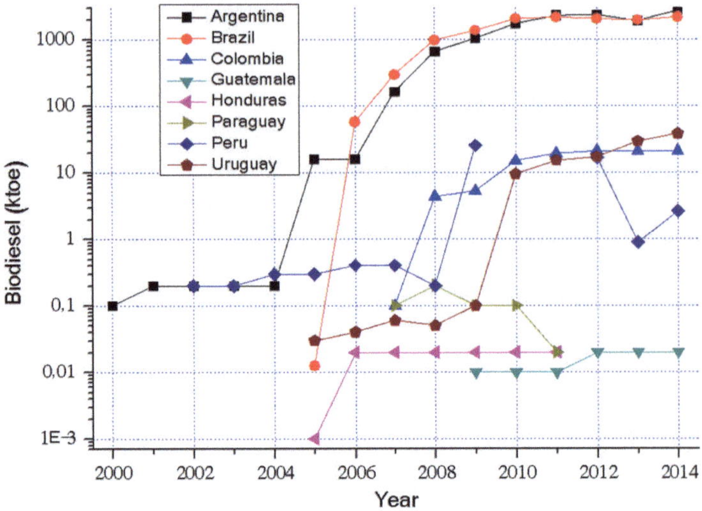

Fig. 11.14 ktoe of biodiesel and country produced in Latin America in the period 2000–2014 [2]

Case Example 11.7

Jatropha adoption by smallholders in Mexico:

This study [35] investigates the allocation of economic subsidies to jatropha cultivation at both the community and household (420 farms) levels in Chiapas. The promotion of jatropha cultivation by the State of Chiapas consisted on providing all interested farmers with seeds and seedings and technical assistance. Also the State of Chiapas supported farmers in applying for the subsidy provided by the Federal Government (USD 486 per ha planted). Quantitative data was complemented with key stakeholder interviews and focus groups. Results show that subsidies tend to be allocated to larger, better-connected communities which have access to better services. Within adopter communities, households that have more resources, better risk-coping strategies, access to information, experiences with similar technologies and positive attitude to risk are more successful in allocating subsidies. Although the biofuel promotion policy has primarily a pro-poor approach, the results show that subsidies have benefited both the least marginalised communities and the most favoured households, and that poor farmers may have been prevented from participating. Efforts should be made in the development of well-adapted promotion strategies, as the poorest farmers cannot afford risk to access the subsidy. But the policies should also avoid raising farmer expectations when the profitability of the crop is unproven [35].

11.4 Conclusions and Future Perspectives

Liquid biofuels (mainly bioethanol and biodiesel) are produced from biomass through chemical and biological processes, and are chiefly used in the transportation sector in Africa and Latin America, and as an alternative cooking fuel option in Africa. In the transportation sector, biofuel production is limited largely to Brazil and to a lesser extent Colombia and Argentina with almost no production ongoing in Africa. Bioethanol can be directly used in some specific internal combustion engines (i.e., flex vehicles), with E85 serving as a standard for Brazilian vehicles. Biodiesel may be burned directly in diesel engines with additives or blended with diesel to promote improved combustion. In both Africa and Latin America, diesel engines are preferred over gasoline even though Brazil is the second largest producer of bioethanol in the world. Efforts to produce ethanol in Africa (largely Ethiopia) have been impacted severely by delays, water limitations, opposition from local communities, variable governmental policies, subsidies to fossil fuels and falling gasoline prices globally.

Ethanol and ethanol gels are emerging as cooking fuel options in Africa because they have an advantage of over liquid ethanol and kerosene. Ethanol gel is manufactured by mixing ethanol with cellulose to produce a clean-burning fuel that does not spill and it is denatured to prevent accidental ingestion. Several ethanol gel stoves are now available, incorporating single or double burners and which accommodate custom pots and Mozambique has developed the first supply chain to distribute ethanol and efficient ethanol stoves.

In conclusion, stable policies, a balance between subsidies to fossil fuels and biofuels and sharing profits from biofuels cultivation and transformation with local communities are key elements to increase the share of biofuels in the transportation sector as an alternative cooking fuel option.

References

1. International Energy Agency database (2015) http://data.iea.org/. Accessed 30 Nov 2015
2. Energy Information Administration. US Department of Energy (2016)
3. Twidell J, Weir T (2006) Renewable energy resources. Taylor & Francis, London
4. Larkin S, Ramage J, Scurlock J (2004) Chapter 4: Bioenergy. Renewable energy. Oxford University Press, Oxford
5. Bloomberg New Energy Finance (2016) Week in review. Available 19 July 2016
6. IEA (2008) Energy technology perspectives
7. Wyman CE (2008) Cellulosic ethanol: a unique sustainable liquid transportation fuel. Harnessing materials for energy. MRS Bull 33:381–383
8. Atsumi S, Higashide W, Liao JC (2009) Direct photosynthetic recycling of carbon dioxide to isobutyraldehyde. Nat Biotechnol 27:1177–1180
9. Wendimu MA (2016) Jatropha potential on marginal land in Ethiopia: reality or myth? Energy Sustain Dev 30:14–20
10. Arevalo J, Ochieng R, Mola-Yudego B, Gritten D (2014) Understanding bioenergy conflicts: case of a jatropha project in Kenya's Tana Delta. Land Use Policy 41:138–148

11. Williams PJ, Laurens LM (2010) Microalgae as biodiésel and biomass feedstocks. Energy Environ Sci 3:554–590
12. Indexmundi: wheat: http://www.indexmundi.com/commodities/?commodity=wheat; soyabeans: http://www.indexmundi.com/commodities/?commodity=soybeans; corn: http://www.indexmundi.com/commodities/?commodity=corn&months=60. Accessed 30 Nov 2015
13. BFAP (2008) Impact analysis of the biofuels industrial strategy on the South African agricultural and biofuel subsector. Bureau for Food and Agricultural Policy, Pretoria
14. IRENA (2015) Africa 2030: roadmap for a renewable energy future
15. Lloyd PDJ, Visagie EM (2008) A comparison of gel fuels with alternative cooking fuels. J Energy South Afr 18:26–31
16. CMEGroup. Ethanol: http://www.cmegroup.com/trading/energy/ethanol/cbot-ethanol.html; biodiesel: http://www.cmegroup.com/trading/energy/refined-products/rme-biodiesel-argus-fob-rdam-red-compliant-swap-futures.html; gasoil: http://www.cmegroup.com/trading/energy/refined-products/european-gasoil-bullet-swap.html; gasoline: http://www.cmegroup.com/trading/energy/refined-products/rbob-gasoline.html. Accessed 30 Nov 2015
17. IEA (2008) 1st- to second-generation biofuel technologies: an overview of current industry and R&D activities
18. IEA (2011) Technology roadmap. Biofuels for transport
19. IEA (2010) Sustainable production of second-generation biofuels
20. Guerrero-Lemus R, Martínez-Duart JM (2010) Updated hydrogen production costs and parities for conventional and renewable technologies. Int J Hydrogen Energy 35:3929–3936
21. US Department of Energy (2015) Quadrennial technology review—an assessment of energy technologies and research opportunities
22. GTM (2010) Transitioning from 1st generation biofuels to advanced biofuels. A White Paper from Enterprise Florida and GTM Research/Josua Kagan GTM Research
23. Melillo JM et al (2009) Indirect emissions from biofuels: how important? Science 326:1397–1399
24. Fargione J et al (2008) Land clearing and the biofuel carbon debt. Science 319:1235–1238
25. Goldemberg J, Guardabassi P (2008) Are biofuels a feasible option? Energy Policy 37:10–14
26. Mathews JA, Tan H (2009) Biofuels and indirect land use change effects: the debate continues. Biofuels, Bioprod Biorefin 3:305–317
27. Tomei J (2016) The sustainability of sugarcane-ethanol systems in Guatemala: land, labour and law. Biomass Bioenergy 82:94–100
28. Globalpetrolprices.com. http://www.globalpetrolprices.com/articles/4/. Accessed 30 Nov 2015
29. World Bank (2015) http://data.worldbank.org/indicator/EP.PMP.SGAS.CD. Accessed 30 Nov 2015
30. World Bank (2015) http://data.worldbank.org/indicator/EP.PMP.DESL.CD/countries?display=map. Accessed 30 Nov 2015
31. Climatescope (2015) The clean energy country competitiveness index. The Multilateral Investment Fund, UK Department of International Development, Power Africa and Bloomberg New Energy Finance
32. Moyo P, Moyo M, Dube D, Rusinga O (2014) Biofuel policy as a key driver for sustainable development in the biofuel sector: the missing ingredient in Zimbabwe's biofuel pursuit. Modern Appl Sci 8:36–58
33. Schut M, Florin MJ (2015) The policy and practice of sustainable biofuels: between global frameworks and local heterogeneity. The case of food security in Mozambique. Biomass Bioenergy 72:123–135
34. IEA (2014) Africa energy outlook
35. Soto I, Achten WMJ, Muys B, Mathijs E (2015) Who benefits from energy policy incentives? The case of jatropha adoption by smallholders in Mexico. Energy Policy 79:37–47

Chapter 12
Waste-to-Energy

Ricardo Guerrero-Lemus and Les E. Shephard

Abstract This chapter shows that municipal and other waste forms make a significant contribution to GHG emissions and have an adverse impact on other environmental factors (e.g., health, water quality, air, etc.) in those regions where waste management systems are rudimentary or non-existent. Waste-to-energy provides a very limited amount of electricity or heat on a global basis and has virtually no contribution in either Africa or Latin America. There is a wide variety of Municipal solid waste (MSW) production rates between countries in Africa related to GDP per capita, and variations in national policies. In Latin America the average values of MSW production rates are significantly higher but are less related to GDP per capita than other socio-economic factors that are difficult to identify. Landfill gas and biogas projects are increasing in frequency within Africa and Latin America. For small communities, biogas digesters using animal and human manure could provide cooking solutions while also providing supplementary benefits to sewage treatment. A number of programmes are in place in Africa to increase the use of biogas in domestic applications. Therefore, instead of carbon capture and storage in landfills, which generally occurs at the beginning of the resource flows, the principle of carbon capture, transformation and reuse adding waste-to-energy paths in the cycle could be further developed and applied throughout the whole socio-industrial metabolism.

Keywords Energy · Africa · Latin America · Electricity · Waste · Waste to energy · Anaerobic digestion · Incineration · Pyrolysis · Hydrothermal carbonization · Biogas · Municipal solid waste (MSW) · Gasification · Feedstock · Combined heat and power (CHP) · Fermentation · Digester

12.1 Introduction

The waste-to-energy cycle is used in many countries around the world, supplying 93,540 GWh of electricity (0.4% of the total) and 399,734 TJ of heat (2.7% of the total) in 2014 (last data available) [1]. The only country obtaining energy from

R. Guerrero-Lemus and L.E. Shephard, *Low-Carbon Energy in Africa and Latin America*, Lecture Notes in Energy 38,
DOI 10.1007/978-3-319-52311-8_12

waste in Africa and Latin America is Mexico which started producing electricity in 2013 (142 GWh) and reaching 77 GWh in 2014.

Waste derived from life activity on Earth provides a large source of methane, a powerful GHG that is particularly impactful in the short-term (it breaks down more quickly than CO_2). These wastes also impact the environment in many other aspects (mainly public health impacts, air pollution and water supply) when the management systems are rudimental or non-existing, as in many areas in Africa and Latin America. Efficient use of carbon containing waste and all other waste materials must be increased to reduce the need for virgin raw materials acquisition, including biomass, and to reduce carbon being emitted to the atmosphere. Although materials are used more and more efficiently, there are no indications that overall material consumption will decline. As a result, it is expected that large amounts of waste will continue to be generated in the future [2].

In accordance to the International Energy Agency, waste can be classified in different types depending on its origin and how the energy is obtained from it:

- Municipal waste: consists of products that are combusted directly to produce heat and/or power and comprises wastes produced by households, industry, hospitals and the tertiary sector that are collected by local authorities for incineration at specific installations. Municipal waste can be split into renewable and nonrenewable.
- Biogas: Biogas is derived principally from the anaerobic fermentation of biomass and solid wastes and combusted to produce heat and/or power. Included in this category are landfill gas, sludge gas and other biogas such as biogas produced from the anaerobic fermentation of animal slurries and of wastes in abattoirs, breweries and other agro-food industries.
- Sludge gas: Covers gas produced from the anaerobic fermentation of sewage sludge.
- Other biogas: Covers gas such as biogas produced from the anaerobic fermentation of animal slurries and of waste abattoirs, breweries and other agro-food industries.
- Industrial waste: Industrial waste of non-renewable origin consists of solid and liquid products (e.g., tyres) combusted directly, usually in specialised plants, to produce heat and/or power. Renewable industrial waste is not included in this concept, but with solid biomass, biogas or liquid biomass.

Thus, animal manure and debris from urban wastewater are a concern and can be a major source of greenhouse gases. In some countries these sources represent a significant percentage of methane emissions.

Worldwide waste management systems are still characterized by a heavy dependence on landfilling, very low levels of recycling and composting, and near absence of waste-to-energy systems. In OECD countries the average generation of municipal solid waste (MSW) reaches 2.20 kg/capita·day but, excluding touristic state-islands, in Africa and Latin America is between 0.09 and 2.00 kg/capita·day, averaging 0.65 kg/capita·day in Sub-Saharan Africa and 1.1 kg/capita·day in MENA, Latin America and the Caribbean [2].

In developing countries the quantity of solid waste generated in urban areas is low when compared to industrialized countries, waste management capacity still remains inadequate [3]. Thus, the percentage of waste collection is close to 100% in developed countries but ranges between 11 and 98% in Africa and Latin America (in touristic islands the waste collection is close to 100% due to territorial limitations and environmental concerns) [2]. Solid waste management is almost always the responsibility of local governments and it is often their single largest budget item, particularly in developing countries. Also, solid waste management and street sweeping is often the city's single largest source of employment.

Case Example 12.1

Landfill gas projects in Africa:

Landfill gas projects are becoming increasingly common in Africa. For example, in South Africa, the Durban municipality has implemented a landfill gas-to-electricity project with an installed power generation capacity of 7.5 MW. Also a number of interesting biogas-generation projects have been initiated in Kenya, such as producing off-grid electricity from biogas generated by manure and sisal, utilising slaughterhouse waste to produce biogas for electricity production, and another producing 20 kW of electricity from vegetable waste [4].

The waste management sector follows a generally accepted hierarchy responding to financial, environmental, social and management considerations: (1) reduce, (2) reuse, (3) recycle, (4) recover (mainly digestion or composing), (5) landfill/incineration (with energy recovery), and (6) controlled dump [2]. The waste-to-energy (WtE) systems are those mainly based in the following technologies: incineration, pyrolysis, gasification, anaerobic digestion and co-combustion. WtE systems are not common, and generally not successful in developing countries because of high capital, technical, and operation costs, high moisture content in the waste, high percentage of inerts, and air pollution control equipments in these countries are not advanced and often by-passed.

Recovering the energy value embedded in waste prior to land disposal is considered preferable to landfilling, assuming pollution control requirements and costs are adequately addressed. Conversely, incineration without energy recovery (or non-autogenic combustion: the need to regularly add fuel) is not a preferred option due to cost and pollution. Open-burning of waste is particularly discouraged due to severe air pollution associated with low temperature combustion.

The use of the combustible fraction of waste as a fuel either in a dedicated combustion facility (incineration) with or without energy recovery, or as RDF (refuse-derived fuel) in a solid fuel boiler, is also an strategy to mitigate worldwide

GHG as the WtE route avoid combustion of fossil fuels. Other strategies to avoid GHG emissions using the WtE path include: (i) compost the organic material after digestion to produce a useful soil conditioner and, consequently, reducing nitrogen requirements in agriculture and avoiding landfill disposal; and (ii) recycle the carbon for use as material feedstock [5]. Currently municipal waste is being energy-valued through combustion processes or anaerobic digestion, producing gas that can be used to generate heat and/or electricity. Thus, worldwide, total primary energy supplied from renewable municipal waste accounts for 0.67 EJ, and 0.65 EJ from nonrenewable municipal waste, a 0.2% of the total (2014), but no municipal waste (renewable and nonrenewable) is supplying energy in Africa and Latin America, according to IEA statistics [1].

Case Example 12.2

Methane production from sanitation improvement in Haiti:

Haiti has no centralized sewage treatment and less than 27% of the population has access to improved sanitation. The 2010 earthquake and ongoing cholera outbreak in Haiti highlighted the consequences of inadequate waste treatment, with 734,134 cholera cases and 8896 confirmed deaths due to cholera as of 28 February 2015. In the U.S., approximately 0.5 kWh of electricity is needed to treat each m^3 of wastewater, accounting for 3% of all U.S. electricity use. In Haiti centralized wastewater treatment would exceed the local capacity to pay. Additionally, the country's electric grid provides irregular coverage to only 10% of the population, with wood and charcoal representing 70% of the energy use within the country. Cooking fuel is estimated to require 25–50% of daily income and resulted in soil erosion and deforestation, leaving only 1.5% of the land forested. Anaerobic digestion that utilizes wastewater for energy generation is proposed and a the wastewater from a hospital (Partners in Health medical complex in Cange, providing healthcare to about 2000 people daily, 100–400 patients and a 1000-student K-12 school) was considered. Low-cost non-mechanized AD designs, such as the plug-flow, Taiwanese-model digester and other similar tubular designs, have successfully treated agricultural manure in Latin America and Asia with minimal energy input. Moreover, in tropical climates, mesophilic (20–45 °C) digestion can be maintained without heating. Separately treating black water results in a decrease in AD reactor size, thereby reducing capital costs and space requirements. Biochemical methane potential (BMP) was calculated as 0.006–0.017 m^3 $person^{-1}$ day^{-1}. The BMP of Haitian latrine waste (blackwater: feces, urine and flush water) was determined (16 $ml \cdot ml^{-1}$) and compared to other waste streams, such as grey water (originating from washrooms, showers and kitchens: 0.09 $ml \cdot ml^{-1}$), septage (0.58 $ml \cdot ml^{-1}$), and dairy manure. A AD system was implemented and began receiving blackwater in October 2013, with dairy manure used as the inoculum source to increase biogas production during the startup phase. The biogas production averaged 0.108 m^3 day^{-1}, with an average of 65% CH_4, and the average use of toilets was 31 uses day^{-1}, resulting in a CH_4 production value of 0.002 m^3 use^{-1} day^{-1}. Also, there was 99.9% reduction in pathogens (coliforms and *Escherichia coli*), 98.5% reduction in chemical oxygen demand (COD) [6].

Also organic waste can serve as feedstock for the production of transportation fuels. Current research with regards to biorefineries emphasizes the utilization of lignocellulosic materials, originating from agriculture and forestry, as second generation feedstock for the production of biofuels, as discussed in Chap. 11.

In this chapter we will focus our attention on different technologies offering WtE routes except for industrial waste because it uses specialized plants and this analysis is beyond the scope of this book.

12.2 Technology State of the Art

There are different approaches to extract energy (electricity and/or heat) from waste, mainly divided in biological and thermochemical processes (Fig. 12.1). Most of these approaches have been briefly described in Chaps. 5 and 11, with an emphasis on the fermentation process (also used for obtaining second generation bioethanol). Thus, the state-of-the-art will be focused on anaerobic digestion, incineration, pyrolysis-gasification and hydrothermal carbonization. No MSW commercial plants applying hydrolysis followed by fermentation of the sugars are operating. In contrast to fermentation, pyrolysis and gasification techniques are widely used for the processing of waste feedstock. Anaerobic digestion of organic waste has advanced mainly in Europe but facilities have been built in other developed countries [7].

A biological drying process is often used as a pretreatment to convert the MSW into SRF (solid refuse fuel—a secondary fuel). This process comprises waste shredding, aerobic drying for reducing the moisture of the wastes to less than 20%, metal recovery, and separation of rejects. The energy required for drying the wastes in the bioreactor is generated by the exothermic aerobic decomposition of a limited fraction of easily biodegradable organics. This process does not biostabilize the

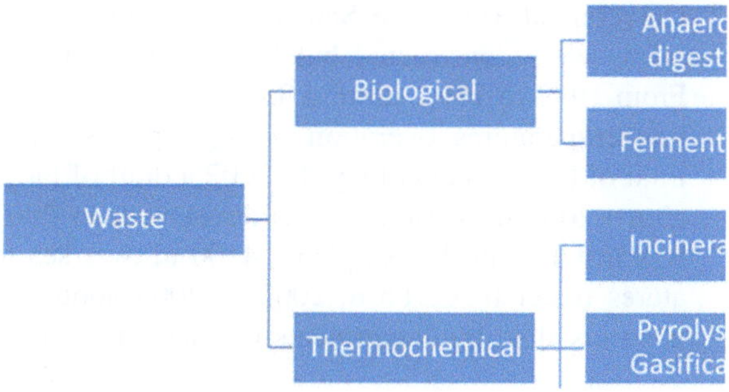

Fig. 12.1 Summary of current waste-to-energy technologies [8]

wastes but is designed to minimize the bio-decomposition so as to preserve the energy content of the wastes [9].

12.2.1 Anaerobic Digestion

Anaerobic digestion of organic wastes can be produced in households, farms, sewage water plants and in landfills. It produces biogas, comprising mainly CH_4 (25–75%) and CO_2 (30–40%), and trace amounts of other gases as N_2, H_2O, NH_3, CO, O_2 and H_2S that escape into the atmosphere and pollute the environment. It has been reported that 1 m^3 of biogas from anaerobic digestion is equivalent to 21–37.5 MJ of energy, it could generate 2.04 kWh of electricity considering a 35% efficiency [10] and it is equivalent to about 5.5 kg of firewood [11].

The anaerobic process consists of essentially 4 sub-processes or stages, namely hydrolysis, acidogenesis (fermentation), acetogenesis/dehydrogenation and methanogenesis. Each of these sub-processes uses different bacterial communities functioning under different temperature range (psychrophilic <20 °C; mesophilic 20–40 °C; thermophilic 40-60 °C). In the case of domestic and community biogas systems, the operating temperature range is often close to ambient temperature. Most of the anaerobic systems today use feedstock diluted with water to a total solids content of around 25%, thermophilic and single-stage type, with retention times ranging from 14 to 20 days. The stabilized residues contain large amounts of water, most of which is removed by filtration. Only a portion of the removed water can be reused so as to maintain the electrical conductivity within proper limits in order to protect the microbial activity in the reactors.

Case Example 12.3

Biogas technology and production of fertilizers in Bolivia:

One important motivation for biogas plants in Bolivia is to use the fermenters for the production of fertilizers from organic materials. Therefore the government decided to develop biogas technology as part of agricultural production, to prevent ecosystems from being destroyed by agricultural activities. In 1986, a biogas cooperation project was started by GTZ and the Universidad Mayor de San Simon (UMSS) in Cochabamba. A total of 27 plants were constructed but in 1988 there was only one still functioning. From 1989 to 1992, around 35 plants were constructed [12]. Because of the low temperatures, operation of biogas plants at high altitude is the theoretical limit of biogas technology. In 2002 a draft of biogas digesters for the regions above 2000 m in Bolivia was developed. In 2003 a tubular biogas digester was installed in the Altiplano, 4100 m over sea level, with ambient temperatures under 0 °C. From 2002 to 2006 about 250 tubular biogas digesters were installed by departments Cochabamba and La Paz [13].

Table 12.1 Feedstock production and gas yields from various common sources [14]

Substrate	Daily production (kg/animal·day)	Dry matter (DM) (%)	Biogas (m^3/kg DM)	Biogas (m^3/animal·day)
Pig mature	2	17	3.60–4.80	1.43
Cow mature	8	16	0.20–0.30	0.32
Chicken mature	0.08	25	0.35–0.80	0.01
Human excreta	0.5	20	0.35–0.50	0.04
Straw/grass		80	0.35–0.40	
Water hyacinth		7	0.17–0.25	
Maize		20	0.25–0.40	
Rice straw		87	0.18	
Rice husk		86	0.014–0.018	
Bagasse			0.165	
Leaf matter			0.60	

The biogas plant for the household and community level can show different configurations, but comprises the following essential components [11]:

- the inlet: for the input of the raw material referred to as feedstock
- the digester: hosts the feedstock and is where the anaerobic process takes place
- the gas holder: stores the gas produced for later use
- the outlet or compensation tank: collects the digested material from the digester
- the pipeline: conveys the gas to the consumption point.

The gas yield depends on the design and the type of feedstock, operating temperatures, pH, hydraulic retention time and agitation. From revised literature, laboratory studies show that the potential gas yield of biomass feedstock ranges between 0.3 and 0.5 m^3 biogas/kg total solid [11]. Table 12.1 shows the gas yield for common feedstocks [14].

Case Example 12.4

Factors affecting household's decisions in biogas technology adoption in northern Ethiopia:

The study involves 179 biogas-user and 179 non-user sample households in Ofla and Mecha, and is based on questionnaires and interviews produced between June, 2013 and January, 2014. Key factors considerd to increase the adoption of biogas technology included empowering females and female-headed households, improving educational levels of the household heads, increasing cattle size (as the primary input for biogas digesters in Ethiopia is cattle dung, and possession of four or more heads of cattle is indicated to be sufficient to feed at least the minimum recommended digester size), raising incoming levels, improving access to credit, and encouraging households to plant more trees (more planted trees means more access to energy, but also better cash to finance biogas installation). Faster and more convenient biogas stoves and brighter biogas lighting were the most important reasons for adopting the technology. Almost all the heads of the non-user sample households have some awareness about biogas technology, but also interest to invest on it in future. The majority of the non-users seemed to have a 'wait and see' attitude. The main reasons provided for not adopting biogas technology are fear to the absence of regular maintenance, user training gaps, shortage of human and/or animal labor power required for the construction of the biogas digester (e.g., transport construction stones and dig the digester pit), financial constraints and the need to improve the existing biogas model with the addition of injera stove [15].

In Brazil, large-scale crop-based digesters using sugar cane residues as feedstock are most common. The main parameter for dimensioning a biogas plant is the retention time (typically 20–30 days), approximately obtained by dividing the digester volume by the daily influent rate, and it is highly dependent of process temperature and digested substrates. For rural biogas plants in developing countries retention time is normally around 150–200 days, sometimes more than a year [13]. Climatic conditions need to be taken into account especially when planning a facility without heating equipment (low ambient temperatures decrease microbiological activity in the digester). Thus, climatic conditions in most areas of Africa and Latin America are optimal for biogas plants, especially in tropical countries [13]. Different pretreatments can enhance the production of biogas, as

(i) physical: mechanical and high pressure
(ii) thermal: microwave and thermochemical liquidation
(iii) chemical: dilute-acid hydrolysis and alkaline addition
(iv) biological: enzymatic hydrolysis [16].

Several types of bioreactors are currently applied, but the three major systems commonly used include batch, continuous one-stage, and continuous two-stage reactors with a variety of methanizers [16]. A disadvantage of anaerobic digestion is that anaerobic bacteria are sensitive to changes in process conditions such as pH, temperature, salts and toxic materials generated during the reaction, requiring stringent process control and optimization operations.

Anaerobic digestion can operate with any type of biomass that can be digested by animals (that is, except biomass from wood). The process is particularly efficient with wet biomass, sludge from sewage treatment plants, as well as in landfill cells for municipal waste storage. If the biomass used as raw material is not contaminated, the solid digestate can be also used as fertilizer, adding value to the process.

Case Example 12.5

Development of biogas technology in Colombia:

Colombia has ideal conditions for the implementation and expansion of biogas technology. In the lowlands the average annual temperatures is over 24 °C. There are 2.5 million pigs and more than 25 million cattle. The main incentive to introduce biogas technology is the severe pollution of surface waters and thus contamination of the water resources through the deposits from animal husbandry. The first project for the construction of biogas plants started in the middle of the 1980s, with the assistance of a German consulting company. Between 1985 and 1992 25 biogas plants have been built. The types of biogas plants and their sizes were varied: a floating-drum plant, a tunnel plant, an upflow anaerobic sludge blanket and some fixed-dome plants with the size from 14 to 115 m^3. Some of the plants were equipped with plastic balloons to include biogas equipment. These plants were designed for medium and large pig and cattle breeders who had between 20 and 2000 animals [13, 17]. However, inadequate training of users, material errors, technical defects, reduced animal holdings, water problems, etc., have rendered many to be non-functional.

Often associated with wastewater treatment facilities, anaerobic digestion will generate methane that can either be flared or used to generate heat and/or electricity. Also, the renewable municipal waste can be stored in impermeable disposal cells in which gases are produced naturally by anaerobic digestion, and later collected through a mesh of interconnected and perforated pipes to depths of about 20 m. The production of landfill gas per ton of waste is estimated, in theory, at about 150–300 m^3, containing a high proportion of methane, and an energy of about 6 GJ per ton of waste material [7]. Also, a new route has been proposed for the bioconversion of syngas from waste gases, into lipids that can be used for biodiesel production [18]. In a first stage, an anaerobic bioreactor converts mixtures of gases of CO_2 and CO and H_2 to acetic acid, using an anaerobic acetogen. In a second bioreactor the acetic acid product is fed as a substrate for aerobic conversion into lipids by an engineered oleaginous yeast.

12.2.2 Incineration

The waste incinerator plants are designed to handle large volumes of MSW with no-pretreatment. These plants include: (i) the incinerator, usually of inclined moving or roller grate type (single-stage chamber), with the presence of a flame for the oxidation of the MSW; (ii) the boiler for energy recovery and electricity/heat production; and (iii) the flue gas treatment system (>850 °C) for the removal of particles, HCl, HF, SO_2 dioxins, furans and heavy metals, including Hg. The target of this complicated physicochemical operation is the evaporation, degradation and inertization of organic elements in MSW and also the reduction of the weight and the volume of MSW. Most of the solid residues are generated from the combustion chamber and the boiler. The incombustible ash usually constitutes a concentrated inorganic waste that has to be properly disposed. Smaller and more toxic quantities are generated from the flue gas treatment system [9]. The major flue gas cleaning systems are dry or wet scrubbers, electrostatic filters, fabric bag filters and cyclones, activated carbon filters and chemicals (NH_3, CaO, $Ca[OH]_2$). The optimum selection of the flue-gas cleaning system is based on the composition of the gaseous pollutants under treatment and the emission limits of the plant. After stabilization and solidification, the fly ash can disposed in sanitary landfill, while the bottom ash is relatively inert and in general can be used for construction applications (roads, earthworks, in mines, etc.) [19].

Incineration of waste (with energy recovery) can reduce the volume of disposed waste by up to 90% [2]. These high volume reductions are seen only in waste streams with a very high amounts of packaging materials, paper, cardboard, plastics and horticultural waste [2].

The municipal renewable raw material is very heterogeneous and polluted, thus corroding and damaging the standard infrastructure, and therefore requiring strict emission controls, as well as a robust technology to enhance energy recovery. For this reason, average conversion efficiencies to produce electricity are only of about 22% [7]. The newly built projects for electricity production seem to be more efficient, compared to older installations (i.e., WtE plant MKW Bremen with efficiency of 30.5%, EVI Laar 30.5%, AEC Amsterdam 34.5%, and AZM Moerdijk 32.5%). In the case of CHP production, the net electrical efficiency is close to 23% and its thermal efficiency is approximately 45%, which is technically possible by using back pressure turbine technology [20].

12.2.3 *Pyrolysis and Gasification*

Pyrolysis and gasification are both thermal processes and work on carbon based wastes. Pyrolysis converts waste, in an oxygen-free environment at temperatures generally higher than 600 °C, into bio-oil (75%, of which the heating value is around 17 mJ/kg [21]) as the major product along with syngas ($CO + H_2$) and solid bio-char. Gasification converts waste into a combustible gas mixture by partially oxidizing waste at high temperature, typically in the range 800–900 °C. The exact temperature of these processes depends on the reactor type and the waste characteristics.

Gasification seems to be favored over biochemical conversion due to the fact that contaminants (alkali metals, halides, sulfur gases, and tars) present in the biodegradable fraction of the waste can (in comparison to pre-sorting and steam-cleaning the biomass itself) be removed from the produced syngas before catalytic conversion to the intermediate products takes place [7]. Considering the environmental concerns of incineration, pyrolysis, gasification, or both combined may serve as alternatives to combustion. Gasification may theoretically produce electricity at an efficiency of about 27% and heat at about 24% [22]. However, in practice, gasification has not been properly proven and only recently has been achieved in some WtE applications [20]. In the report of the Thermoselect project in Karlsruhe [23] it is stated that no more than 0.56 MWh_e/tMSW may be achieved even in optimized future realizations (by assuming highly efficient gas engines). This performance indicates an electrical efficiency of about 20%, which has to be proven in practice.

The most important characteristics of wastes for gasification are elemental composition, lower heating value, ash content and composition, moisture content, volatile matter content, other contaminants (N, S, Cl, alkalies, heavy metals, etc.), bulk density and size. Some of these characteristics are so crucial that most or prevailing gasification technologies generally use pre-processed waste, or refuse-derived fuel as feedstocks rather than the waste as it is [16].

The quality of the waste as a feedstock for fermentation or gasification is important in terms of pre-treatment and conversion facility design. The composition of MSW varies significantly among countries as well among regions within individual countries (e.g., urban vs. rural areas). For example, these variations are caused by differences in consumer habits, diet and disposal patterns and relate to the level of affluence and development of the country. Generally, the thermochemical platform, using gasification, will be superior to the biochemical platform if an organic waste fraction with low water content is used, whereas the biochemical conversion generally works better if waste with high water content is utilized [24]. Thus, one of the biggest challenges faced by developers of waste conversion facilities is the heterogeneity of the feedstock. System performance will depend on their flexibility to cope with varying MSW composition over time.

Case Example 12.6

Planning waste-to-energy integration in the Venezuelan grid:

The authors considered Venezuela because it represents a developing country with a lack of interest on improving waste management, delayed implementation of renewable energies and an additional barrier based on the existence of a very significative fuel subsidy (about a 4% of the GDP in 2009). The electricity sector has been nationalized under one corporation, and it is about 66% based on hydroelectricity and 34% on traditional thermal units using natural gas, diesel and fuel oil. The current waste collection service rate is considered at 88% (2010), reaching 1.24 kg/day·capita. There exists an informal recycling sector, mostly for aluminium, glass and paper, but the exact amount of waste collected by this sector is unknown. The composition of the waste collected (different during different seasons) was estimated to be 17.26% paper and cardboard, 12.45% plastic, 7.4% glass, 3.17% ferrous metals, 0.99% non-ferrous metals, 2.79% fabrics, 14.15% food, 9.39% garden, 0.32% wood and straw and 33.25% others. The disposal sites are 32% sanitary landfills, 43% controlled disposal (without leachate or other emissions control) and 24% non-controlled disposal. Only 30–40% of toxic wastes from hospitals and clinics are treated. The scenarios analysed consider the current collection rate, improved collection rate to 94%, fuel subsidies, subsidies to WtE, carbon tax policy and carbon reduction policy. The thermal conversion technologies considered for electricity generation were gasification with gas turbines and incineration with steam turbines. The success for the introduction of WtE integration mainly depends on the elimination of the fuel subsidies and the introduction of taxes carbon emissions [25].

12.2.4 Hydrothermal Carbonization

Hydrothermal carbonization (HTC) is attracting increasing attention especially for waste streams with high moisture content (80–90%). During HTC, wet feedstocks undergo a series of simultaneous reactions, including hydrolysis, condensation, dehydration, and decarboxylation. A result of this process is the formation of a highly carbonized and energy densified material (often termed hydrochar) that has been reported to have composition equivalent to that of lignite coal. Thus, HTC is a wet process that converts food wastes to a valuable energy-rich resource under autogenous pressures and relatively low temperature (180–350 °C). Compared to other waste-to-energy conversion methods using biological processes, this technology has various advantages including smaller treatment footprints, greater waste volume reductions, no process-related odors, the reaction takes only a few hours and the high temperature helps to eliminate pathogens and inactivates other potential.

However, (i) the process heat requirements may not be offset by the energy generated from HTC products as well as byproducts from process water; (ii) a proper catalyst is needed to lower the reaction temperature and pressure; (iii) further research is needed to digest food waste biomass by HTC to produce functional carbonaceous hydrochar materials for their possible energy applications; and (iv) the efficiency of the process can be low when the waste is composed by highly mixed materials, as it is usual in MSW [16].

12.2.5 Costs

For an initial cost analysis, input data of MSW plants collected in the literature are considered [20]. The results indicate the advantage of CHP over solely electricity generation, and the most attractive technology among the ones examined proves to be incineration, mainly due to its higher power production efficiency, lower investment costs and lower emission rates (Table 12.2) [20]. However, incinerators must keep being fed rubbish for decades to be economical, and poorly run incinerators can be worse for the environment than none at all.

Cost functions for MSW treatment plants with specific configurations are available that allow estimates of initial capital investment (Fig. 12.2) and annual operating cost versus annual quantity of MSW processed, based on data collected in the literature [19]. The results show that aerobic mechanical and biological treatments represent the processes with lower initial capital investment but also offer no WtE treatment. On the other hand, biodrying and incineration or gasification shows

Table 12.2 Electricity, thermal energy and carbon emissions calculated from MSW plant's input data collected in the literature [20]

	$MWh_e/tMSW$	$MWh_{th}/tMSW$	$tCO_2/tMSW$	MW_e
Incineration	0.8	0	0.28	135
Incineration (CHP)	0.6	1.2	0.28	102
Gasification	0.53	0	0.28	90
Gasification (CHP)	0.33	0.7	0.28	56
Landfill biogas	0.2	0	0.35	26
Landfill biogas (CHP)	0.2	0.3	0.35	25
Anaerobic digestion	0.5	0	0.31	95
Anaerobic digestion (CHP)	0.35	0.5	0.31	60

No comparable information about hydrothermal carbonization has been detected in the literature

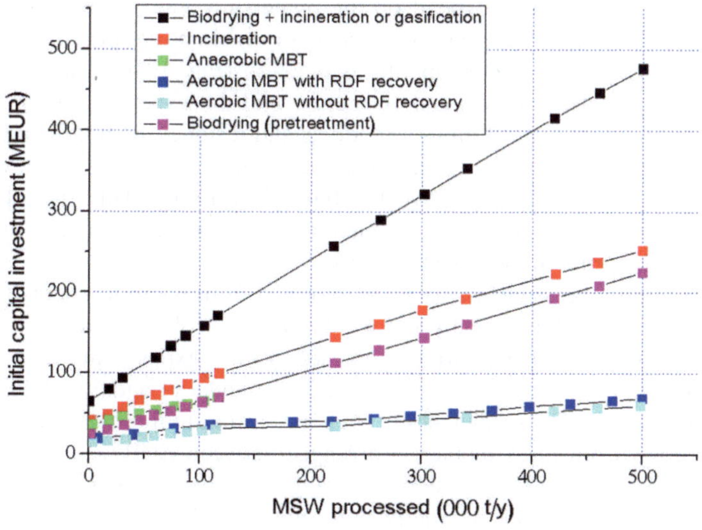

Fig. 12.2 Initial capital investment versus mass of MSW per year for different processing technologies [19]

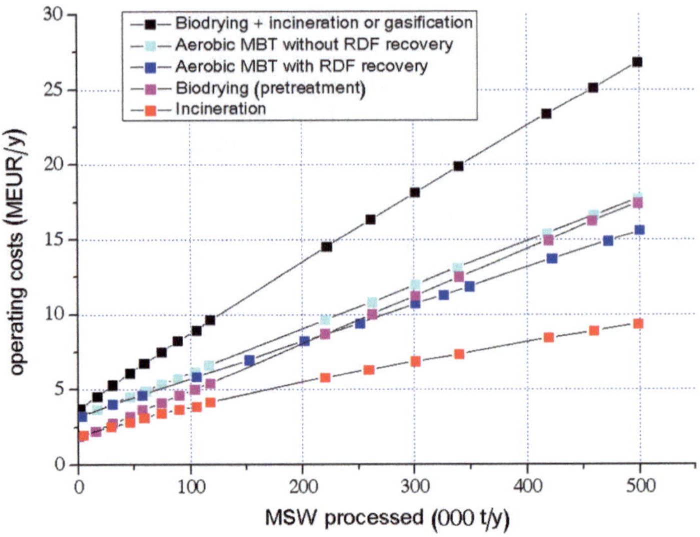

Fig. 12.3 Annual operating costs versus mass of MSW per year for different processing technologies [19]

the highest up front costs but this treatment also includes the most efficient WtE treatment compared to direct incineration.

However, in terms of annual operation costs, the most cost effective treatment is the direct incineration (Fig. 12.3) as it is expected to produce revenues from the sale of electricity and heat.

The operating costs also depend on the ownership and operation entity of the MSW plant. Thus, the treatment costs per mass of MSW are provided for a plant owned and operated by a municipality association (Fig. 12.4). The revenue from the sale of recyclable materials and/or energy, and the expenditures for the transportation and management of unused products and residues are not included. These costs can be substantially increased if significant annual internal return on investment and gate fees are included [26]. The biological drying plants do not offer any significant economy of scale due to their modular construction and the use of multiple parallel units in large installations.

The amount of excess electricity to be exported from waste-to-energy treatments depends on many factors, mainly the technology considered and the waste composition. As an approximation we can consider values exposed in [26] for Greece: 464.6 kWh/tMSW for mass incineration (20% efficiency) and 384.1 kWh/tMSW for SRF incineration (27% efficiency). Then, according to these data and Fig. 12.4, electricity costs from mass incineration ranges between EUR 132–212/MWh and electricity cost from SRF incineration ranges between EUR 160–256/MWh. Furthermore, aerobic MBT internally consumes 32 kWh/t MSW and biological drying 140 kWh/t MSW.

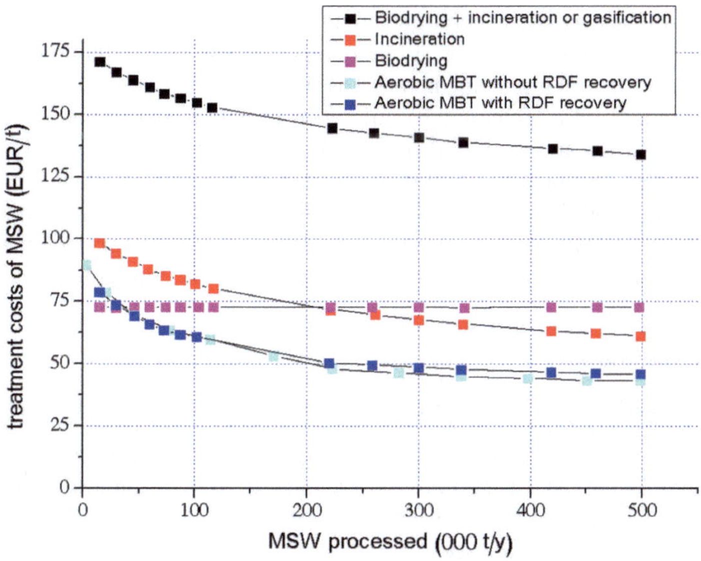

Fig. 12.4 Annual treatment costs versus mass of MSW per year for different processing technologies [19]

Table 12.3 Electricity, efficiency, fixed costs, running costs for different MSW technologies simulated [20]

	Electricity (MWh_e/tMSW)	Efficiency (%)	Fixed costs (EUR/kW/y)	Running costs (EUR/tMSW)	Learning rates
Incineration	0.8	30	4.5	42	0.01
Incineration (CHP)	0.6	23/45	5		
Gasification	0.53	20	3.2	60	0.02
Gasification (CHP)	0.33	12/26	4		
Landfill biogas	0.2	6	1.4	15	0.05
Landfill biogas (CHP)	0.2	5/9	2		
Anaerobic digestion	0.5	22	1.2	40	0.05
Anaerobic digestion (CHP)	0.35	13/19	2.1		

If we consider input values available in the literature and simulate different technologies, results show that CHP-incineration and incineration for only electricity production constitute the optimal strategies (Tables 12.2 and 12.3 [20]) because of higher power generation efficiency and lower emission rates. On the other hand, gasification is the least competitive technology, although the learning rates can vary the ranking as incineration is the most mature technology. However, waste gasification is a technology unlikely to be either appropriate or financially affordable in Africa and Latin America in the short-term. This is simply because a modern gasifier designed for high-heating value waste in developed countries is likely to require additional fuel inputs to gasify a typical high-organic and relatively wet waste in a developing country, and citizens have lower incomes and are therefore not able to pay as much for waste management and carbon recycling [7].

Case Example 12.7

Anaerobic digestion in a beef cattle feedlot in Brazil and GHG emissions:

The manure generated from 140 beef Red Angus heifers (20 months of age) averaging 345 ± 50 kg confined for 90 days in a representative feedlot pen (840 m^2) was considered. Measurements of CH_4 and N_2O fluxes originating from the feedlot pen surface were made 17 times during the period. The feedlot produced an average of 1.6 kWh of electricity with 1 m^3 of biogas (55% CH_4) and 0.26 m^3 biogas per kg of volatile solid. The manure accumulated on the floor surface was fluxed every 15 days. Then, the diluted manure, containing an average of 1,945.2 kg of C and 113.0 kg of N, was sieved; the solid (3.0 m^3) were stockpiled and the liquid part (15.0 m^3) was delivered to the AD. It was estimated a C- and a N-manure recovery of 87.3 and 43.4%, respectively (leaching, runoff and volatilization as some sources of manure losses). The offset generated by the AD system through the replacement of conventional sources of nitrogen fertilizer and electricity was evaluated. Results showed that direct GHG emissions accounted for 0.14 ± 0.06 of carbon dioxide equivalent (CO_2eq) per kg of animal live weight gain (lwg), with ~80% originated from field application. These emission values does not differ from the conventional manure management (without AD). However, 2.4 MWh and 658 kg of N-manure were estimated with AD utilization, potentially offsetting 0.13 ± 0.01 kg of CO_2eq per kg lwg or 95% (±45%) of total direct emissions from the manure management (considering the power mix of the country, where hydropower represents a 75.2% of the total and, consequently, 0.2 kg of CO_2 per kWh). The crop grain (maize) productivities achieved in this experiment were 4.0 ± 0.5 and 5.5 ± 0.3 t·ha^{-1} using sieve and digested manure, respectively [27]. Related studies for the production of swine meat in Brazil, and evaluating the life cycle assessment of four different manure management systems (liquid manure storage in slurry tanks, the biodigester by flare, the biodigester for energy purposes, and composting) shows that the biodigester for energy purposes had the best environmental performance for almost all the environmental impacts, mainly due to the biogas capture and the potential of energy saves [28].

12.3 Regional and National Perspectives on Technology

It is very difficult to find reliable data of MSW on many African and Latin American countries [2], but we will consider the data available from the World Bank as the most credible source of information.

12.3.1 MSW Production Rates

There is a wide variety of MSW production rates between countries in Africa (Fig. 12.5), in part related to GDP per capita, as it can be observed in South Africa and African MENA countries, but also to the policies each country is developing [2]. Moreover, Africa shows the global lowest MSW production rates, meaning that there are huge opportunities to design more efficient and advanced waste management strategies to minimize the environmental impact of human activity, mainly carbon and methane emissions, but also increasing from zero the share of waste in total primary energy supply in this region.

In Latin America the average values of MSW production rates are significantly higher than in Africa and they are not so much related to GDP per capita than to other factors far from being fully identified (Fig. 12.6).

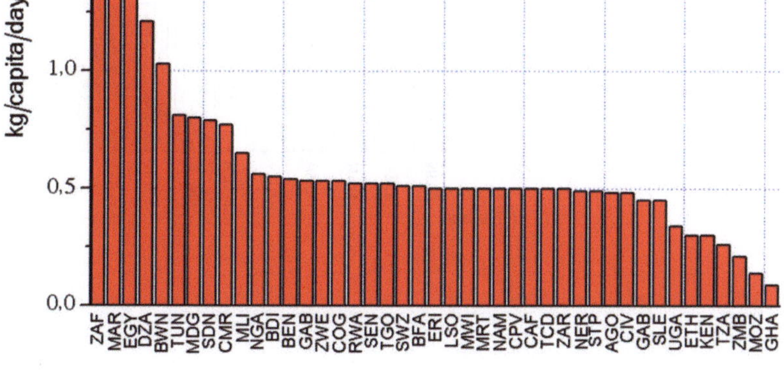

Fig. 12.5 MSW produced per capita and day in different African countries (2012, last data available) [2]

12.3.2 MSW Collection Rates

There is a lack of data in many African countries related to the collection rates of MSW, and most of the countries offering collection rates are only referred to urban areas (Fig. 12.7), as in most African rural areas no collection services are expected. Again, the highest collection rates are mostly related to countries from the Maghreb area.

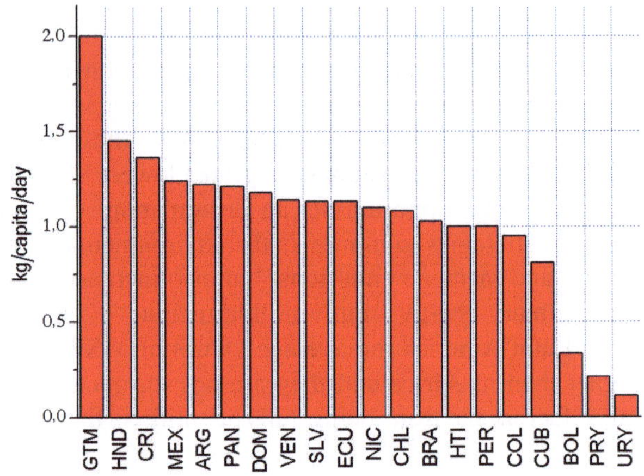

Fig. 12.6 MSW produced per capita and day in different Latin American countries (2012, last data available) [2]

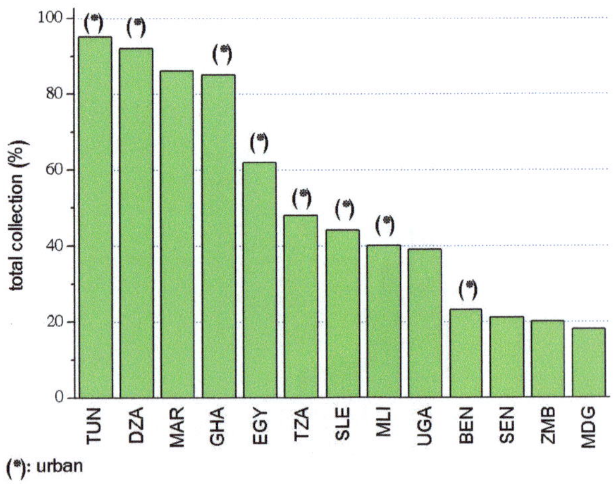

Fig. 12.7 Available MSW collection rates in different African countries (2012, last data available) [2]

Case Example 12.8

Overview of solid waste in Libya:

The work considers if solid waste can be used as a source of energy in Libya. Organic matter is considered the primary category as it represented 59% of the waste collected (12% paper-cardboard, 8% plastic, 7% metals, 4% glass, 2% wood and 8% miscellaneous). Elimination is the solution applied to 95% of waste produced in Libya, either thrown into one of approximately 2,300 open dumps (67%), most of them nearly saturated, or in landfill mode (30%). Composting represents only 2% of all waste produced, and recycling a 1%. However, preliminary estimates for Libyan cities revealed that 25% of the waste can be recycled. The authors propose the implementation of landfill disposal techniques for the valorization of biogas [29].

On the contrary, in Latin America all data are related to total collection rates, not only urban collection activity (Fig. 12.8), and also data from most countries are available. The availability of collection data and the fact that most collection rates are above 70% can be considered as an indicator of the higher development of the waste collection in Latin America compared to Africa.

Case Example 12.9

The potential of biogas production from waste in Uruguay:

Uruguay's intensive use of land for agriculture and livestock, and a significant agro-industrial activity, make this country attractive for a national program to exploit the energy from biogas production. The study was based on a national survey of agricultural and agro-industrial production, and the use of organic solid/liquid waste for producing biogas by anaerobic digestion. There is a wide variety of industrial wastewaters that have been successfully treated in anaerobic systems: malting, brewing, drinks, distilleries, pulp and paper, food processing (i.e., slaughterhouses, dairy), pharmaceuticals, yeast, leachate, etc. The industrial and urban waste treatment systems were developed with the primary aim of reducing pollution levels. There are systems based in anaerobic digesters in three malting companies, two dairy industries, the main wool scouring factory, etc. The domestic wastewater is mainly aerobically treated or discharged directly to the sea. Digesters in rural areas have not been sustained because of the imported technology and because rural electrification is broadly extended. Centralised biogas plants that process organic waste from diverse sources can offer an adequate solution for disperse and variable waste production, mainly derived from the agricultural and forest activity. However, the author considers that the main barrier to implement the anaerobic technology is the lack of investment to improve collection of wastes and update the treatment plants. Estimations suggest that 1.3–2.1% of current total primary energy could be achieved [30].

12.4 Conclusions and Future Perspectives

Municipal and other waste forms make a significant contribution to GHG emissions and have an adverse impact on other environmental factors (e.g., health, water quality, air, etc.) in those regions where waste management systems are rudimentary or non-existent. Waste-to-energy provides a very limited amount of electricity or heat on a global basis and makes virtually no contribution in either Africa or Latin America except in Mexico, which has begun producing electricity in small amounts (~ 140 GWh) annually. Organic wastes (discussed in Chap. 11) also serve as feedstock for the production of transportation fuels.

There is a wide variety of MSW production rates between countries in Africa related to GDP per capita, and variations in national policies (some nations have very few policies on MSW and very limited data on MSW collection rates) to minimize the environmental impact of human activity, mainly carbon and methane emissions, or to utilize MSW as a potential energy resource in the future. In Latin America the average values of MSW production rates are significantly higher but are less related to GDP per capita than other socio-economic factors that are difficult to identify. Total collection rates for both urban and rural areas are available for most of Latin America indicating collection rates exceed 70%. Landfill gas and biogas projects are increasing in frequency within Africa (e.g., South Africa and Kenya) and Latin America (e.g., Colombia, Brazil and Bolivia).

Fig. 12.8 Available MSW collection rates in different African countries (last data available) [2]

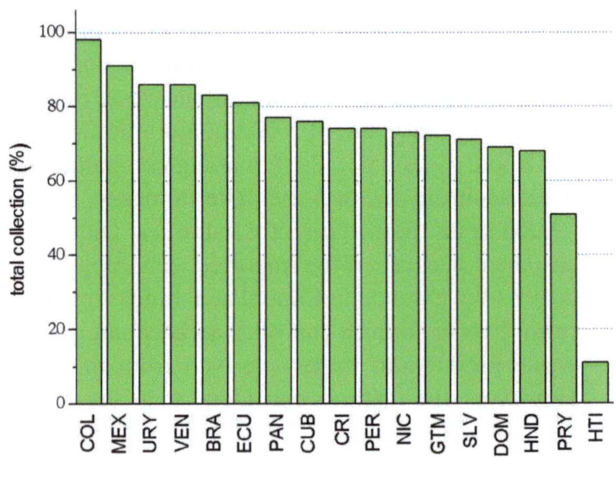

For small communities, biogas digesters using animal and human manure could provide cooking solutions while also providing supplementary benefits to sewage treatment. A number of programmes are in place in Africa to increase the use of biogas in domestic applications. The Africa Biogas Partnership programme has already installed 46,000 digesters (Africa Biogas Partnership Programme 2015) and intends to extend the programme to reach another 100,000 households by 2017, in Kenya, Ethiopia, Tanzania, Uganda and Burkina Faso (Africa Biogas Partnership Programme 2014). Cameroon and Rwanda have initiated national programmes for the implementation of biogas.

For building the future of waste management in Africa and Latin America, ecosystems can provide the best example of a system that works in a sustainable fashion. One of the central principles in industrial ecology is the vision that industrial systems can use materials extracted and metabolized in a cyclical manner, driven by renewable energy which is used in a cascading manner [8]. Thus, instead of carbon capture and storage in landfills, which generally occurs at the beginning of the resource flows, the principle of carbon capture, transformation and reuse adding waste-to-energy paths in the cycle could be further developed and applied throughout the whole socio-industrial metabolism.

References

1. International Energy Agency database (2015) http://data.iea.org/. Accessed 30 Nov 2015
2. Hoornweq D, Bhada-Tata P (2012) What a waste. A global review of solid waste management. Urban Development Series. Knowledge Papers. World Bank
3. Henry RK, Yongsheng Z, Jun D (2006) Municipal solid waste management challenges in developing countries-Kenyan case study. Waste Manage 26:92–100
4. IRENA (2015) Africa 2030: roadmap for a renewable energy future
5. Nuss P, Bringezu S, Gardner KH (2012) Waste-to-materials: the longterm option. In: Karagiannidis A (ed) Waste-to-energy: opportunities and challenges for developing and transition economies. Green Energy and Technologies Series. Springer-Verlag London Ltd. ISBN 978-1-4471-2305-7
6. Lansing S, Bowen H, Gregoire K, Klavon K, Moss A, Eaton A, Lai Y-J, Iwata K (2016) Methane production for sanitation improvement in Haiti. Biomass Bioenergy 91:288–295
7. IEA (2009) Bioenergy—a sustainable and reliable energy source. IEA Bioenergy
8. Ayres RU, Ayres L (2002) A handbook of industrial ecology. Edward Elgar Publishing, Cheltenham
9. Economopoulus AP (2012) Planning tools and procedures for rational municipal solid waste management. In: Karagiannidis A (ed) Waste-to-energy: opportunities and challenges for developing and transition economies. Green Energy and Technologies Series. Springer-Verlag London Ltd. ISBN 978-1-4471-2305-7
10. Murphy JD, McKeogh E, Kiely G (2004) Technical/economic/environmental analysis of biogas utilisation. Appl Energy 77:407–427
11. Mapelli F, Mungwe JN (2013) Modern energies services for cooking: from improved cook-stoves to domestic and community biogas based systems. In: Colombo S, Bologna S, Masera D (eds) Renewable energy for unleashing sustainable development. Springer International Publishing, Switzerland

12. Schwarz W (2006) Landwirtschaftliche Nutzung der Biogas-Technology in Lateinamerika. Master thesis, University of Stuttgart. Institute of Sanitary Engineering, Water Quality and Solid Waste Management
13. Kornau J, Albers H (2012) Modelling waste characteristicas and WtE plants as a tool for optimal operation conditions. In: Karagiannidis A (ed) Waste-to-energy: opportunities and challenges for developing and transition economies. Green Energy and Technologies Series. Springer-Verlag London Ltd. ISBN 978-1-4471-2305-7
14. Bond T, Templeton MR (2011) History and future of domestic biogas plants in the developing world. Energy Sustain Dev 15:347–354
15. Mengistu MG, Simane B, Eshete G, Workneh TS (2016) Factors affecting households' decisions in biogas technology adoption, the case of Ofla and Mecha Districts, northern Ethiopia. Renew Energy 93:215–227
16. Pham TPT, Kaushik R, Parshetti GK, Mahmood R, Balasubramanian R (2015) Food waste-to-energy conversion technologies: current status and future directions. Waste Manage 38:399–408
17. GTZ-GATE (1999) Biogas digest, vol 4, biogas country report, information and advisory service on appropriate technology. Eschborn, Germany
18. Hu P, Chakraborty S, Kumar A, Woolston B, Liu H, Emerson D, Stephanopoulos G (2016) Integrated bioprocess for conversion of gaseous substrates to liquids. Proc Natl Acad Sci 113:3773–2778
19. Kalogirou E, Bourtsalas A, Klados M, Thelemis NJ (2012) Waste management in Greece and potential for waste-to-energy. In: Karagiannidis A (ed) Waste-to-energy: opportunities and challenges for developing and transition economies. Green Energy and Technologies Series. Springer-Verlag London Ltd. ISBN 978-1-4471-2305-7
20. Tolis A, Rentizelas A, Aravossis K, Tatsiopoulos I (2012) Decisions under uncertainty in municipal solid waste cogeneration investments. In: Karagiannidis A (ed) Waste-to-energy: opportunities and challenges for developing and transition economies. Green Energy and Technologies Series. Springer-Verlag London Ltd. ISBN 978-1-4471-2305-7
21. Digman B, Kim D-S (2008) Review: alternative energy from food processing wastes. Environ Prog 27:524–537
22. Murphy JD, McKeogh E (2004) Analysis of energy production from municipal solid waste. Renew Energy 29:1043–1057
23. Hesseling WFM (2002) Thermoselect Facility report, TNO R2002/126
24. Kamm B, Gruber PR, Kamm M (2006) Biorefineries—industrial processes and products: status quo and future directions, vol 2. Wiley-VCH, London
25. Balda MC, Furubayashi T, Nakata T (2016) Integration of WTE technologies into the electrical system for low-carbon growth in Venezuela. Renew Energy 86:1247–1255
26. Economopoulos AP (2010) Technoeconomic aspects of alternative municipal solid wastes treatment methods. Waste Manage 30:707–715
27. Junior CC, Cerri CEP, Pires AV, Cerria CC (2015) Net greenhouse gas emissions from manure management using anaerobic digestion technology in a beef cattle feedlot in Brazil. Sci Total Environ 505:1018–1025
28. Cherubini E, Zanghelini GM, Alvarenga RAF, Franco D, Soares SR (2015) Life cycle assessment of swine production in Brazil: a comparison of four manure management systems. J Clean Prod 87:68–77
29. Hamad TA, Agll AA, Hamad YM, Sheffield JW (2014) Solid waste as renewable source of energy: current and future possibility in Libya. Case Stud Therm Eng 4:144–152
30. Moreda IL (2016) The potential of biogas production in Uruguay. Renew Sustain Energy Rev 54:1580–1591

Chapter 13
Natural Gas

Ricardo Guerrero-Lemus and Les E. Shephard

Abstract This chapter shows that Africa and Latin America have substantial proven reserves of natural gas, and quite homogeneously located in both regions. Natural gas is playing an increasingly prominent role in the development of electricity generation systems, as it is versatile, supports both baseline and peaking power generation demands, provides proven reserve generation capacity that supplements non-dispatchable power (i.e., wind and solar), has lower GHG emissions and uses less water than coal-based generation and has lower capital costs and shorter new plant construction times than nuclear. Unconventional natural gas production is associated with a number of socio-political, environmental and economic issues that limit drilling in some areas globally and that have created local concerns about impacts on water supply, air quality and enhanced methane emissions. Natural gas plays three prominent roles in the development of the future low-carbon energy future in Africa and Latin America: complementing deloyment of renewables by increasing the flexibility of non-dispatchable power systems, reducing emissions by displacing other fossil fuel generation and reducing health concerns related to household biomass burning. Africa consumes much less gas than it produces on a country-by-country basis in part because of limited infrastructure investment. In Latin America, Mexico and most South American countries have considerable natural gas reserves, and the gas pipeline infrastructure is more highly developed with gas being transported between countries and with all countries producing natural gas using it to produce electricity. Then, natural gas can play a key role in the development of the future low-carbon energy future in Africa and Latin America if a mitigation of social and environmental impacts and the extension of the power grid are achieved.

Keywords Energy · Africa · Latin America · Electricity · Natural gas · Gas turbines · CHP · Trigeneration · Combined cycle · Internal combustion engines · Energy efficiency · Gas reserves · Carbon emissions · Dispatchable power · Carbon capture

R. Guerrero-Lemus and L.E. Shephard, *Low-Carbon Energy in Africa and Latin America*, Lecture Notes in Energy 38,
DOI 10.1007/978-3-319-52311-8_13

13.1 Introduction

Natural gas is playing an increasingly prominent role in the development of electricity generation systems on a global scale. In 2014, 21.6% of the world electricity was produced from natural gas [1]. Growing global gas supplies have increased availability and moderated prices, electricity generation from gas is versatile and can support both baseline and peaking power generation demands and can also serve as a proven reserve generation capacity that supplements non-dispatchable power generating capabilities (mainly wind and solar) [2]. This role of supporting continued expansion of non-dispatchable power generation to assure system reliability is likely to continue well into the future until utility-scale storage options are integrated into the electricity sector at scale. Electricity generation from natural gas has lower GHG emissions and uses less water than coal-based generation and the capital costs and new plant construction times are significantly less than nuclear.

The recent growth in unconventional gas production in the US in particular, but elsewhere around the world, is a direct result of technology innovation that combines enhanced drilling technology with hydraulic fracturing practices to produce oil and gas from shale formations. The increase in natural gas production from hydraulic fracturing has also contributed to a number of socio-political, economic and environmental issues that have limited or precluded drilling in some US states and in many nations around the world [3]. Many of these issues are related to the potential impact of natural gas production on local and regional water resources through fresh groundwater contamination during drilling [4], on-site surface spills, inadequate well completion and closure, and the disposal of large volumes of water extracted as part of the gas production process. Other impacts include excessive road traffic, accelerated deterioration of local and regional roads and impacts on air quality. Finally methane emissions during the production, transportation, storage, delivery and end-use of natural gas continues to require additional consideration if natural gas use is to continue to expand globally. Historically, a major driver of natural gas-fired generation technology has been increasing efficiency. In the future other technical objectives include part-load efficiency, ramp rate, turndown ratio and start-up times, among others.

Natural gas is a fossil fuel used mainly for primary energy supply to power plants, heating systems and CHP plants, industry (e.g., cement, glass, fertilizer and chemical production) and a critical feedstock for a variety of products including plastics. Natural gas plays three important roles in the development of the future low-carbon energy future in Africa and Latin America: (i) complementing deployment of renewables by increasing the flexibility of non-dispatchable power systems, (ii) reducing emissions by displacing other fossil fuel generation sources (i.e., coal- and oil-fired power generation), and (iii) reducing health concerns related to household biomass burning. As of January 1, 2015, there were an estimated 6973 trillion cubic feet (Tcf) of total world proved reserves of dry natural gas, 606 in Africa and 295 in Latin America [5]. Projections on proven global reserves will vary regionally across the globe depending on available data, existing wells and production history, among other factors [6] but in many countries or regions within Latin America and Africa, natural gas will likely become more abundant to support electricity production [7].

Natural gas plays a central role in the generation capacity of a power system. Generation is a mix of baseline, peaking, spinning reserve, offline reserve, and non-dispatchable (mainly wind and solar) power plants [2]. Baseline refers to coal, nuclear, oil and gas-fueled generators that run continuously except for scheduled maintenance. These are the largest generators and offer the cheapest electricity but can require long start up times to achieve full capacity. Nuclear and coal generators are usually considered baseline. On the contrary, peaking generators are smaller, can be started quickly and inject energy to the grid in a relatively short time to cover peak daily or seasonal demand and are responsive to market opportunities when electricity costs are most expensive. Peaking generators are usually fueled with oil or natural gas. Spinning reserves operate at low power output in anticipation of ramp up from dispatched conditions to full load matching non-programmed power demands. Any generator that is running at low output is not as economical as at full load and, consequently, spinning reserves are avoided as much as possible by system operators. Offline reserve generators are operated seasonally or during scheduled maintenance times for baseline and peaking reserves.

Natural gas occurs deep beneath the earth's surface and consists mainly of methane and small amounts of hydrocarbon liquids and non-hydrocarbon gases. The gross calorific value (GCV) of natural gas can vary depending on the composition. The top-ten producers in 2014 based on GCV varies from 37.89 MJ/m^3 in Turkmenistan to 39.57 MJ/m^3 in Algeria [8] (to calculate the net calorific value, the gross calorific value is multiplied by 0.9). From the 126,779 Bcf of natural gas produced globally in 2015, 5.8% was produced in Africa and a 6.9% in Latin America [5].

13.2 Technology State of the Art

With the advent of hydraulic fracturing and the sustained development of unconventional natural gas resources natural gas prices have been reduced significantly. As a result, the state of the art has experienced significant change on a global scale. This short state of the art summary is focused on those changes that are likely to have a greater potential impact on the energy future of Africa and Latin America.

13.2.1 Shale Gas and Hydraulic Fracturing

Natural gas is produced using conventional drilling technology in high permeability rock formations, typically sandstones and limestones, where the gas has migrated from finer grained, organic rich source rocks. Unconventional drilling technology that includes horizontal drilling and hydraulic fracturing in shales, low permeability sandstones and limestones, and coal bed methane, today accounts for more than 65% of the total US natural gas production. Unconventional drilling technologies were pioneered by Mitchell Energies in the early to mid-1990s in the Barnett Shale and has continued to expand throughout the last decade and has expanded globally to many nations around the world.

Natural gas production from shales and other low porosity rock that contain gas in pores is produced by the process of hydraulic fracturing or "fracking". The hydraulic fracturing process begins by drilling a vertical well to the "shale" horizon where the well is turned horizontal drilled in laterals roughly parallel with the producing horizon often for distances that exceed a mile. Water, chemicals, and sand is pumped down the well and pressurized to levels that exceed the confining pressure and strength of the reservoir rock resulting in the fracturing event. Then, sand is injected into the fractures to maintain permeabilities that allow the gas (and oil) to flow to the wellbore. On average, about 15,000 m^3 of water mixed with chemicals are typically required for the 'fracking" process [9].

Concerns about the environmental impacts from unconventional shale gas production range from surface impacts on forest fragmentation, stream sedimentation, surface spills or leakage of contaminated hydraulic fracturing fluids or reservoir flowback waters from unlined pits [10]. In the subsurface, contamination of freshwater aquifers has been documented as a result of water withdrawal at times of low water availability, injection of water into wells with inadequate mechanical integrity, and leakage through poorly designed well casings. Induced seismicity has been observed in many areas where large quantities of water recovered from the hydraulic fracturing process has been re-injected into the subsurface for disposal [11]. Often, low magnitude earthquake swarms occur in regions where deep well injection is taking place. Other concerns are related to naturally occurring radioactive material (NORMs) that occur in shales and may become more concentrated from the drilling process as fluids move across the site and pose some potential risk to oil field workers. Diminishing air quality related to emissions from the field development process often increase as does the potential for methane emissions as a result of leakage around valves and fittings associated with gas production onsite. Each of these concerns are the basis for scientific research activities directed at improving resource evaluation techniques and alternative drilling technologies [9].

13.2.2 Gas Turbine Power Plants

Gas turbines produce power by utilizing the energy created from combustion of gas and air at high temperature and pressure that drives a series of rings and fixed blades. Also water is used in the combustion process to lower temperature and, subsequently, for reducing NO_x gases. Gas turbine power plants are classified into two groups depending on the cycle of operation: (i) open cycle turbine plants that are most commonly used in the power sector and whose rating is limited by atmospheric backpressure; and (ii) close cycle, where backpressure can be increased with a working fluid in a close cycle. Generally, the thermal efficiency of the simple open cycle gas turbine is only about 16–23% unless the waste heat is re-utilized rather than exhausted. The cycle efficiency depends directly upon the temperature of the inlet gases to the turbine that are typically limited to less than 1000° C due to metallurgical limitations that are temperature dependent. This efficiency handicap can be overcome by incorporating thermal refinements in the simple open cycle, e.g., regeneration and reheating, raising the plant efficiency over 30%, but these refinements can increase system complexity. In regeneration, the heat energy from the exhaust gases is transferred to the compressed air before it enters the combustion chamber, saving fuel and utilizing waste heat. In a reheat cycle, the combustion gases are expanded using two turbines: the exhaust of the high-pressure turbine is reheated in a reheater and then expanded in a low-pressure turbine. Improvements in the design and application of gas turbines have increased the unit capacities of the gas turbine unit, improved their overall efficiency and decreased their capital cost.

Case Example 13.1
Gas flaring and its impact on electricity generation in Nigeria:
A major cause of irregular power supply in Nigeria is believed to be shortfall in gas supply to power plants. About 100 billion cubic meters (bcm) of gas is being flared globally on annual basis with Russia and Nigeria leading at 35.5 and 15.2 bcm, respectively. For ethical and security reasons the Nigerian gas and electric companies of the case study are not identified. The site is composed of 7 gas wells, producing 7.2 million cubic meters per day (mcmpd), and supplying 3.0 mmcmpd to the electric company, 3.6 mmcmpd to the gas utility, and the reminder is flared. The electric company constitutes of 20 gas turbines, of which only one is actively in service or in good conditions; 6 units are unavailable due to one minor or major mechanical problem; and 13 units have been technically decommissioned. If the demand of the end users decreases, there is a regulatory system, and the production is minimized to reduce losses (flaring). However, whenever there is a sudden reduction in consumption from final users, it brings excessive pressure to the inlet valve. Then, for safety reasons the excess gas is channelled to the flare stack. Emergency shutdown of the plant also happen due to loss of control (e.g., fire outbreak in the facility), and involves instant shutting down of the plant and then the subsequent result is gas flare. Planned shutdown can be also arranged, where the gas is flared too. When loss of electric power is produced (e.g., power failure or decrease in power demand) the gas excess is also flared. The electricity generated by the electric company depends to a larger extent on the availability and working conditions of the gas turbines than on the availability of gas. Then, the acquisition of new gas turbines and routine maintenance of the old and new turbines will boost electricity generation and minimize gas flaring [12].

13.2.3 Cogeneration and Trigeneration

Cogeneration, also called combined heat and power (CHP) production, is based on the utilization of the heat energy of combustion with direct heating in different applications (e.g., district heating), resulting in a very flexible and efficient way of utilizing fossil and renewable fuels, mainly natural gas and biomass. The interest in CHP has its origin in increasing the total efficiency of the energy supply systems (85–95% [13]) but also getting rid of an increasing surplus of residue products: (i) oil refineries and WtE systems can fuel CHP systems with hydrocarbon gases; and (ii) agriculture and forestry can supply biomass residues. Natural gas cogeneration can be based on gas turbines, gas engines, and stirling engines. Gas turbines have been described above. Gas engines are gas-fired internal combustion engines, with sizes from 7 kW power to several MW and the power efficiency is about a 20% for the small engines and over 40% for the largest [13]. The Stirling engine is a very simple hot-air engine, nearly noiseless, and very stable in operation with a low power range (typically, 1–150 kW) [13].

Trigeneration is cogeneration with the additional use of low-grade heat to drive an absorption chiller. Absorption refrigeration chillers have become more popular because they use environmentally friendly working fluids with zero global warming potential (e.g., a mixture of water-lithium bromide or ammonia-water) [14]. The absorption chillers can also be used with renewable energy sources. The current development of trigeneration systems is focused on the selection of working fluids as refrigerant/absorbent (e.g., n-butane/n-octane [15] or ionic-liquid/mixed refrigerant) [16], and on combined system configurations (e.g., ejector-double effect absorption refrigeration systems [17], ammonia-water refrigeration systems powered by industrial waste heat or gas turbine exhaust gas [18]).

13.2.4 Combined Cycle Power Plants

The combined cycle (CC) power plants are based on the utilization of the heat energy of the exhaust gases of the gas turbine to produce work, thereby rejecting a minimum of energy. Traditionally, CC power plants have been associated with base load supply, while open cycle gas turbines provide flexible response and peaking electricity. There may be various combinations on combined cycles depending upon the place or country requirements although the typical scheme is a combination of an open cycle gas turbine and steam turbine. The exhaust of the gas turbine is used as the inlet gas to a steam generator. For any given total power output, the energy input is reduced (i.e., saving in fuel) and the installed cost of gas turbine per unit of power output is about one-fourth of that of a steam turbine [13], reaching total efficiencies above 50%. A modern CC gas turbine has an efficiency of around 60% LHV (lower heating value basis), and emits around 350 CO_2 kg/MWh when operating at full load. Manufacturers aim to bring CC gas turbines full-load efficiency toward 65%

LHV by 2020 through materials improvements (using a 1700 °C-class gas turbine could raise its efficiency to around 63% but also results in increased NO_x production) , and through combustion and compression processes. Dry cooling (a wet cooling tower consumes and withdraws around 700 liters per MWh) can reduce water use by two orders of magnitude, but adds cost and efficiency penalties [19].

The greater disadvantages include the complexity of the plant, different fuel requirements and possible loss of flexibility and reliability. There are also some technologies being tested that add other systems to CC power plants to increase some features: (i) adding carbon capture (e.g., oxyfuel combustion) and storage systems for lowering CO_2 emissions; (ii) adding solid oxide fuel cells for increasing the efficiency; (iii) integrating concentrated solar thermal power, lowering CO_2 emissions; and (iv) adding humid air as working fluid to the cycle for increasing efficiency, reducing start-up time and minimum load, and lower NO_x emissions, but increasing water consumption [19].

Case Example 13.2
Toward the hybridization of gas-fired power plants in Algeria:

Algeria offers a strong combination of natural gas and solar resources that can increase the yield of gas-fired power plants if they are hybridized. A feasibility study of potential solar thermal hybridization with parabolic trough technology was performed in 21 Algerian OCGT and CCGT, which represent about 97% of electricity generation [20]. The aim of the work is the use of the exhaust gases from the gas turbines for supplying heat to the CSP systems. Their performance was analyzed in relation to the meteorological parameters, obtaining the yield of each turbine. Gains in terms of additional electricity production and carbon emissions avoided are reported, but depend on the amount of new CSP systems integrated and the ratio gas turbine/steam turbine. Thus, optimal number of trough loops per turbine are determined. Solar steam is used only when atmospheric conditions made the steam turbine work partially loaded. The internal return rate for the CCGT with CSP obtained was negative while for OCGT with CSP was positive. This result can be explained by the higher cost of electricity for OCGT compared to CCGT and the higher potential of the higher temperature of exhaust gases from an open gas turbine to run a steam cycle compared to CC turbines. However, the different daily production patterns that both technologies provide to the power grid should be considered in a more defined analysis. On the other hand, the low price of electricity in Algeria and the fact that the existing regulatory framework only supports solar shares higher than 5% currently lowers the profitability of investments for the hybridization of gas-fired power plants with CSP.

13.2.5 Flexibility for Non-dispatchable Generation

Electricity supply and demand must balance at every instant and, therefore, electricity systems will increasingly require technologies that can provide electrical energy, power and other ancillary services that can respond to changes in non-dispatchable energy generation (mainly wind and solar energy). Thus, thermal generators are increasingly called upon to operate more flexibly. Natural gas power plants can ease the integration of non-dispatchable energy generation by increasing the flexibility of the electricity system and reaching what the IEA calls a "low carbon electricity system" (<100 $kgCO_2/MWh$) [19].

Flexibility has four main sources: (i) generation; (ii) electricity storage; (iii) interconnection with other electrical systems; and (iv) demand response [21]. Curtailment of non-dispatchable generation can help to ensure system stability but it is likely to remain an exceptional measure to maximize endogenous resources and CO_2 reductions. Generation flexibility refers to the extent to which generators across a given system can respond to the variability (expected or otherwise) in the residual load on a time scale of a few minutes to several hours, and in the short to medium term. Generation flexibility is likely to remain the critical resource to deliver flexibility to the power system.

Natural gas generation can provide flexibility to the power system, but continuous cycling (hot start, warm start, cold start, ramps up and down, and shut downs) reduces material lifetime because of temperature variations (Fig. 13.1) [19], which increases outage rate and fault risk and, consequently, drives up operation and maintenance costs.

Until recently, research activities have emphasized boosting the base load efficiency of gas turbines. Today, because of the need for flexibility, there is growing emphasis on the need to provide high efficiency over a variety of operating conditions. The most important indicators are the full-cycle efficiency of the thermal plant over each year, part-load efficiency, start-up time, ramping capability and turndown ratio. Thermal plants with high full-load efficiency but low flexibility can generate less revenue than plants with higher part-load efficiency and, consequently, the latter have a higher potential to be dispatched. However, at part-load efficiency the cost and carbon emissions per unit of electricity generated are higher than at full-load efficiency.

Short start-up times allow plants to be called upon nearer to actual dispatch, which increases the power capacity available within a given time window and avoids unnecessary spinning to provide reserve capability. Depending on the preceding shutdown duration, start-up can be classified as hot (within 8 h), warm (8–60 h) and cold (more than 60 h). Hot start-up times for open cycle gas turbines and internal combustion engines can be below 30 min, and also CC gas turbines can have hot start-up in about 1 h if the steam cycle is decoupled [19]. Turndown ratio represents the minimum stable operational load and is expressed as a percentage of full load. Efforts are increasing to develop an "overnight parking mode" for CC gas turbines to keep plants warm during the night at minimum fuel use while also minimizing

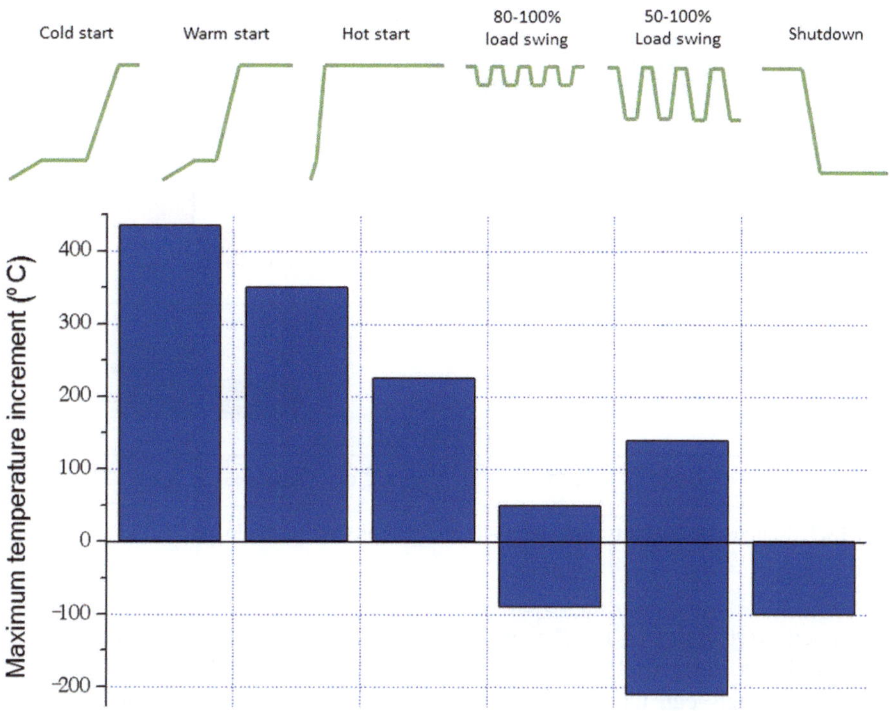

Fig. 13.1 Maximum temperature increments in natural gas turbines due to different cycling conditions [19]

start-up cost the next day. Most of the main flexibility parameters are optimal for open cycle gas turbines, but also for internal combustion engines (Fig. 13.2).

New technologies and upgrades improve flexibility with modest efficiency penalties and roughly identical capital costs: (i) variable pitch-guide vanes and inlet pre-heaters in the compressor to improve part-load efficiency and to lower minimum load; (ii) improved heat recovery steam generator to avoid thermal stress increases ramping capability and to reduce start-up time; (iii) advanced new materials and sensing to reduce cycling impact; and (iv) higher degree of modularity into the gas turbine platform to exchange components relatively quickly if market conditions require different technology characteristics [19].

Case Example 13.3

Natural gas engines to power a gold mine and the grid in the Dominican Republic:
In 2011 the Canadian gold mining company Barrick Gold Corporation signed a turnkey contract with Wärtsilä, a Finnish corporation, for building two 215-MW Flexicycle plants at Pueblo Viejo comprising engines of 16.5 MW

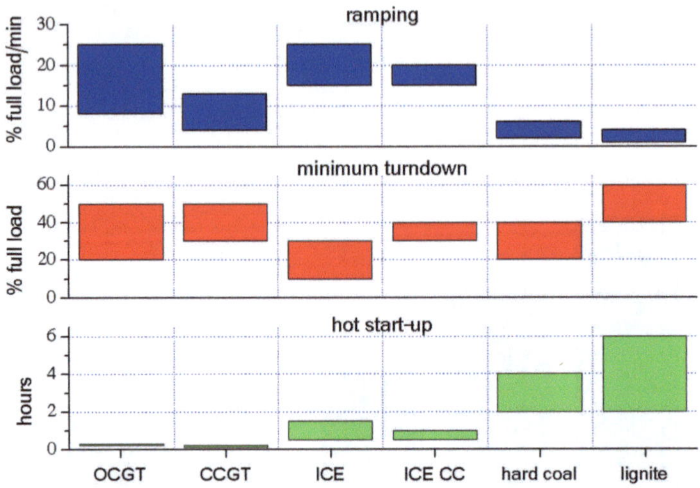

Fig. 13.2 Ranges of flexibility parameters for thermal electric technologies [19]

capacity (Quisqueya I and II power plant). The engines can operate on either natural gas, light fuel oil (LFO), or heavy fuel oil (HFO), and switching between fuels can take place seamlessly during operation, without loss of power or speed. The engine exhaust is directed through a heat-recovery steam generator, and the steam is used to drive a steam turbine. The combined cycle can be shut off and on individually for each generating set. When operating on gas, it offers a low emissions profile and it can achieve efficiencies approaching those of gas turbine combined cycle plants and can maintain that high efficiency across their full operating range (unlike gas turbines). The estimated cost of each combined plant is about USD 700 million. The plants have separate owners (Barrick and EGE Haina, the nation's largest generator by installed capacity) but are operated from one control room. Quisqueya I powers the Pueblo Viejo mine with excess electricity flowing to the grid, while Quisqueya II delivers power solely to the grid. Quisqueya is also capable of providing frequency support to the national grid. The plants have been initially run on fuel oil, but they plan to switch over to natural gas as soon as reliable supplies are available (from a liquified natural gas import terminal being proposed for San Pedro de Macorís). The startup of the Quisqueya plants had an immediate impact on the Dominican energy market, causing a 30% drop in spot prices for electricity, from USD 20/kWh to USD 14/kWh [22].

13.2.6 Internal Combustion Engines (ICE) and Off Grid Power Supply

Some counter-intuitive relationships between unit size and efficiency can exist in natural gas: (i) smaller unit sizes can be attractive in smaller electricity grids for providing ancillary services and because larger units may be difficult to finance; and (ii) internal combustion engine plants, although less efficient than CC gas turbines for large units, can offer higher efficiencies at small scales and also higher efficiencies than a large CC gas turbine operating at partial load [19].

ICE is a very mature technology used in many areas, and large natural gas engines have efficiencies of up to 48% [19]. ICE can be used as single-unit plants (<20 MW), stacked in so-called "bank" or "cascade" electricity plants (20 MW–200 MW), or operated with a combined steam cycle (>250 MW). In island systems or weakly interconnected electricity systems, the modularity of ICE banks allows a more cost effective management of the "N−1" reliability criterion compared with large-scale turbine-based plants. ICEs can execute a cold start in under 15 min and ramps from 80 to 100% within seconds, making them strong flexibility resources. Natural gas ICEs can burn a broad range of gaseous (e.g., landfill gas and biogas) and petroleum-based fuels. However, gas turbines can deliver more energy per unit of land, have considerably lower operating noise levels and deliver higher full- and part-load efficiency than ICE plants, even in combination with a steam cycle [19].

Case Example 13.4

Developing compressed natural gas for vehicles in Africa and Latin America:

In this work the selected countries in Africa and Latin America are Argentina, Brazil and Nigeria. This is based on three criteria: presence of NG reserves, experience with NG fuelled vehicles (NGVs) and the stage of the market. In Argentina, the share of NGVs is 12.6%, and it is due to the fact that Argentina is a net importer of oil, has codes and standards, the government offered incentives, the price of CNG is at 45% of premium gasoline price by energy equivalency, and has the fourth largest network of stations in the world because of the pipelines system attached to. In Brazil, the share of NGVs is 9.8%, mainly because of the search of alternative fuels to oil, financial incentives, NG priced at 50% of gasoline, the development of tri-fuel cars (ethanol, petrol and NG), and the fifth largest CNG refuelling station network. In Nigeria, the share of NGVs is 0.012% because there are no legal provisions nor specific agencies to promote NGVs, there is a disproportionate focus by gas suppliers on LNG export, the differential price of CNG compared to gasoline is marginal, and there are only six refuelling stations, located in Benin City [23].

13.2.7 Natural Gas Appliances

There are many natural gas appliances ready for consumers if proper supply is available: (i) water heaters and boilers; (ii) natural gas range; (iii) natural gas dryers; (iv) natural gas furnaces; (v) natural gas fireplaces; (vi) snow melting systems; and (vii) outdoor lighting. Main research activities are focused on efficiency (combustion analysis, greenhouse gases, fine particle emissions, etc.) and total costs of the appliances, but also on real life operation (consumer behavior, automation, safety procedures, etc.), influence of ventilation systems on domestic use and integration of appliances.

Case Example 13.5
LPG's policies in Brazil:

In Brazil, 98% of households (including 93% of rural households) were reported to have access to LPG—a situation attributed to government policy that has promoted the development of an LPG delivery infrastructure in all regions, including rural regions, and subsidies to LPG users. Until the late 1990s, the rise in LPG use was accompanied by a sharp decline in residential wood consumption. During the period 1973–2001, retail LPG prices were set at the same level in all regions and the average level of the subsidy amounted to 18% of the retail price. In May 2001, end-user prices were liberalised, as part of a process of deregulating the petroleum sector. At the same time, the government introduced an Auxilio-gas ("gas assistance") programme to enable qualifying low-income households to purchase LPG. The total programme cost in 2002 was about half that of price subsidisation. This programme now forms part of the Bolsa Familia, a conditional cash transfer programme. Subsequent LPG price increases, however, appear to have led to a reversal of the trend towards lower residential biomass consumption [24]. Compressed natural gas has also become common in taxi fleets in cities of Rio de Janeiro and Sao Paulo [25].

13.2.8 Costs

The price of natural gas in energy markets (Fig. 13.1) [26] and industry is variable and can be volatile (from USD 15.98/MWh_{gcv} in Canada to USD 73.55/MWh_{gcv} in Switzerland). Household costs also vary (from USD 34.92/MWh_{gcv} in Canada to USD 153.95/MWh_{gcv} in Sweden) depending on the country [8]. Moreover, when natural gas is liquefied (LNG), mainly for shipping exports, an extra-cost is expected (Fig. 13.3) [26].

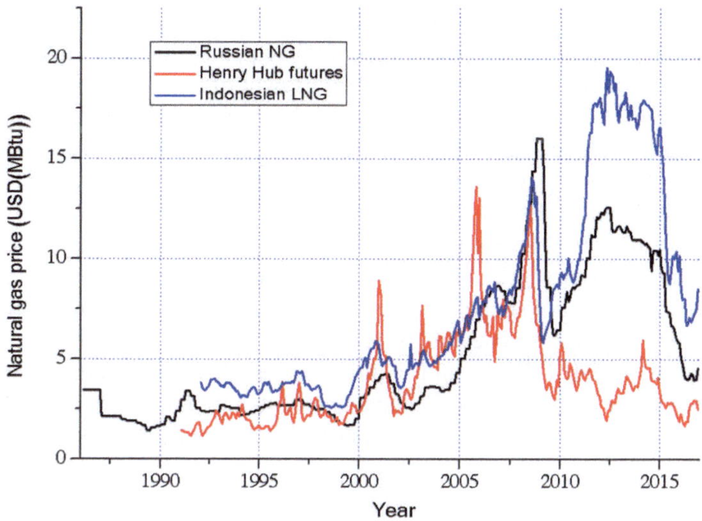

Fig. 13.3 Natural gas price in different global markets [26]

The volatility of natural gas prices in international markets (Fig. 13.3) has a measurable impact on electricity costs based on natural gas. There are regional events that affect the price of natural gas in different markets: (i) Asia's heavy dependence on imports of LNG and increasing competition for LNG cargoes in the region; and (ii) the shale gas boom in USA have led to the lowest natural gas prices in the world.

Case Example 13.6

Effect of low natural gas prices in Bolivia:

According to Petrolíferos Fiscales Bolivianos (YPFB), Bolivia's oil and gas company, the country exported the bulk of its gas production in 2014 to Argentina and Brazil. Additionally, the price of natural gas for power plants has been fixed since 2001 at well below international prices. Such low prices make it difficult for renewable projects to compete [27].

Long-run marginal costs (LRMC), generally presented as the LCOE, are often quoted in the literature that compare electricity generation options. However, for installed conventional capacity (coal, oil or natural gas based), it is short-run marginal costs (SRMC) that determine whether a plant operates and the revenues it receives. SRMC is the change in total cost resulting from a one-unit change in the output of an existing production facility. It ignores adjustments in the capital stock, and includes only fuel costs and variable O&M costs. Theoretically, plants with the lowest SRMC are dispatched first and all generators receive the price paid to the dispatched plant that is last in the merit of order (i.e., with the highest SRMC). SMRC explains recent coal-to-natural-gas fuel switching in USA (shale gas boom) and natural-gas-to-coal in Europe (sustained growth of renewable energy and lower cost of coal partly due to temporary increased exports of US coal) [19].

Providing flexible, dispatchable capacity to complement the increasing flexibility requirements of future power systems has cost impacts as power systems are not at the highest efficiency. Estimates suggest 2 to 5% cost increases on average for fossil-fuel plants when levels up to 33% non-dispatchable energy generation are added to the electricity grid [28]. Moreover, the additional costs of cycling are likely to be lower than the benefits of reduced CO_2 emissions and reduced fuel costs associated with variable renewables [28]. From a macro perspective, ICEs are cost-competitive with OCGTs and could become increasingly attractive in future electricity systems. From a system management standpoint, the challenge will be to share the benefits of renewables in a manner that covers the costs of supporting them [19].

Cost Example 13.7

Diesel versus natural gas fuelling for power distribution in Nigeria:

A life cycle cost model was developed to compare the electricity cost of diesel generators and compressed natural gas (CNG) fuelling. Generally most diesel powered generators can be configured to run on natural gas. Thus, at the prevailing diesel pump price of USD 3.7 per gallon the unit cost of electricity varies between USD 0.28–0.33/kWh depending mostly on the generation capacity. With CNG fuelling, the unit cost, at the prevailing USD 2 per diesel gallon equivalent (DGE), varies between USD 0.14–0.16 kWh. The study concludes that Nigeria needs to extend the natural gas distribution network to take advantage of the nation's natural gas resources. With abundant gas reserves and the heavy flaring of associated gas by oil companies, encouraging the use of CNG for electricity can offer an opportunity to extend the consumption of electricity at a more reduced cost in Nigeria [29].

The natural gas emission factor (56 kg of CO_2 per GJ) is about half of the coal emission factor (96 kg of CO_2 per GJ) [30] due to fewer carbon atoms per unit of energy relative to coal. However, it is not substantially lower than the average global CO_2 emissions per unit of energy: 68 kg of CO_2 per GJ [31]. Consequently, even if natural gas were to substitute for the entire global energy supply, CO_2 emissions would decline by a maximum of 20%. Upstream emissions from natural gas leakage from wells, pumps, pipelines and storage tanks [32] can significantly increase GHG emissions from natural gas. Methane is about 25 times more potent as a cause of global warming than carbon dioxide, but its 12 year lifetime in the atmosphere is significantly less (carbon dioxide lifetime in the atmosphere is over 500 years) [33].

13.3 Regional and National Perspectives on Technology

13.3.1 Production, Consumption and Reserves

Africa has relatively high natural gas proven reserves (606 Tcf) in 2014. This value is below the proven reserves for top global natural gas producers (Fig. 13.4), but as the African consumption is currently very low, their reserves can have a significant impact on transforming much of the power sector to low carbon energy technologies. Most proven reserves today are in the Maghreb, potentially in part because this is where the natural gas industry is best developed. Other large undiscovered but expected reserves are likely in some Sub-Saharan countries. Thus, there are countries in the Sub-Saharan region where natural gas reserves are localized but not exploited. With the exception of Senegal, all African countries producing and consuming natural gas have substantial gas reserves compared to the level of consumption. This fact can attract investment in gas pipelines in the future, mainly to the most populated areas in Africa.

Case Example 13.8

The impact of natural gas consumption in Tunisia's output:

The work analyses the period 1980–2010, and the findings show a relationship between natural gas consumption and real GDP. Tunisia is a net-importer of hydrocarbon products, and two-thirds of natural gas reserves are located

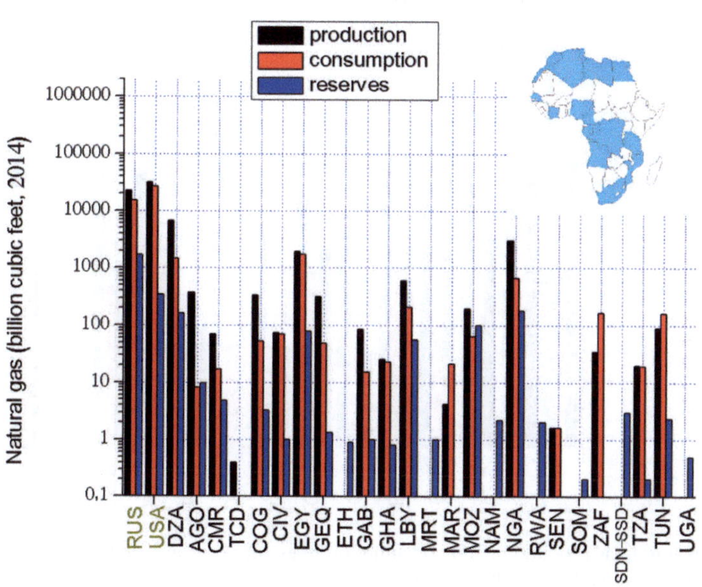

Fig. 13.4 Natural gas production and consumption rates per year, and natural gas proven reserves in African countries in 2014 [34]

offshore. The relationship between natural gas consumption and trade is found to be bidirectional, and also bidirectional causality is found between capital use and natural gas consumption. Also, in the short-term, the relationship between natural gas consumption and economic growth is bidirectional. Then, the authors suggest that the feedback between natural gas consumption and domestic output in Tunisia makes necessary energy exploring (natural gas) policies to sustain long-term economic growth. On the opposite, the adoption of energy conservation policies will not only limit the Tunisian domestic output but also reduce demand for natural gas in return, lowering also exporting potential and trading. The authors also consider that the promotion of renewable energy sources can also be positive to the country [35].

In Latin America the proven reserves are lower (295 Tcf) than in Africa and are mostly concentrated in Venezuela (Fig. 13.5). However, Latin America produces and consumes more natural gas than in Africa. Then, if we add the information shown in chapters devoted to hydropower and, to a lesser extent, to other renewable energy sources, we can consider that Latin America is more advanced in the transition to low carbon energy technologies than Africa. Also in Latin America natural gas consumption is linked to proven reserves in the different countries with the exception of Dominican Republic and Uruguay, showing the advantage of gas pipelines from Venezuela to Central America and from Brazil and Argentina to Paraguay and Uruguay.

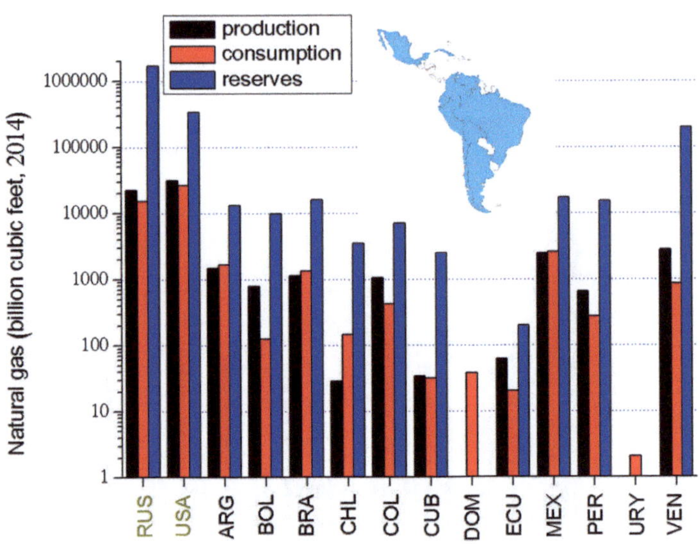

Fig. 13.5 Natural gas production and consumption rates per year, and natural gas proven reserves in Latin American countries in 2014 [34]

Case Example 13.9

Decision-making tool for a LNG regasification plant in Argentina:

Argentina has natural gas reserves that substantially exceed the demands in this country (Fig. 13.5). Nevertheless, Argentina's government has decided to import natural gas from overseas due to the delayed development of local sources and the increasing energy demand. Thus, Argentina began to import LNG to be regasified and injected into the existing gas network in the two injection sites that this country has (Ingeniero White and Escobar ports) and using floating storage and regasification units (regasification ships moored at specially designed docks) for quickly satisfying the increasing demand. However, to satisfy long term demand, new projects of onshore plants are being considered. LNG plants siting analysis and regulations are especially concerned with issues related to safety zones, maritime hazards, near shore navigation, environmental impact, near populated areas, presence of other industrial facilities, evacuation routes, weather conditions, dredging operations, and distance to be traveled by the regasified LNG in the ducts. There are several studies focussed on the manner to take into account these aspects in siting LNG terminals, but only few articles consider simultaneously safety and environmental issues. D'alessandro et al. [36] use this simultaneous approach, consider two alternative operation sites in the Bahia Blanca estuary and conclude that Rosales port seems to be the most suitable.

13.3.2 Electricity Share from Natural Gas

Comparing Figs. 13.4 and 13.6, the number of African countries consuming natural gas is 21, but the number of African countries producing electricity from natural gas is 15. This means that electricity based on natural gas is underdeveloped in Africa compared to the role this resource is currently playing. However, there are countries, most of them in the Maghreb, where the power mix is almost completely based in natural gas.

On the contrary, in Latin America all countries consuming natural gas also produce electricity with natural gas (Figs. 13.5 and 13.7) and the amount of natural gas contributing to the power mix is on average much lower. These facts also suggest a higher degree of maturity in the use of natural gas in Latin America compared to Africa, and also to the role of hydropower in many Latin American countries.

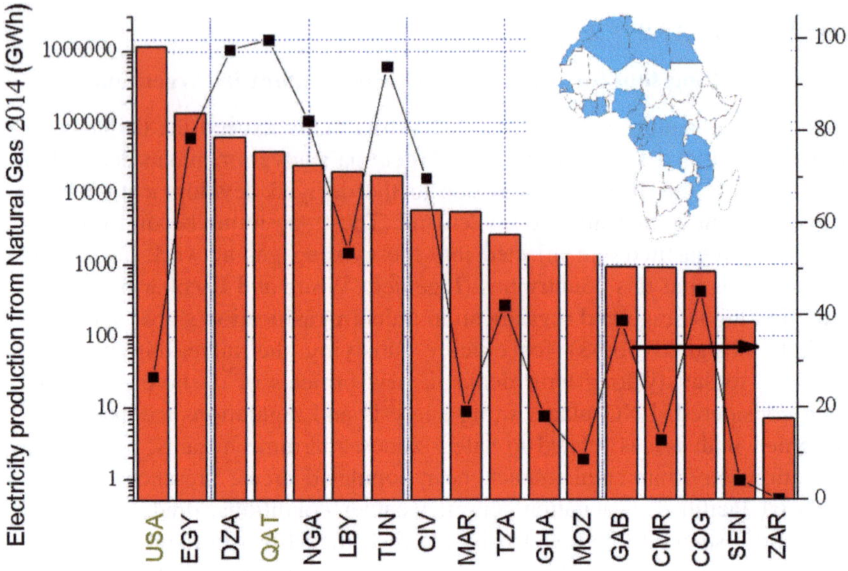

Fig. 13.6 Natural gas electricity output in African countries [1]

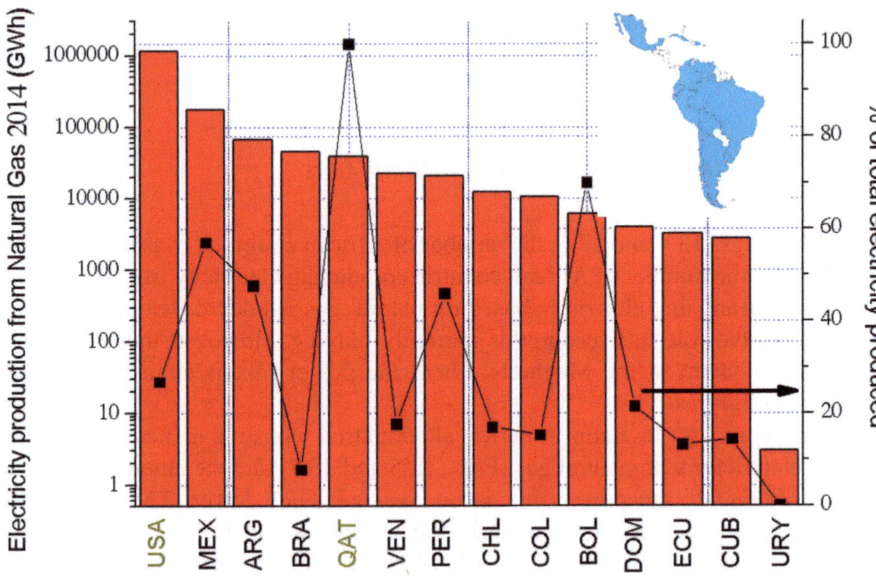

Fig. 13.7 Natural gas electricity output in Latin American countries [1]

13.4 Conclusions and Future Perspectives

Africa and Latin America have an estimated 13% of the total global proven reserves (Africa ~606 Tcf (8.5%); Latin America ~295 Tcf (4%)). Africa's proven reserves are primarily located in the Maghreb but development in unexplored regions is expected to intensify in Sub-Saharan countries. Latin America proven reserves are mostly located in South America and Mexico, and some areas in the Caribbean.

Unconventional production of natural gas is associated with a number of socio-political, environmental and economic issues that have limited drilling in some areas globally and created local concerns about impacts on water supply, excessive road traffic, accelerated deterioration of local and regional roads, impacts on air quality and enhanced methane emissions. As a result it is difficult to start extracting this unconventional resource in areas of Africa and Latin America where local communities are concerned about such impacts and previous ones.

Natural gas is playing an increasingly prominent role in the development of electricity generation systems on a global scale. Electricity generation from natural gas is versatile, supports both baseline and peaking power generation demands, provides proven reserve generation capacity that supplements non-dispatchable power (i.e., wind and solar), has lower GHG emissions and uses less water than coal-based generation and has lower capital costs and shorter new plant construction times than nuclear. However, Africa consumes much less gas than it produces on a country-by-country basis (6 of 21 countries that produce natural gas do not use it for electricity production) in part because of limited infrastructure investment outside of the Maghreb region where the power mix is almost entirely based on natural gas. In Latin America the gas pipeline infrastructure is more highly developed with gas being transported between countries and with all countries producing natural gas using it to produce electricity.

In conclusion, natural gas can play three important roles in the development of the future low-carbon energy future in Africa and Latin America: (1) complementing deployment of renewables by increasing the flexibility of non-dispatchable power systems, (2) reducing emissions by displacing other fossil fuel generation sources (i.e., coal and oil fired power generation), and (3) reducing health concerns related to household biomass burning. The mitigation of social and environmental impacts and the extension of the power grid are essential to achieve this position for natural gas.

References

1. International Energy Agency database (2015). http://data.iea.org/. Accessed 30 Nov 2015
2. Price GD (2014) Power systems and renewable energy. Design, operation, and system analysis. Power generation collection. Momentum Press. ISBN-13: 978-1-60650-570-0
3. Jackson RB, Vengosh A, Carey JW, Davies RJ, Darrah TH, O'Sullivan F, Petron G (2014) The environmental costs and benefits of fracking. Annu Rev Environ Resour 39:327–62

4. US Environmental Protection agency (2016). EPA final report hydraulic fracturing for oil and gas: impacts from the hydraulic fracturing water cycle on drinking water resources in the United States, EPA-600-R16-236 Fa
5. Energy Information Administration. US Department of Energy. http://www.eia.gov/cfapps/ipdbproject/IEDIndex3.cfm?tid=5&pid=57&aid=6. Accessed 30 Nov 2016
6. Inman M (2014) The fracking fallacy. Nature 516:28–30
7. McKinsey (2015) Brighter Africa—McKinsey Report. http://www.mckinsey.com/industries/electric-power-and-natural-gas/our-insights/powering-africa. Accessed 20 Dec 2016
8. IEA (2015) Key world energy statistics
9. Malakoff D (2014) The gas surge. Science 344:1464–1467
10. U.S. EPA (2016) Hydraulic Fracturing for Oil and Gas: Impacts from the Hydraulic Fracturing Water Cycle on Drinking Water Resources in the United States (Final Report). U. S. Environmental Protection Agency, Washington, DC, EPA/600/R-16/236F
11. Frolich CB, Brunt M (2013) Two-year survey of earthquakes and injection/production wells in the Eagle Ford Shale, Texas, prior to the MW4.8 20 October 2011 earthquake. Earth Planet Sci Lett 379:56–63
12. Ojijiagwo E, Oduoza CF (2014) The economics of gas flaring in oil and gas processing environments: a case study of electric power station in a developing country. In: FAIM 2014—Proceedings of the 24th international conference on flexible automation and intelligent manufacturing: capturing competitive advantage via advanced manufacturing and enterprise transformation, pp 143–151
13. Raja AK, Srivastava AP, Dwivedi M (2006) Power plant engineering, New Age International Publishers, Daryaganj. ISBN-10:81-224-2333-7
14. Usóna S, Kostowski WJ, Stanek W, Gazdab W (2015) Thermoecological cost of electricity, heat and cold generated in a trigeneration module fuelled with selected fossil and renewable fuels. Energy 92:308–319
15. Chekir N, Bellagi A (2011) Performance improvement of a butane/octane absorption chiller. Energy 36:6278–6284
16. Kim YJ, Kim S, Joshi YK, Fedorov AG, Kohl PA (2014) Thermodynamic analysis of an absorption refrigeration system with ionic-liquid/refrigerant mixture as a working fluid. Energy 44:1005–1016
17. Farshi LG, Mahmoudi SMS, Rosen MA (2013) Exergoeconomic comparison of double effect and combined ejector-double effect absorption refrigeration systems. Appl Energy 103:700–711
18. Liua M, Zhanga N (2007) Proposal and analysis of a novel ammonia–water cycle for power and refrigeration cogeneration. Energy 32:961–970
19. IEA (2014) Energy technology perspectives
20. Antonanzas J, Alia-Martinez M, Martinez-de-Pison FJ, Antonanzas-Torres F (2015) Towards the hybridization of gas-fired power plants: a case study of Algeria. Renew Sustain Energy Rev 51:116–124
21. IEA (2011) Harnessing variable renewables: a guide to balancing challenge
22. Overton TW (2014) Quisqueya I & II, San Pedro de Macorís, Dominican republic. Power 158:9
23. Ogunlowo OO, Bristow AL, Sohail M (2015) Developing compressed natural gas as an automotive fuel in Nigeria: lessons from international markets. Energy Policy 76:7–17
24. IEA (2006) World energy outlook. Chapter 15: energy for cooking in developing countries
25. IEA (2015) Energy efficiency market report 2015. Market trends and medium-term prospects
26. Indexmundi. www.indexmundi.com. Accessed 30 Nov 2015
27. Climatescope (2015) The clean energy country competitiveness index. The Multilateral Investment Fund, UK Department of International Development, Power Africa and Bloomberg New Energy Finance
28. NREL (2015) The western wind and solar integration study phase 2, Technical report, NREL, Golden, CO, US. www.nrel.gov/docs/fy13osti/55588.pdf. Accessed 30 Nov 2015
29. Oladokun VO, Asemota OC (2015) Unit cost of electricity in Nigeria: a cost model for captive diesel powered generating system. Renew Sustain Energy Rev 52:35–40

30. IPCC (2006) Guidelines for national greenhouse gas inventories. http://www.ipcc-nggip.iges.or.jp/public/2006gl/. Accessed 30 Nov 2015
31. McJeon H et al (2014) Limited impact on decadal-scale climate change from increased use of natural gas. Nature 514:482–485
32. The Economist (2016) Tunnel vision. Methane leaks. Available 23 July 2016
33. The Economist (2016) A dirty little secret. Methane leaks. Available 23 July 2016
34. EIA. US Department of Energy. http://www.eia.gov/energyexplained/index.cfm?page=natural_gas_home. Accessed 30 Nov 2015
35. Farhani S, Shahbaz M, Arouri M, Teulond F (2014) The role of natural gas consumption and trade in Tunisia's output. Energy Policy 66:677–684
36. D'alessandro AA, Izurieta EM, Tonelli SM (2016) Decision-making tools for a LNG regasification plant siting. J Loss Prev Process Ind 43:255–262

Chapter 14
Nuclear Energy

Ricardo Guerrero-Lemus and Les E. Shephard

Abstract This chapter shows that there are nine nuclear reactors located in Africa and Latin America with two in South Africa (1830 MW), three in Argentina (1627 MW), two in Mexico (1620 MW), and two in Brazil (1891 MW). South Africa began its nuclear program as part of the Atoms For Peace Program in the mid-1960s that also included the development of a nuclear weapons program in the early 1970s that was dismantled in 1989. In 1984, South Africa completed construction of two operating reactors and recently has deferred decisions on additional expansion of nuclear power. Africa has several other nations contemplating the use of nuclear, particularly those in North Africa and others that have produced uranium ore for fuel production. In Latin America, Argentina is contemplating further expansion of nuclear power and has ongoing dialogue with China, Russia and South Korea on developing new civilian nuclear power capability. Brazil is the only nation currently producing uranium for nuclear fuel manufacturing. Limited electrical grid capacity and connectivity in Africa (and potentially some parts of Latin America) and financing limit the future development of traditional nuclear power. Several nations are considering small modular reactors (SMRs) as an option (modular capacity nominally 25–250 MW). Argentina has constructed the world's first SMR (25 MW) near the existing Atucha facility that is undergoing testing and is expected to go critical in 2017. However, there are many reasons to expect no nuclear future in most African and Latin American countries in the near future.

Keywords Energy · Africa · Latin America · Electricity · Nuclear energy · Nuclear reactors · Small modular reactors · Nuclear waste · Uranium · Thorium · Spent nuclear fuel · Enrichment · Fission · Neutrons · Fuel rods · Pressurized water reactor (PWR) · Boiling water reactor (BWR) · Fissile materials · Enriched uranium · Plutonium · Nuclear fuel cycle · Centrifuge

R. Guerrero-Lemus and L.E. Shephard, *Low-Carbon Energy in Africa and Latin America*, Lecture Notes in Energy 38,
DOI 10.1007/978-3-319-52311-8_14

14.1 Introduction

Nuclear power has a long, sustained history of providing energy for a variety of applications worldwide that began back in the 1950s. Nuclear power provides over 10% of the world's electricity (2014) as continuous, reliable baseload power without the production of carbon emissions [1]. There are nominally 440 commercial nuclear power reactors operable in 31 countries around the world with over 380,000 MW_e of total generation capacity. Nine nuclear reactors are located in Africa and Latin America with two in South Africa (1830 MW) and three reactors in Argentina (1627 MW), two in Mexico (1620 MW), and two in Brazil (1891 MW). An additional 60 nuclear reactors are under construction worldwide. In addition, there are a number of countries that rely on imported nuclear power as a substantial share of their national electricity supply (e.g., Italy and Denmark ~10%).

Fifty-six nations operate a total of 240 research reactors used primarily for the production of medical and industrial isotopes and as a neutron source for scientific research in a multitude of applications. An additional 180 reactors are used for propulsion of surface ships and submarines worldwide. Eight countries are known to have a nuclear weapons capability.

There are number of attributes of nuclear power which make it particularly beneficial apart from its actual generation cost per unit—MWh or kWh. Fuel is a low proportion of power cost, giving power price stability, its fuel is onsite (not depending on continuous delivery), it is dispatchable on demand, it has fairly quick ramp-up, it contributes to clean air and low-carbon objectives and provides steady voltage support for grid stability. These attributes are mostly not monetised in merchant markets, but have great value which is increasingly recognized where dependence on intermittent sources has grown. Similarly, challenges for nuclear power are well-documented and include safety of operations, security of nuclear technology and materials and the management and disposal of spent nuclear fuel. The power capacity of most conventional nuclear reactors is nominally 1 GW and over, beyond the electricity demands of most African and Latin American countries. However, small modular reactors (SMR) with capacities on the order of 25–250 MW have been proposed as alternative low carbon energy technologies for these regions [2].

The first electricity from nuclear power in Latin America and Africa was produced by Argentina in 1974 and subsequently one decade later in South Africa (Fig. 14.1). Today there are also limited amounts of nuclear power produced by Brazil and Mexico relative to their overall national power mix (Fig. 14.2). In Africa, Egypt and Nigeria have used small test reactors for limited amounts of nuclear power production [3].

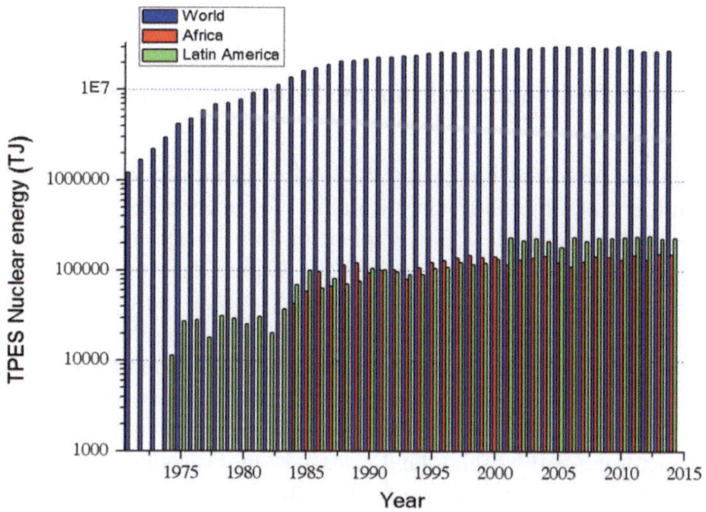

Fig. 14.1 Evolution of the total primary energy supply from nuclear energy in the World, Africa and Latin America from 1971 to 2014 [4]

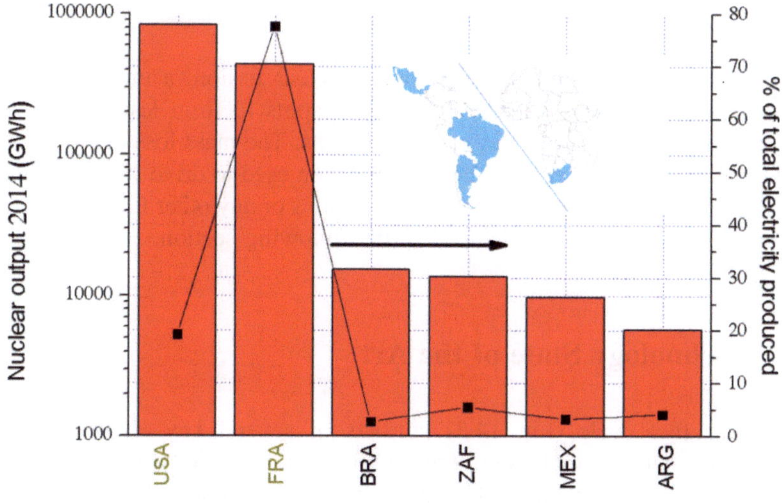

Fig. 14.2 Nuclear power output and share of total electricity produced in 2014 in African and Latin American countries and the country which produces the larger amount of nuclear energy in absolute terms (USA) and the larger share of nuclear energy in the power mix (France) [4]

Case Example 14.1

Multicriteria decision analysis for the location of a nuclear power plant in Egypt:

The study is due to the increasing demand of electricity and fresh water in Egypt and, on the other side, the availability of fewer sites, as time goes by. The analysis of the nuclear power plant optimal location has been defined with the participation of personnel from the Siting Department of the Nuclear Power Plant Authority of Egypt and based on a multi-criteria decision analysis using 22 factors: geology, seismicity, oceanography (flooding, depth of the sea, turbidity and water temperature), meteorological phenomena (wind direction and rainfall), ecology (terrestrial environments and aquatic environment), sociology (tourism, archeology and population distribution), land uses (agriculture and special land uses), economy (site topography, grid connection and proximity to hazardous facilities), construction and operating costs. Four sites were chosen as candidates, all located along the North western and the Red Sea coast. The El Dabaa site was found the most suitable due to less seismicity, less environmental impact and proximity to major load centers, followed by the East El Negila site on the Mediterranean Sea [5].

Nuclear energy is largely derived from nuclear fission, a nuclear reaction or radioactive decay process in which one atom splits to form less massive atoms, while releasing free neutrons and gamma photons. The mass loss that occurs in the process is converted into energy (heat) that is then used to drive steam turbines that produce electricity without direct carbon emissions or any other GHG. An extended description of the process is included in the following section.

14.2 Technology State of the Art

The nuclear fuel cycle (Fig. 14.3) consists of several key elements including mining, conversion, enrichment, fuel fabrication, electricity production, reprocessing and waste management. The maturity and distribution of these technologies vary by nation on a global scale and is limited in extent to a few nations within either Africa or Latin America.

14.2.1 Mining

Conventional uranium resources are defined as resources from which uranium is recoverable as a primary product, a co-product or an important by-product.

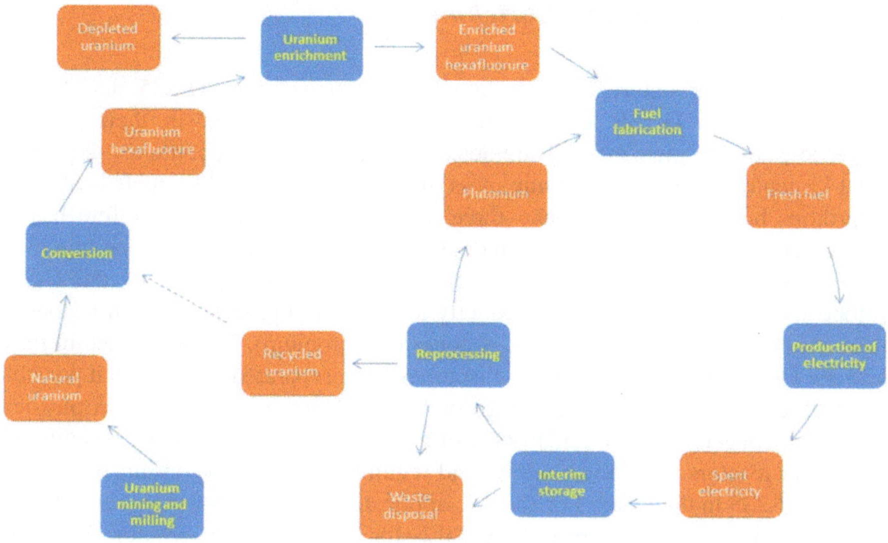

Fig. 14.3 Diagram of the nuclear fuel cycle illustrating both open and closed fuel cycle processes

Unconventional resources are resources from which uranium is only recoverable as a minor by-product, such as uranium associated with phosphate rocks, non-ferrous ores, carbonatite, black shale and lignite. Other unconventional sources may include reprocessed uranium (RepU), depleted uranium (DU) and mixed oxide (MOX) [6]. Uranium production methods vary depending on the characteristics of the resource and the host country but include open-pit mining, underground mining, in situ leaching, acid, in situ leaching alkaline and co-product/by-product. Open pit and underground mining extraction processes employ standard mining practices whereas in situ leaching involves dissolving naturally occurring uranium in sub-surface host rock in the ground and pumping the solution to the surface where the ore minerals can be recovered. Long lead times (typically in the order of ten years or more in most producing countries) are required to develop production facilities that can turn resources into refined uranium ready for nuclear fuel production.

The International Atomic Energy Agency defines *identified conventional resources* as reasonably assured resources (RAR) and inferred resources (IR) recoverable at a cost of less than USD 260/kgU (USD 100/lbU_3O_8). Estimated global uranium resources approached 7.6 million tU in 2013 which is sufficient to support the existing global nuclear fleet for 128 years based on an annual demand of 59,170 tU in 2013 (last data available) [6]. Ten countries accounted for 97% of the world production in 2013 (58,816 tU) with Africa providing 18% of the total and Latin America about 0.5% [6]. The remainder of uranium has been provided or derived from secondary sources including stockpiles of natural and enriched uranium, blending down weapons grade uranium, reprocessing of spent fuel and the re-enrichment of depleted uranium tails [6].

14.2.2 *Conversion and Enrichment*

Uranium found in nature consists largely of two isotopes, U-235 and U-238. Uranium-235 and U-238 are chemically identical, but differ in their physical properties, notably their mass. The nucleus of the U-235 atom contains 92 protons and 143 neutrons, giving an atomic mass of 235 units. The U-238 nucleus also has 92 protons but has 146 neutrons—three more than U-235—and therefore has an atomic mass of 238. Natural uranium contains 0.7% of the U-235 isotope, the main fissile isotope of uranium, with the remaining 99.3% comprised of the U-238 isotope which does not contribute directly to the fission process (though it does so indirectly by the formation of fissile isotopes of plutonium). The difference in mass between U-235 and U-238 allows the isotopes to be separated and makes it possible to increase or "enrich" the percentage of U-235. Isotope separation is a physical process to concentrate ('enrich') one isotope relative to others. Most reactors are light water reactors (described below) and require uranium to be enriched from 0.7 to 3–5% U-235 in their fuel. For specific applications and reactor types, enrichment levels may exceed 5%, approaching 20% for special power reactor fuels (for comparison, uranium used for nuclear weapons would have to be enriched in plants specially designed to produce at least 90% U-235). Also, some reactors use natural uranium as their fuel (e.g., Canadian-designed Candu and the British Magnox reactors).

Case Example 14.2

South Africa—Moving Full Cycle on Non-proliferation:

South Africa is the only country to have built nuclear weapons and then voluntarily dismantled them. Under the umbrella of the "Atoms for Peace Program" South Africa acquired a nuclear research reactor and highly enriched uranium from the United States in 1965. In 1971, under the auspices of the Ploughshares Peaceful Nuclear Explosion program, South Africa began to pursue the use of nuclear explosions for supporting the mining industry and extended their nuclear efforts to include the development of a limited nuclear deterrent capability reported to have six nuclear weapons using highly enriched uranium in the mid-1970s. The program continued until 1989 when South Africa ended their nuclear weapon program in conjunction with ending Apartheid. In 1991 South Africa signed the Non Proliferation Treaty, became a nuclear free zone in 1998, and ratified the Comprehensive Test Ban Treaty in 1999. Recently, South Africa has terminated their research program on Pebble Bed Reactors and are re-considering additional investment in the next generation of nuclear power [7].

Uranium enrichment processes transform the uranium oxide obtained from the mining process into uranium hexafluoride (UF_6) at low temperatures. Nearly 90% of the global enrichment production capacity occurs in the five nuclear weapons states (i.e., USA, Russia, UK, France and China) with additional enrichment production capability in 13 other nations. Large commercial enrichment plants are in operation in France, Germany, Netherlands, UK, USA, and Russia, with smaller plants elsewhere. New centrifuge plants are being built in France and USA, and other nations (e.g., China) are extending existing capacity.

A number of enrichment processes have been demonstrated historically or in the laboratory but only two, the gaseous diffusion process and the centrifuge process, have operated on a commercial scale. In both of these, UF_6 gas is used as the feed material. Molecules of UF_6 with U-235 atoms are about one percent lighter than with U-238, and this difference in mass is the basis of both processes.

The feedstock for enrichment consists of UF_6 obtained from the conversion plant. Following enrichment, two streams of UF_6 are formed: (1) the enriched 'product' containing a higher concentration of U-235 which will be used to make nuclear fuel, and the 'tails' containing a lower concentration of U-235, and known as depleted uranium. Feedstock will have varying concentrations of U-235, depending on the source. Natural uranium will have a U-235 concentration of approximately 0.7%, while recycled uranium will be around 1% and tails around 0.25–0.30%. The capacity of enrichment plants is measured in terms of 'separative work units' or SWUs. The SWU is a complex unit which indicates the energy input relative to the amount of uranium processed, the degree to which it is enriched (i.e., the extent of increase in the concentration of the U-235 isotope relative to the remainder) and the level of depletion of the remainder (tails).

Enrichment accounts for almost half of the cost of nuclear fuel and about 5% of the total cost of the electricity generated. Enrichment costs are largely related to electrical energy used. The gaseous diffusion process consumes about 2500 kWh (9000 MJ) per SWU, while modern gas centrifuge plants require only about 50 kWh (180 MJ) per SWU. In the past, enrichment has accounted for the main greenhouse gas impact from the nuclear fuel cycle because of the large electricity demand much of which produced from coal.

Commercial uranium enrichment using the diffusion process was first performed in the USA and subsequently in Russia, UK, France, China and Argentina. It is a very energy-intensive process, requiring about 2400 kWh per SWU. United States Enrichment Corporation (USEC) says that electricity accounts for 70% of the production cost at its Paducah plant, which is the last operating gas diffusion plant. The diffusion process involves forcing uranium hexafluoride gas under pressure through a series of porous membranes or diaphragms. As U-235 molecules are lighter than the U-238 molecules they move faster and have a slightly better chance of passing through the pores in the membrane. The UF_6 which diffuses through the membrane is thus slightly enriched, while the gas which did not pass through the membrane is depleted in U-235. This process is repeated many times in a series of diffusion stages called a cascade. Each stage consists of a compressor, a diffuser and a heat exchanger to remove the heat of compression. The enriched UF_6 product is withdrawn from one end of the cascade and the depleted UF_6 is removed at the other end. The gas must be processed through some 1400 stages to obtain a product with a concentration of 3–4% U-235. Diffusion plants typically have a small amount of separation through one stage (hence the large number of stages) but are capable of handling large volumes of gas. In recent years the gaseous diffusion process has accounted for about 25% of the world's enrichment capacity. Today, gas diffusion plants have reached the end of their design life and most nations are becoming more reliant on centrifuge enrichment technology.

The centrifuge process uses UF_6 gas as its feed and also makes use of the slight difference in mass between U-235 and U-238. The gas is fed into a series of vacuum tubes, each containing a rotor several metres tall and tens of centimeters in diameter depending on country. When the rotors are spun rapidly, at 50,000–70,000 rpm, the heavier molecules with U-238 increase in concentration towards the cylinder's outer edge. There is a corresponding increase in concentration of U-235 molecules near the centre. The countercurrent flow set up by a thermal gradient enables the enriched product to be drawn off axially, heavier molecules at one end and lighter ones at the other. The enriched gas forms part of the feed for the next stages while the depleted UF_6 gas goes back to the previous stage. Eventually enriched and depleted uranium are drawn from the cascade at the desired assays. To obtain efficient separation of the two isotopes, centrifuges rotate at very high speeds, with the outer wall of the spinning cylinder moving at between 400 and 500 m/s to give a million times the acceleration of gravity.

Although the volume capacity of a single centrifuge is much smaller than that of a single diffusion stage, its capability to separate isotopes is much greater. Centrifuge stages normally consist of a large number of centrifuges in parallel. Such stages are then arranged in cascade similarly to those for diffusion. In the centrifuge process, however, the number of stages may only be 10–20, instead of a thousand or more for diffusion. Centrifuges are designed to run for about 25 years continuously, and cannot simply be slowed or shut down and restarted according to demand.

An alternative option, laser enrichment processes are continuing to evolve as a potential third-generation technology promising lower energy inputs, lower capital costs and lower tails assays, hence significant economic advantages. Laser enrichment uses either an atomic or molecular process to enhance enrichment. Atomic vapor processes work on the principle of photoionization, whereby a powerful laser is used to ionise particular atoms present in a vapor of uranium metal (the laser uses frequencies which are tuned to ionize a U-235 atom, but not a U-238 atom). The positively-charged U-235 ions are then attracted to a negatively-charged plate and collected. Atomic laser techniques may also separate plutonium isotopes. Most molecular processes emphasize the application of photo-dissociation of UF_6 to solid UF_5^+, using a tuned laser radiation as above to break the molecular bond holding one of the six fluorine atoms to a U-235 atom. This enables the ionized UF_5 to be separated from the unaffected UF_6 molecules containing U-238 atoms, hence achieving a separation of isotopes.

14.2.3 Fuel Fabrication

Fuel assembly performance has improved since the 1970s to allow increased burn-up of fuel from 40 GW·day/tU to more than 60 GW·day/tU [8]. This is correlated with increased enrichment levels and the use of advanced burnable absorber designs for pressurized water reactors (PWRs) that incorporate gadolinium. Real time reactor core monitoring available today has considerably improved core performance. There are three main stages in the fabrication of nuclear fuel structures used in the most common light water reactors (LWR) and pressurized heavy water reactors: (i) producing pure uranium dioxide (UO_2) from incoming UF_6 or UO_3; (ii) producing high-density, accurately shaped ceramic UO_2 pellets; and (iii) producing the rigid metal framework for the fuel assembly and loading the fuel pellets in the fuel rods, sealing them and assembling the rods into the final fuel assembly structure [8]. There is considerable variation among fuel assemblies designed for the different types of reactors, meaning that utilities have limited choice in suppliers of fabricated fuel assemblies, especially for PWRs.

The standard fuel form comprises a column of ceramic pellets of uranium oxide, clad and sealed into zirconium alloy tubes. For light water reactor (LWR) fuel, the uranium is enriched to various levels up to about 4.8% U-235. Pressurised heavy water reactor (PHWR) fuel is usually unenriched natural uranium (0.7% U-235), although slightly-enriched uranium is also used. Uranium arrives at a fuel manufacturing plant in one of two forms, uranium hexafluoride (UF_6) or uranium trioxide (UO_3), depending on whether it has been enriched or not. The uranium needs to be converted to uranium dioxide (UO_2) prior to pellet fabrication. Most fabrication plants have their own facilities for effecting this chemical conversion (some do not, and acquire UO_2 from plants with excess conversion capacity). Chemical conversion to and from UF_6 are distinct processes, but both involve the handling of aggressive fluorine compounds and plants may be set up to do both.

For the conversion of UO_3 to UO_2, water is added to UO_3 so that it forms a hydrate. This solid is fed (wet or dry) into a kiln operating with a reducing atmosphere and UO_2 is produced. The UO_2 powder may require additional homogenization to ensure uniformity, additives to tailor the microstructure and density of the pellets and other fuel additives such as lubricants, burnable absorbers (e.g., gadolinium) and pore-formers before it can be formed into pellets. Conditioned UO_2 powder is fed into dies and pressed biaxially into a cylindrical pellet form using a load of several hundred MPa (this is done in pressing machines operating at high speed). These 'green' pellets are then sintered by heating in a furnace at about 1750 °C under a precisely controlled reducing atmosphere (usually argon-hydrogen) in order to consolidate them. This also has the effect of decreasing their volume. The pellets are then machined to exact dimensions—the scrap from which being fed back into an earlier stage of the process. Rigorous quality control is applied to ensure pellet integrity and precise dimensions.

For most reactors, pellets are just under one centimetre in diameter and a little more than one centimetre long. A single pellet in a typical reactor yields about the same amount of energy as one tonne of steaming coal. Burnable absorbers (or burnable 'poisons') such as gadolinium may be incorporated (as oxide) into the fuel pellets of some rods to limit reactivity early in the life of the fuel. Burnable absorbers have a very high neutron absorption cross-section and compete strongly for neutrons, after which they progressively 'burn-out' and convert into nuclides with low neutron absorption leaving fissile (U-235) to react with neutrons. Burnable absorbers enable longer fuel life by allowing higher fissile enrichment in fresh fuel, without excessive initial reactivity and heat being generated in the assembly.

The fabrication of nuclear fuel assemblies is the last stage of the front end of the nuclear fuel cycle and represents less than 20% of the final cost of the fuel. The process for uranium-plutonium MOX fuel fabrication is essentially the same—notwithstanding some specific features associated with handling the plutonium component.

Nuclear fuel designs dictate that the pellet-filled rods have a precise physical arrangement in terms of their lattice pitch (spacing), and their relation to other features such as water (moderator) channels and control-rod channels. The physical structures for holding the fuel rods are therefore engineered with extremely tight tolerances. They must be resistant to chemical corrosion, high temperatures, large static loads, constant vibration, fluid and mechanical impacts. Yet they must also be as neutron-transparent as possible.

Assembly structures comprise a strong framework made from steel and zirconium upon which are fixed numerous grid support pieces that firmly hold rods in their precise lattice positions. These are made from zirconium alloy and must permit (and even enhance) the flow of coolant water around the fuel rod. The grid structures grip the fuel rod and so are carefully designed to minimise the risk of vibration-induced abrasion on the cladding tube—called 'fretting' wear.

All fuel fabricators have highly sophisticated engineering processes and quality control for the timely manufacture of their assembly structures. Pellets are loaded into tubes made from an appropriate zirconium alloy, referred to as the 'cladding'. The filled tube is flushed with helium and pressurized with tens of atmospheres (several MPa) of this gas before the ends are sealed at each end by precision welding. A free space is left between the top of the pellet stack and the welded end-plugs—this is called the 'plenum' space and it accommodates thermal expansion of the pellets and some fission product gases. A spring is usually put into the plenum to apply a compressive force on the pellet stack and prevent its movement. The completed fuel rods are then fixed into the prefabricated framework structures that hold the rods in a precisely defined grid arrangement depending on the type of reactor.

In order to maximize the efficiency of the fission reaction the cladding and, indeed, all other structural parts of the assembly must be as transparent as possible to neutrons. Different forms of zirconium alloy, or zircaloy, are therefore the main materials used for cladding. This zircaloy includes small amounts of tin, niobium, iron, chromium and nickel to provide necessary strength and corrosion resistance. Hafnium, which typically occurs naturally with zirconium deposits, needs to be removed because of its high neutron absorption cross-section. The exact composition of the alloy used depends on the manufacturer and is an important determiner in the quality of the fuel assembly. Zircaloy oxidizes in air and water, and therefore it has an oxidized layer which does not impair function.

14.2.4 *Types of Nuclear Fuel Assemblies for Different Reactors*

There is considerable variation among fuel assemblies designed for the different types of reactors. This means that utilities have limited choice in suppliers of fabricated fuel assemblies, especially for PWRs.

14.2.4.1 PWR Fuel

Pressurised water reactors are the most common type of nuclear reactor accounting for two-thirds of current installed nuclear generating capacity worldwide. A PWR core uses normal water as both moderator and primary coolant. Water is kept under considerable pressure (about 10 MPa) to prevent it from boiling, and its temperature rises to about 330 °C after its upward passage past the fuel. It then goes through massive pipes to a steam generator.

Fuel for western PWRs is built with a square lattice arrangement and assemblies are characterized by the number of rods they contain, typically, 17 × 17 in current designs. A PWR fuel assembly stands between four and five metres high, is about 20 cm across and weighs about half a tonne. The assembly has vacant rod positions (space left for the vertical insertion of a control rod). Not every assembly position requires fuel or a control rod, and a space may be designated as a "guide thimble" into which a neutron source rod, specific instrumentation, or a test fuel segment can be placed.

A PWR fuel assembly comprises a bottom nozzle into which rods are fixed through the lattice and to finish the whole assembly it is crowned by a top nozzle. The bottom and top nozzles are heavily constructed as they provide much of the mechanical support for the fuel assembly structure. In the finished assembly most rod components will be fuel rods, but some will be guide thimbles, and one or more are likely to be dedicated to instrumentation. PWR fuel assemblies are rather uniform compared with boiling water reactors (BWR) ones, and those in any particular reactor must have substantially the same design.

An 1100 MW_e PWR core may contain 193 fuel assemblies composed of over 50,000 fuel rods and some 18 million fuel pellets. Once loaded, fuel stays in the core for several years depending on the design of the operating cycle. During refueling, every 12–18 months, some of the fuel—usually one third or one quarter of the core—is removed to storage, while the remainder is rearranged to a location in the core better suited to its remaining level of enrichment.

Russian PWR reactors are usually known by the Russian acronym VVER. Fuel assemblies for these are characterized by their hexagonal arrangement, but are otherwise of similar length and structure to other PWR fuel assemblies. Most is made by TVEL in Russia, but Westinghouse in Sweden also fabricates it. TVEL is instigating using erbium as a burnable poison in fuel enriched to about 6.5% in order to prolong the intervals between refueling to two years.

14.2.4.2 BWR Fuel

Boiling water reactors are the second most common nuclear reactor type accounting for almost one-quarter of installed nuclear generating capacity. In a boiling water reactor, water is turned directly into steam in the reactor pressure vessel at the top of the core and this steam (at about 290 °C and 7 MPa) is then used to drive a turbine.

BWRs also use fuel rods comprising zirconium-clad uranium oxide ceramic pellets. Their arrangement into assemblies is again based on a square lattice, with pin geometries ranging from 6×6 to 10×10. Fuel life and management strategy is similar to that for a PWR. However, BWR fuel is fundamentally different from PWR fuel in certain ways: (i) Four fuel assemblies and a cruciform shaped control blade form a 'fuel module'; (ii) each assembly is isolated from its neighbours by a water-filled zone in which the cruciform control rod blades travel (they are inserted from the bottom of the reactor); (iii) each BWR fuel assembly is enclosed in a zircaloy sheath or channel box which directs the flow of coolant water through the assembly and during this passage it reaches boiling point; and (iv) BWR assemblies contain larger diameter water channels—flexibly designed to provide appropriate neutron moderation in the assembly.

The zircaloy tubes are allowed to fill with water thus increasing the amount of moderator in the central region of the assembly. Different enrichment levels are used in the rods in varying positions—lower enrichments in the outer rods, and higher enrichments near the centre of the bundle. A BWR reactor is designed to operate with 12–15% of the water in the top part of the core as steam, and hence with less moderating effect and, thus, efficiency there.

For many BWR models, control of reactivity to enable load-following can be achieved by changing the rate of circulation inside the core. Jet pumps located in the annulus between the outer wall of the vessel and an inner wall called the shroud increase the flow of water up through the fuel assembly. At high flow rates steam bubbles are removed more quickly, and hence moderation and reactivity is increased. When flow rate is decreased, moderation decreases as steam bubbles are present for longer and hence reactivity drops. This allows for a variation of about 25% from the maximum rated power output, enabling load-following more readily than with a PWR.

Control rods are used when power levels are reduced below 75%, but they are not part of the fuel assembly as in a PWR. They are bottom-entry (pushed upwards so that rods intercept the lower, more reactive, zone of the fuel assemblies first).

14.2.5 Nuclear Plants and Electricity Production

A nuclear power plant uses controlled nuclear fission chain reactions with neutron-absorbers, as the number of fissions caused by the neutron population determines the energy released. Power reactors use the heat from fission to produce steam, which turns turbines to generate electricity. In this respect, they are very similar to plants fueled by coal. There are several different designs for nuclear reactors, but the common components include fuel assembly, control rods, a coolant, a pressure vessel, a container structure, and an external cooling facility.

About 85% of the energy released initially is from the kinetic energy of the fission fragments. However, in solid fuel they can only travel a microscopic distance, so their energy becomes converted into heat. The balance of the energy comes from gamma rays emitted during or immediately following the fission process and from the kinetic energy of the neutrons. Some of the latter are immediate (so-called prompt neutrons), but a small proportion (0.7% for U-235, 0.2% for Pu-239) is delayed, as these are associated with the radioactive decay of certain fission products. The longest delayed neutron group has a half-life of about 56 s.

About 6% of the heat generated in the reactor core originates from radioactive decay of fission products and transuranic elements formed by neutron capture, mostly the former. This must be allowed for when the reactor is shut down, since heat generation continues after fission stops. Even after one year, typical used fuel generates about 10 kW of decay heat per tonne, decreasing to about 1 kW/t after ten years.

Nuclear fission is based in uranium, the most popular fissile element. Natural uranium is 99.3% ^{238}U and 0.7% ^{235}U. As each ^{235}U that undergoes fission produces an average of 2.5 neutrons and some ^{238}U nuclei capture neutrons becoming ^{239}U, and subsequently emit two β particles to produce ^{239}Pu, also a fissile material [9]. Approximately one third of the energy produced by a thermal power reactor comes from fission of this plutonium [10].

$$^{235}U_{92} + {}^{1}n_{0} \rightarrow \text{fission products} + (2.4)\,{}^{1}n_{0} + 192.9\,\text{MeV} \tag{14.1}$$

$$^{238}U_{92} + {}^{1}n_{0} \rightarrow {}^{239}U_{92} \tag{14.2}$$

$$^{239}U_{92} \rightarrow {}^{239}Np_{93} + \beta^{-1} \quad t_{1/2} = 23.5\,\text{min} \tag{14.3}$$

$$^{239}Np_{93} \rightarrow {}^{239}Pu_{94} + \beta^{-1} \quad t_{1/2} = 2.33\,\text{days} \tag{14.4}$$

$$^{239}Pu_{94} + {}^{1}n_{0} \rightarrow \text{fission products} + (2.9)\,{}^{1}n_{0} + 198.5\,\text{MeV} \tag{14.5}$$

Thorium can also be used as nuclear fuel. Although not fissile itself, ^{232}Th, when loaded into a nuclear reactor, absorbs neutrons to produce ^{233}U, which is fissile. Then ^{233}U can be chemically separated from the thorium and used as fuel in a nuclear reactor [6].

Fission of U-235 nuclei typically releases 2 or 3 neutrons, with an average of about 2.5. One of these neutrons is needed to sustain the chain reaction at a steady level of controlled criticality; on average, the others leak from the core region or are absorbed in non-fission reactions. Neutron-absorbing control rods are used to adjust the power output of a reactor. These typically use boron and/or cadmium (both are strong neutron absorbers) and are inserted among the fuel assemblies. When they are slightly withdrawn from their position at criticality, the number of neutrons available for ongoing fission exceeds unity (i.e., criticality is exceeded) and the power level increases. When the power reaches the desired level, the control rods are returned to the critical position and the power stabilises.

The ability to control the chain reaction is entirely due to the presence of the small proportion of delayed neutrons arising from fission. Without these, any change in the critical balance of the chain reaction would lead to a virtually instantaneous and uncontrollable rise or fall in the neutron population. It is also relevant to note that safe design and operation of a reactor sets very strict limits on the extent to which departures from criticality are permitted. These limits are built into the overall design.

While fuel is being burned in the reactor, it is gradually accumulating fission products and transuranic elements which cause additional neutron absorption. The control system has to be adjusted to compensate for the increased absorption. When the fuel has been in the reactor for three years or so, this build-up in absorption, along with the metallurgical changes induced by the constant neutron bombardment of the fuel materials, dictates that the fuel should be replaced. This effectively limits the burn-up to about half of the fissile material, and the fuel assemblies must then be removed and replaced with fresh fuel. Fuel life can be extended by use of burnable poisons such as gadolinium, the effect of which compensates for the build-up of neutron absorbers.

In naval reactors used for propulsion, where frequent fuel changes are inconvenient, the fuel is enriched to higher levels initially and burnable poisons—neutron absorbers—are incorporated. Hence as the fission products and transuranic elements accumulate, the 'poison' is depleted and the two effects tend to cancel one another out. To maximise the burn-up of commercial reactor fuel, burnable poisons such as gadolinium are increasingly used, along with increasing enrichment towards 5% U-235.

Nuclear power plants are classified in terms of the reactor types. There are three types of classifications: (i) by moderator material; (ii) by coolant material; and (iii) by reaction type. The speed of the neutrons in the chain reaction determines the reaction type (Fig. 14.4). Thermal reactors use slow neutrons to maintain the reaction, as ^{235}U undergoes fission more readily with slow neutrons than with fast ones. Consequently, they require a moderator to reduce the speed of neutrons produced by fission. Light water (H_2O), heavy water (D_2O), and carbon in the form of graphite are the most common moderators. Fast neutron reactors, also known as fast breeder reactors (FBR), use high speed, unmoderated neutrons to sustain the chain reaction [10].

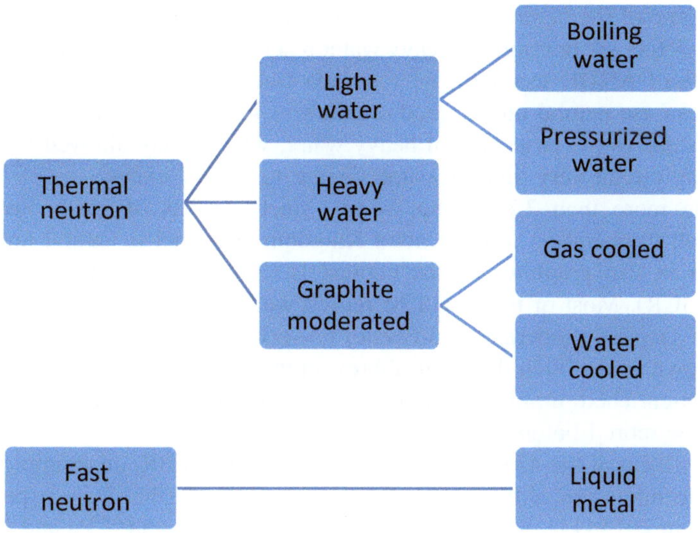

Fig. 14.4 Reactors classification in terms of moderator, coolant and reaction type

Light water reactors use fuel assemblies containing either natural uranium (0.7% ^{235}U) or slightly enriched uranium (0.9–2.0% ^{235}U) fuel. Rods composed of neutron absorbing material such as cadmium or boron are inserted into the fuel assembly in defined locations that determines the rate of the fission chain reaction. For example, a typical 1100 MW PWR core may contain 193 fuel assemblies composed of over 50,000 fuel rods and some 18 million fuel pellets [8]. The fuel assemblies remain in the core of the power reactor for up to 3 years to maximize the energy produced.

The coolant is a liquid or gas that removes the heat from the core and produce steam to drive the turbines. In reactors using either light water or heavy water, the coolant also serves as the moderator. The pressure vessel, made of heavy-duty steel, holds the reactor core containing the fuel assembly, control rods, moderator, and coolant. The containment structure, composed of thick concrete and steel, inhibits the release of radiation in case of an accident and secures components of the reactor from potential intruders [10].

The average capacity factor of a nuclear power reactor currently is very high (about 90% in USA). Between 1980 and 2000 the world median capacity factor increased from 68 to 86% for reactors that exceed 150 MW_e capacity and since then has been maintained near 85%. At the end of December 2014 nearly 82% of operable nuclear reactors were LWRs, 63% of which are PWRs and 19% boiling water reactors (BWRs). The PWR is unique in that in a primary loop water passes through the reactor core to act as moderator and coolant and produces steam in a secondary loop that drives the turbine, avoiding leaks of radioactive material and increasing the Carnot efficiency but also increases the complexity of the system. BWR light water reactors use low enriched uranium as fuel because natural water absorbs some of the neutrons, reducing the number of nuclear fissions. 11% of the world's reactors are pressurised heavy water reactors (PHWRs), operating mainly in Canada (the CANDU technology ["CANada Deuterium Uranium"]) and in India. PHWRs can use natural un-enriched uranium. A great disadvantage of this type of reactors is due to the large cost of heavy water, which is not allowed to boil, so in the primary circuit very high pressure, similar to PWRs, exists.

A little more than 3% of the world's fleet consists of gas-cooled reactors (GCRs), all in operation in the United Kingdom. The fuel is natural uranium, the moderator is also graphite and the coolant is CO_2 (helium is used in high temperature GCR). Most of these will be retired within the next decade. Another 3% consist of graphite-moderated light water gas-cooled reactors (LWGR), which are better known under their Russian abbreviation RBMK. These reactors may use natural un-enriched uranium, are today only in operation in Russia and will probably be retired before the end of the next decade.

Finally, 1 out of the 438 reactors is a sodium-cooled FBR, an example of one of the main technologies of future Gen IV reactors, and a further 2 are expected to be connected in 2015 [11]. Sodium cooled FBRs use plutonium (^{239}Pu and ^{241}Pu from recycled fuel) as their basic fuel, or sometimes high-enriched uranium to start them off. The liquid metal, mostly sodium offers a large temperature working range (liquefies at 98 °C and does not boil until 892 °C) and it is used as coolant, which makes the reactor safer to use. They are called FBRs because they have a conversion ratio above 1 (i.e., more fissile nuclei are created than fissioned). Fast neutron reactors are 60% more efficient than normal reactors as they extract much more energy from recycled nuclear fuel, minimizing the risk of weapon proliferation and markedly reducing the time nuclear waste must be isolated (a few hundred years compared to current tens of thousands of years) [8].

Finally, it is important to mention that the US Department of Energy uses a different classification of reactors developed as part of the Generation IV reactor concept. This system includes: (i) Gen I refers to the early prototypes and civil nuclear reactors from the 1950s and 1960s currently decommissioned; (ii) Gen II refers to commercial reactors designed to be economical, reliable, for a typical operational lifetime of 40 years and comprise the bulk of the world's 400+ commercial PWRs and BWRs; (iii) Gen III are essentially Gen II reactors with evolutionary, state-of-the-art design improvements, potentially to greatly exceed 60 years of operation; and (iv) Gen IV reactors are two-to-four decades away, include the goal of full actinide recycling and on-site fuel-cycle facilities [12].

14.2.6 Thorium as an Alternative Fuel

As discussed above, compared to uranium, thorium is more abundant and harder to divert to nuclear weapons production and it yields less radioactive waste. ^{232}Th is not fissile but when loaded into a nuclear reactor absorbs neutrons to produce ^{233}U, which is fissile and long lived. Much of the ^{233}U will then fission in the reactor and the used fuel can then be unloaded from the reactor. The remaining ^{233}U can be chemically separated from the thorium and used as fuel in a nuclear reactor [6]. Basic research and development, as well as operation of reactors with thorium fuel from 1964, has been conducted in Canada, Germany, India, Japan, the Russian Federation, the United Kingdom, Norway and the United States [6].

Current R&D is being carried out on several concepts for advanced reactors including: high-temperature gas-cooled reactor (HTGR); molten salt reactor (MSR); CANDU-type reactor; advanced heavy water reactor (AHWR); and fast breeder reactor (FBR) [6]. India, with the world's largest known thorium reserves, has been running since 1986 the Kamini research reactor, the only one in the world that uses uranium-233 [13]. Thorium-fueled MSRs could also works for neutering stockpiled plutonium and high-level nuclear waste, which could transmute plutonium and waste into radioactive elements with shorter half-lives and potentially generate power as well.

14.2.7 Reprocessing

Uranium recovery through reprocessing of spent fuel has been conducted in the past in several countries, including Belgium and Japan. It is now routinely produced only in France and the Russian Federation, principally because the production of RepU is a relatively costly industry, in part due to the requirement for dedicated conversion, enrichment and fabrication facilities. Available data indicate that it represents less than 1% of projected annual world requirements [6].

DU stocks represent a significant source of uranium that could displace primary production. However, the re-enrichment of depleted uranium has been limited since it is only economic in centrifuge enrichment plants with spare capacity and low operating costs [6]. Successful development of laser enrichment could potentially result in an additional supply of uranium to the market in the longer term.

14.2.8 Small Modular Reactors (SMR)

Small modular reactors (SMR) are reactors (~ 25–250 MW) that could respond to a demand in rapidly growing countries that have small power grids, or address specific non-pure-electric applications such as district heating or desalination. SMRs are proposed to be built in factories and then transported (by truck, train or barge) to sites for installation. For example, the Russian KLT-40S (for electricity generation, heat processing and possibly desalination) under construction is mounted on the Lomonosov barge and shipped to isolated coastal regions or islands. SMRs require much less water for cooling than conventional reactors so they need not to be located by the sea or large rivers. The economics of SMRs are often cited as a distinct advantage but to date have not been demonstrated [11, 14].

Table 14.1 SMRs under construction [11]

Vendor	Country	Design	Type	Net capacity (MW)
CNES	Argentina	CAREM-25	PWR	25
CNEC	China	HTR-PM	HTR	210
OKBM	Russia	KTL-40S	Floating PWR	2 × 35
NuScale Power	US	NuScale SMR	PWR	50

Certification of SMR designs is still pending by the regulatory community to verify performance, safety and security design criteria.

Many of today's SMR plans have their roots in naval research technology (e.g., Westinghouse reactors that powered the first US nuclear submarines). There are different types of SMRs, some already under construction in Argentina, China or Russia (Table 14.1), others with near-term deployment potential, and still others with longer-term deployment prospects (liquid metal-cooled reactor technologies) including designs of dedicated burner concepts for countries having to dispose of plutonium stockpiles. In total, some 40 SMRs are either under construction or have conceptual or detailed designs around the world [14].

Unless governments and industry work together in the next decade to accelerate the deployment of the first SMR prototypes that can demonstrate the benefits of modular design and construction, the market potential of SMRs may not be realized in the short to medium term. For SMRs, CNEA announced that the CAREM unit has an estimated capital cost of USD 17,840/kW [14], and the US NuScale's design is backed by MUSD 217 from the US Department of Energy and its licensing process is expected to start in 2016 [15].

Case Example 14.3

Argentina builds the world's first SMR:

It has been built by Argentina's National Atomic Energy Commission (CNEA), a 25 MW unit, known as CAREM. This SMR is placed some 60 miles northwest of Buenos Aires at the site of the country's two-unit Atucha nuclear plant and it is based on a pressurized-water reactor. SNEA says the CAREM unit, with a capital cost of USD 446 million, should begin cold testing in 2016 and go critical in 2017 [16].

14.2.9 Nuclear Waste and Management

Nuclear power has unresolved the challenge of long-term management of radioactive wastes. The nuclear fuel cycle starts with the mining of uranium and ends with the disposal of nuclear waste. The fuel requires replacement when only a few percent of the total fissile species have been consumed. With the reprocessing of used fuel as an option for nuclear energy, the stages form a true cycle. The spent fuel assemblies are stored under water for cooling and radiation shielding for a few years and transferred to an interim storage (wet or dry) facility. After 40 years in storage, the fuel radioactivity will be about a thousand times lower than when it was removed from the reactor. The spent fuel contains uranium (96%, less than 1% fissile), plutonium (1%) and high-level waste products (3%). If spent fuel is not reprocessed, the fuel cycle is referred to as an 'open' or 'once-through' fuel cycle; if spent fuel is reprocessed, and partly reused, it is referred to as a 'closed' nuclear fuel cycle [10].

Currently, in the closed cycle the spent fuel is partitioned into U and Pu suitable for MOX fabrication and recycled back to the reactor. The rest of the spent fuel is treated as high-level waste (HLW) which remains highly radioactive and hot, hence it continues to require cooling and shielding. In the future additional isotopes are planned to be separated from the spent fuel to be used to transmute them for reducing the long term radioactivity of the waste. There are also: (i) very low level waste (VLLW) containing radioactive materials at a level which is not considered harmful to people or environment, also produced in other industries, therefore disposed of with domestic refuse; (ii) low level waste (LLW, 90% in volume) containing small amounts of mostly short-lived radioactivity, it does not require shielding during handling and transport and is suitable for shallow land burial (commonly after compacting or incinerating to reduce its volume); and (iii) intermediate level waste (ILW, 7% in volume) requiring shielding, non-solids may be solidified in concrete or bitumen for disposal.

Storage of HLW is mostly onsite in reactor spent fuel pools or in dry cask storage. After storage for about 40 years, deep geologic disposal of the spent fuel or vitrified waste in underground facilities or boreholes is the preferred option for most nations. HLW disposal is more contentious than disposal of lower-level wastes and no country today has an operating disposal site for high-level waste [10]. A pending question is if spent nuclear fuel should be retrievable from geologic repositories in the future or if permanent closure is a preferred design.

14.2.10 Costs

Although overall uranium resources have increased in recent years, the costs of production have also increased, leading to reductions in lower cost category resources. Market prices will determine how and when at least some of this supply is brought to market. Currently the largest identified resource base is in the USD 80–130/kgU category [6]. In the lowest cost category (<USD 40/kgU) underground mining is the dominant production method for RAR. In the <USD 80/kgU category,

Table 14.2 Global reasonable assured resources (RAR) and inferred resources (IR) recoverable at different costs in 2013 [6]

1000 tU	RAR	IR	Total
<USD 260/kgU	4587.2	3048.0	7635.2
<USD 130/kgU	3698.9	2204.0	5902.9
<USD 80/kgU	1211.6	745.1	1956.7
<USD 40/kgU	507.4	175.5	682.9

resources produced by underground mining and in situ leaching methods makes the largest contribution. There is now a more even distribution of resources associated with the open-pit, underground and co/by-product categories in both the <USD 130 kgU and <USD 260/kgU categories, but the underground mining still dominates. Identified resources can be classified in terms of cost (Table 14.2).

Some national and international authorities (Australia, the United States and the ESA), publish price indicators to illustrate uranium price trends for both long-term and short-term (spot price) contract arrangements (Fig. 14.5) [17]. The overproduction of uranium, which lasted through 1990, combined with the availability of secondary sources (nuclear disarmament), resulted in downward cost trends until 2002 and then a more volatile cost trend driven in part by the temporal dependence on a few critical facilities in the supply chain, changes in the value of the USD, an increased sense of declining inventories, the expected expansion of nuclear power generation in countries as China and the recognition of the positive effects of nuclear energy in GHG emissions [6].

The unit costs per kWh for nuclear energy are now comparable to or lower than the unit costs for coal in most parts of the world. China's average overnight cost (include cost of site preparation, construction and contingency costs) of

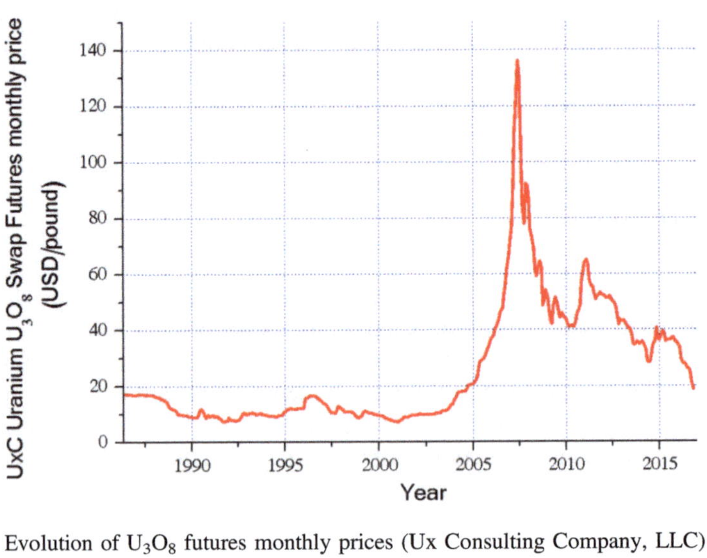

Fig. 14.5 Evolution of U_3O_8 futures monthly prices (Ux Consulting Company, LLC) [17]

approximately USD 3500/kW is less than two-thirds of the EU's cost of USD 5500/kW and USD 4950/kW in USA [11]. Higher costs in the EU and USA can be attributed to the lack of recent experience in building nuclear plants compared to Asia, as well as to the higher labor costs for engineering and construction.

14.3 Regional and National Perspectives on Technology

14.3.1 Resources

RAR and uranium resources are present and quantified in 16 African countries (Fig. 14.6) [6] but today only four nations are producing uranium in Africa: Namibia, Niger, South Africa, and Malawi. In Africa, most of the uranium production in 2012 was owned by private non-domestic companies (41%) or non-domestic, government-owned companies (37%). In addition, there is uranium production in government-owned domestic mining companies (17%) and only in South Africa there is uranium production owned by private domestic mining companies (5%) [6].

South Africa, an early participant in the Atoms for Peace program, has two operating reactors with 290 tU requirements in 2012 and providing 1.8 GW to the power grid. Although no other countries in Africa have nuclear power plants at this time, several have expressed interest in developing nuclear power for electricity generation and desalination in recent years, including Algeria, Egypt, Ghana, Kenya, Morocco, Namibia, Niger, Nigeria, Tunisia and Uganda.

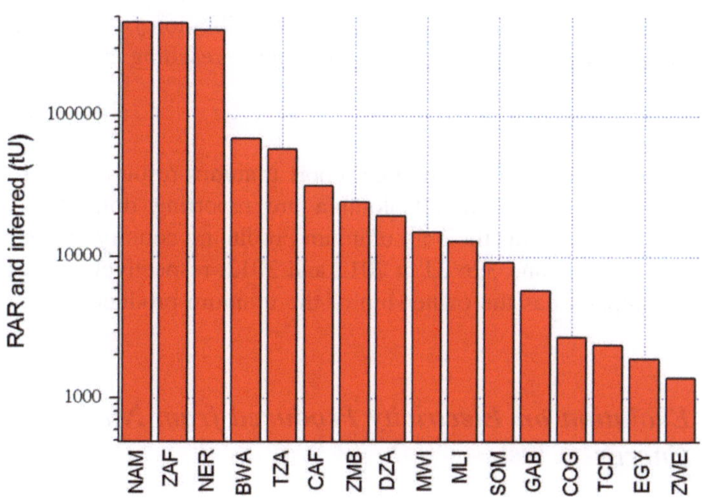

Fig. 14.6 Reasonably assured resources (RAR) and inferred resources in Africa [6]

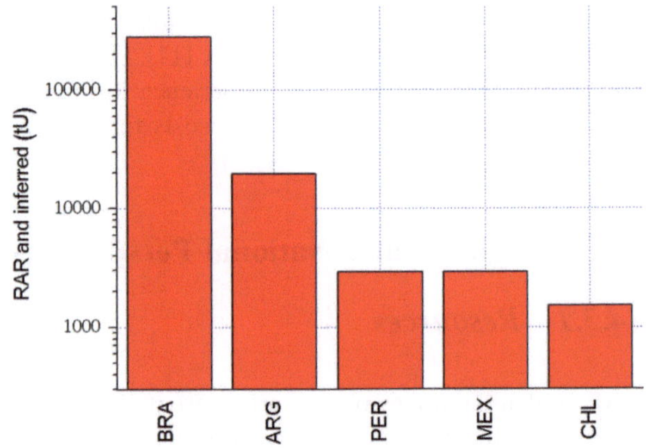

Fig. 14.7 Reasonably assured resources (RAR) and inferred resources in Latin America [6]

Case Example 14.4

Calibration of the Nigeria Research Reactor-1 (NIRR-1):

Nigeria Research Reactor-1 is a low-power, tank-in-pool miniature neutron source reactor with a nominal thermal power of 30 kW under steady conditions, located at the Centre for Energy Research and Training, Ahmadu Bello University, in Zaria. The reactor was designed by the China Institute of Atomic Energy, and was first critically accomplished on February 2004. The current reactor core is a 230 × 230 mm square cylinder and fueled by U-Al_4 enriched to 90% in Al alloy cladding. It has 347 fuel pins and 3 Al dummies in the fuel lattice. The reactor water temperature varies between 23 and 43 °C, and the core of the reactor is located 4.7 m under water close to the bottom of the water light-reactor vessel. The power calibration of this reactor has been just reported to the scientific community reaching a full power of 30.7 ± 2.5 kW [18].

In Latin America only five countries report uranium resources (Fig. 14.7) [6], and only Argentina, Brazil and Colombia are reporting domestic exploration expenditures [6]. Brazil was the only uranium producing country in Latin America with production of 265 and 326 tU in 2012 and 2013, respectively, and a Brazilian government company has the ownership of the uranium production.

14.3.2 Evolution on Electricity Produced from Nuclear Energy

Argentina has two operating reactors (0.34 GW Atucha 1 and 0.6 GW Embalse) and one under construction (0.75 GW Atucha 2), requiring 120 tU in 2012.

Argentina's government is also considering the construction of another two reactors in 2017 and 2020. With the license for the Embalse reactor due to expire in 2014, the government intends to conduct the necessary work to upgrade equipment, increase power output and extend the life of the reactor by 25 years. In early 2013, the Development Bank of Latin America issued a USD 240 million loan in support of this project.

Brazil has two operating reactors (0.5 GW Angra 1 and 1.3 GW Angra 2) and one under construction (1.2 GW Angra 3), requiring 400 tU in 2012. The national long-term electricity supply plan includes a total of 4 GW_e nuclear generating capacity installed by 2030 in order to help meet rising energy demand.

Mexico has two operating reactors (1.4 GW) in Laguna Verde and required 140 tU in 2012. After a refurbishment and uprate program completed in 2011, the power of the two units was increased by about a 20% and extended the operating life to 2029 and 2034. No plan to add additional nuclear generating capacity has been announced, although it has been reported that the government is still considering adding more units in the longer term.

Other countries in the region, currently without nuclear power plants, have been considering the development of such facilities, including Bolivia, Chile, Cuba, and Uruguay. Current legislation in Uruguay promotes development of renewable energy sources and has deferred nuclear development plans at present.

The IAEA has updated total thorium resources, regardless of resource category or cost category and identifies thorium resources in many countries in Africa (Egypt, South Africa, Morocco, Nigeria, Madagascar, Angola, Mozambique, Malawi, Kenya and Democratic Republic of the Congo) and in Latin America (Brazil, Venezuela, Peru, Uruguay and Argentina) [6].

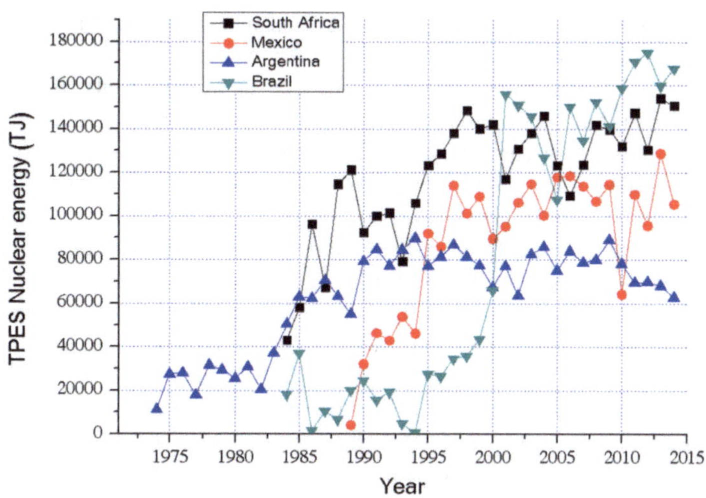

Fig. 14.8 Evolution of the total primary energy supply from nuclear energy per country in Africa and Latin America [4]

Finally, the share of nuclear energy in the power mix of the African and Latin American countries which possess nuclear reactors is very small (2.9–5.1%). However, production of electricity from nuclear energy is reported since 1974, when Argentina started the first nuclear reactor connected to the grid (Fig. 14.8) [4]. Brazil and South Africa started operations in 1984 and Mexico in 1989. Production is almost stagnant in Argentina since 1990, in Mexico and South Africa since 1996 and in Brazil since 2001.

14.4 Conclusions and Future Perspectives

Nuclear power has a long, sustained history of providing energy for a variety of applications worldwide that began back in the 1950s. Today nuclear power provides over 11% of the world's electricity as continuous, reliable base load power without the production of carbon emissions. There are nominally 440 commercial nuclear power reactors operable in 31 countries around the world with an additional 60 under construction and over 380,000 MW_e of total generation capacity. Nine of these reactors are located in Africa (South Africa with two [1830 MW]) and Latin America (three reactors in Argentina [1627 MW], two in Mexico [1600 MW], and two in Brazil [1891 MW]). Nuclear power has a number of favorable attributes including electricity generation cost, available fuel supply, dispatchable on demand, no carbon emissions and well-documented challenges including operational safety, materials security, nuclear waste management and disposal.

Today there are limited amounts of power produced in Africa and Latin America using nuclear energy in absolute terms and relative to their overall national power mix. Africa also has several nations that have produced uranium ore for fuel production (e.g., Namibia, Niger and South Africa, and Malawi). In Latin America, Argentina has two operating reactors (0.34 GW Atucha 1 and 0.6 GW Embalse) and one under construction (0.75 GW Atucha 2) and is contemplating further expansion. Brazil has two operating reactors (0.5 GW Angra 1 and 1.3 GW Angra 2) and one under construction (1.2 GW Angra 3). Mexico has two operating reactors (1.62 GW) in Laguna Verde. Brazil is the only nation currently producing uranium for nuclear fuel manufacturing.

The capacity and connectivity of the electric grid in Africa (and potentially some parts of Latin America) limits the further development of nuclear power using traditional power plants (nominally 0.5 GW or larger). There are many reasons that will likely limit nuclear power expansion in most African and Latin American countries in the near future: (i) it is widely regarded as a more expensive source than other available options; (ii) it implies environmental, safety, and political issues strongly difficult to manage in these countries and would likely face strong resistance from the local public and international community; (iii) most large countries have access to less expensive energy sources with considerably shorter lead times, while smaller countries do not need the large scale that nuclear power offers; (iv) this technology requires the largest amount of upfront capital investment compared to other conventional or renewable technologies; and (v) a lack of skilled engineers and other professionals.

Several nations are considering small modular reactors (SMRs) as an option (modular capacity is variable—nominally 25–250 MW). Argentina has constructed the world's first SMR (25 MW) near the existing Atucha facility that is undergoing testing and is expected to go critical in 2017. Thus, SMRs could be a viable alternative for African and Latin American countries. However, for many of the reasons stated above, and others, it is unlikely that SMRs will become a significant contributor to the energy future of most African and Latin American countries.

References

1. WNA (2016) Nuclear power in the world today. http://www.world-nuclear.org/information-library/current-and-future-generation/nuclear-power-in-the-world-today.aspx. Accessed 30 Nov 2015
2. Kessides ON (2014) Powering Africa's sustainable development: the potential role of nuclear energy. Energy Policy 74:S57–S70
3. McKinsey & Company (2015) Brighter Africa. Electric Power & Natural Gas
4. International Energy Agency database (2015) http://data.iea.org/. Accessed 30 Nov 2015
5. Abudeif AM, Moneim AAA, Farrag AF (2015) Multicriteria decision analysis based on analytic hierarchy process in GIS environment for siting nuclear power plant in Egypt. Ann Nucl Energy 75:682–692
6. IAEA (2014) Uranium 2014: resources, production and demand. https://www.iaea.org/OurWork/ST/NE/NEFW/Technical-Areas/NFC/uranium-production-cycle-redbook.html. Accessed 30 Nov 2015
7. Nuclear Threat Initiative—South Africa—Recent Developments and Current Status, 2015. Nuclear Threat Initiative.Org
8. Nuclear fuel fabrication. World Nuclear Association. http://www.world-nuclear.org/information-library/nuclear-fuel-cycle/conversion-enrichment-and-fabrication/fuel-fabrication.aspx. Accessed 30 Nov 2015
9. Duderstadt J, Hamilton L (1976). Nuclear reactor analysis. Wiley, Inc. ISBN 0-471-22363-8
10. Zohuri B, McDaniel P (2015) Thermodynamics in nuclear power plant systems. Ed. Springer. ISBN 978-3-319-13418-5
11. IEA (2015) Nuclear energy technology roadmap
12. Goldberg SM, Rosner R (2011) Nuclear reactors: generation to generation. American Academy of Art & Sciences (2011). ISBN 0-87724-090-6. http://www.amacad.org/pdfs/nuclearreactors.pdf. Accessed 30 Nov 2015
13. Bagla P (2015) Thorium seen as nuclear's new frontier. Science 350:726–727
14. Malze K (2015) Small modular reactors speaking in foreign tongues. Power 159:30–34
15. The Economist (2016) Is smaller better? Nuclear energy. Available 23 Apr 2016
16. Maize K (2015) Small modular reactors speaking in foreign tongues. Power 159:12. http://www.powermag.com/small-modular-reactors-speaking-in-foreign-tongues/. Available 23 Apr 2016
17. Indexmundi.com. http://www.indexmundi.com/commodities/?commodity=uranium&months=360. Accessed 30 Nov 2015
18. Agbo SA, Ahmed YA, Ewa IOB, Jibrin Y (2016) Analysis of Nigeria research reactor-1 thermal power calibration methods. Nucl Eng Technol 48:673–683

MIX
Papier aus verantwortungsvollen Quellen
Paper from responsible sources
FSC® C105338

If you have any concerns about our products,
you can contact us on
ProductSafety@springernature.com

In case Publisher is established outside the EU,
the EU authorized representative is:
Springer Nature Customer Service Center GmbH
Europaplatz 3, 69115 Heidelberg, Germany

Printed by Libri Plureos GmbH
in Hamburg, Germany